Birkhäuser Advances in Infectious Diseases

Series Editors:

Stefan H. E. Kaufmann
Department of Immunology
Max Planck Institute for Infection Biology
Berlin, Germany

Olaf Weber
Bonn, Germany

More information about this series at http://www.springer.com/series/5444

Jaime Marcelo Altcheh • Hector Freilij
Editors

Chagas Disease

A Clinical Approach

Springer

Editors
Jaime Marcelo Altcheh
Servicio de Parasitología y Enfermedad de Chagas, Hospital de Niños "Ricardo Gutiérrez"
Buenos Aires, Argentina

Instituto Multidisciplinario de Investigación en Patologías Pediátricas (IMIPP-GCBA)
Consejo Nacional de Investigaciones Científicas y Técnicas (CONICET)
Buenos Aires, Argentina

Hector Freilij
Servicio de Parasitología y Enfermedad de Chagas, Hospital de Niños "Ricardo Gutiérrez"
Buenos Aires, Argentina

ISSN 2504-3811 ISSN 2504-3838 (electronic)
Birkhäuser Advances in Infectious Diseases
ISBN 978-3-030-00053-0 ISBN 978-3-030-00054-7 (eBook)
https://doi.org/10.1007/978-3-030-00054-7

Library of Congress Control Number: 2018965747

This Springer imprint is published by the registered company Springer Nature Switzerland AG
The registered company address is: Gewerbestrasse 11, 6330 Cham, Switzerland

To my wife Ada, my children Marina and Pedro, and my parents Chuchi and Enrique.

—Jaime Marcelo Altcheh

To my family, Anabela, Tomas, Ignacio, Facundo, Katy, Mikel, and Lore, and my pets Uvita and Violeta.

—Héctor Freilij

Preface I

Chagas disease was discovered over a hundred years ago, affecting several million people, mainly in Latin America, and for decades, it was ignored and neglected. It is primarily transmitted by insect vectors which carry the parasite *Trypanosoma cruzi*, the agent of the disease. Currently, in areas under vector control and in non-endemic countries, the main route of infection is the congenital route. Subsequent cardiac and gastrointestinal complications are common in untreated infected individuals. There are two available drugs (nifurtimox and benznidazole), but only few people are receiving adequate treatment due to the lack of information about the disease and its epidemiology.

Chagas disease is a potentially life-threatening illness that was historically mainly endemic to Latin America. Over the last decade, however, the disease has spread to, and is increasingly prevalent in, other continents such as North America and Europe, with an estimated seven million people infected worldwide. This migration to non-endemic countries has brought Chagas disease to areas where it had never been before, putting patients at risk of not being diagnosed if asymptomatic or misdiagnosed in symptomatic cases.

The Parasitology Service, at the Ricardo Gutierrez Children's Hospital, Buenos Aires, Argentina, is a reference center for the diagnosis and treatment of infants and children with Chagas disease. During many years, I have been working with Dr. Hector Freilij sharing the experience of improving the medical assistance of infants and children. He transferred to me the passion of the pediatric profession, and I will be grateful to him all my life. Many researchers and healthcare professionals from Latin America and other continents have worked with us, and several collaborative studies were developed during the last 30 years. The main idea for writing this book was to share and disseminate the information about the disease in order to improve the diagnosis and treatment of the affected population.

Many chapters were written by authors from Latin America, where the disease has been studied in depth, including experts working with patients on a day-to-day basis. The authors chosen come from different fields, and therefore the scope of their experience ranges from the asymptomatic disease in children to seriously ill adults, covering all aspects of the disease.

In addition to clinical information, the book includes chapters on epidemiology in the Americas and Europe and basic information about the parasite and immune response to the infection. There are chapters covering new information about diagnosis and new developments in biomarkers of treatment response.

This book gives a complete overview of Chagas disease, including its vectorial, congenital, and oral transmission, as well as practical advice about dealing with infants, children, and adults suffering from the infection.

I would like to thank all the staff at Parasitology Service for the support to my work and also to patients and their families who understand that our work was carried out to improve the diagnosis and treatment of Chagas disease.

Buenos Aires, Argentina Jaime Marcelo Altcheh

Preface II

If we think that parasitic diseases are owned by the poorest people in poor countries, Chagas disease is a sad and valid example.

It is a complex disease due to the multiplicity of determinants that cause it, but not to doubt that all of them have poverty as a backdrop, but not only the economic one. The saddest thing is the poor interest of some health policy-makers and part of the different members of the health system to deal with this pathology.

The vectorial aspects are complex to solve, such as the characteristics of the rural domiciles, the climatic conditions, the ecosystem, the resistance of the vectors to the insecticides, but it is not impossible. In fact, there are several countries that have achieved the cutting of vector home transmission of *Trypanosoma cruzi*.

Some aspects of care, especially regarding the parasiticide treatment of patients, are controversial, as well as so many other diseases. However, there are many concepts in which there is consensus with strong evidence of benefit to patients. Despite this, the actions with clear and proven benefits for neonates, children, and adults are not carried out in the magnitude of the needs.

These are the shadows of this disease; but there are also lights and hopes. In the last decades, important advances have been achieved for the population, for example, the cutting of the transfusion transmission of this protozoan in almost all countries. In addition, several countries and some regions of other countries have certified home vector interruption. In general terms, we can affirm that the area of domiciliary vectors has decreased in the Americas, which has led to a reduction in the total number of infected persons.

In a number of healthcare centers, adequate patient care is carried out. Some health authorities and several NGOs carried out important proactive studies; this allowed that thousands of patients have already received the parasiticide treatment to avoid the serious consequences that this protozoan produces.

Our book speaks a bit of all this and other topics and is aimed at members of various branches of the health system which will update their knowledge and generate behavior in front of patients. Hopefully it will also be useful to those responsible for public health to develop health plans that aim to avoid the generation of new patients and fully assist those who already have it. There are several chapters that tell us about the scientific contributions related to its pathogenesis, the epidemiological aspects of countries located outside the endemic area of the Americas, and the recent approaches of new and old drugs with a more pharmacological vision.

Although each of the authors, excellent referents in their respective specialties, wrote the latest information on each topic, we also know that when this book comes out, some of the concepts that have been spilled will be less valid; this is a product of the permanent concern and interest of many researchers from different branches of science to increase knowledge. Fortunately, the human being is very curious and eager for new knowledge. My thanks go to each and every one of the collaborators.

I would like to thank Dr. Jaime Altcheh for the invitation to share the creation of this book but above all, for one of its virtues, the great vocation for research that allowed our beloved Hospital Service to grow. Also I would like to thank the rest of the members of the Service and the many professionals from various countries with whom we share tasks that allowed us to know many facets of this disease and have acquired great friendships and also to the *patients*, who were the ones who allowed us to really unravel this disease, and they generated the enormous gratification when they reached their cure.

Very different would be the reality if everything that is already known about this disease will be carried out!

We must go on and on; it is not easy, but it is possible.

Buenos Aires, Argentina Héctor Freilij

Contents

Part I
Overview

Chagas Disease: Past, Present, and Future

Héctor Freilij

Abstract In this chapter we present an introduction to diverse aspects of Chagas disease. It is especially targeted to members of the health system in need for a deeper insight on the disease.

Chagas disease is a silent, silenced, and very complex disease. Silent because it may take 20 years since the parasite ingresses the organism until the subject develops a pathology. Silenced, because governments have managed to keep it without a common denominator. It is highly associated to poverty. It is included within the so-termed neglected tropical diseases by the WHO, given the little concern it represents to health officers from many countries and the pharmaceutical industries.

It is caused by the protozoan *Trypanosoma cruzi*, which has a wild life cycle, with over 100 animal species involved, and a human cycle. Its vector is a triatomine hematophagous insect, varying according to the geographic area. The parasite is transmitted through the vector feces. It is the most important endemic disease in the Americas, but due to Latin American population migratory movements, it has reached different countries in several continents. It affects approximately eight million people.

About 30% of the infected subjects develop cardiac disease with severe consequences and mortality. It also produces digestive disease, with a smaller load of morbidity.

We produced this chapter as a review of the history of the disease, and we describe a panoramic view of the infection ways, points relevant to diagnosis, treatment, phases of the disease, criteria of cure, and new research.

H. Freilij (✉)
Servicio de Parasitología y Enfermedad de Chagas, Hospital de Niños "Ricardo Gutiérrez", Buenos Aires, Argentina

J. M. Altcheh, H. Freilij (eds.), *Chagas Disease*, Birkhäuser Advances in Infectious Diseases, https://doi.org/10.1007/978-3-030-00054-7_1

We may say that Chagas disease is an infectious pathology, since it is certainly produced by a protozoan, *Trypanosoma cruzi*. We may add that the spectrum of pathological findings includes subclinical infection, cardiac, digestive, neurologic syndromes, and death. It is also known as American trypanosomiasis.

But in fact, the most important point to stress is that it is included within the 20 neglected tropical diseases, as designated by the World Health Organization (WHO) [1, 2], with all that implies. To only mention an example, this disease is not economically interesting to the pharmaceutical industry, and therefore new drugs are not being developed [3]. Currently we rely on only two anti-parasitic drugs developed in the decade of the 1970s in the twentieth century. Their production was discontinued in several occasions by their manufacturers. To this we must add the lack of appropriate accessibility to these treatments for the patients; several countries do not include them in their official list of approved medicines.

We may add some other characteristics of this disease: it is *silent, silenced, hidden,* and *very complex*. We term it silent because a patient may take 15–30 years since the protozoan enters their organism to develop cardiac or digestive lesions [4–6]. It is the main cause of infectious myocarditis worldwide [7]. Silenced and hidden, because patient care and actions against vectors, likewise, are insufficient in several countries. Governments have committed to keep its visibility low among the public. We also say it is complex, since it involves multiple factors: the protozoan, the age and immunological status of the infected subject, the vector, the housing conditions in endemic rural areas, the insecticide spraying actions against the vector, the post-spraying surveillance, the adequate control at blood banks, the scarce knowledge among the population, and, for worse, among the workers of the healthcare system, the limited amount of time dedicated to the disease in university courses and the little concern of politicians and health administration officers.

We can also term it a *pediatric disease*. Even when an inhabitant of a rural endemic area can acquire the infection by vector-borne transmission at different ages, most of the patients are infected by this means during early childhood. If we also consider children who are infected by *T. cruzi* across the placenta, we arrive to this conclusion [3, 8–14]. It is a pediatric disease that, if undiagnosed and untreated during childhood, has consequences that physicians will have to deal with in the adult patient.

The risk of being infected has several outcomes. Natural history of the disease shows that about 30% of the infected people may develop cardiac and/or digestive disease [3] (Fig. 1). Women can transmit the protozoan during one or many pregnancy terms [7]. In the case of an individual going through a condition of immunodeficiency (HIV infection, organ transplantation, autoimmune disease, or oncological treatments), the infection can be reactivated and generate great morbidity and mortality [2, 13]. The infection bans people from being blood or organ donors and may become a factor of employment discrimination.

We can say of it that it is an essentially biological disease, riding on a social situation of poverty, backwardness, and lack of development [15].

It is the most relevant parasitosis in the Americas, where it is endemic to 21 countries spanning Central and South America, as well as Mexico and the south of

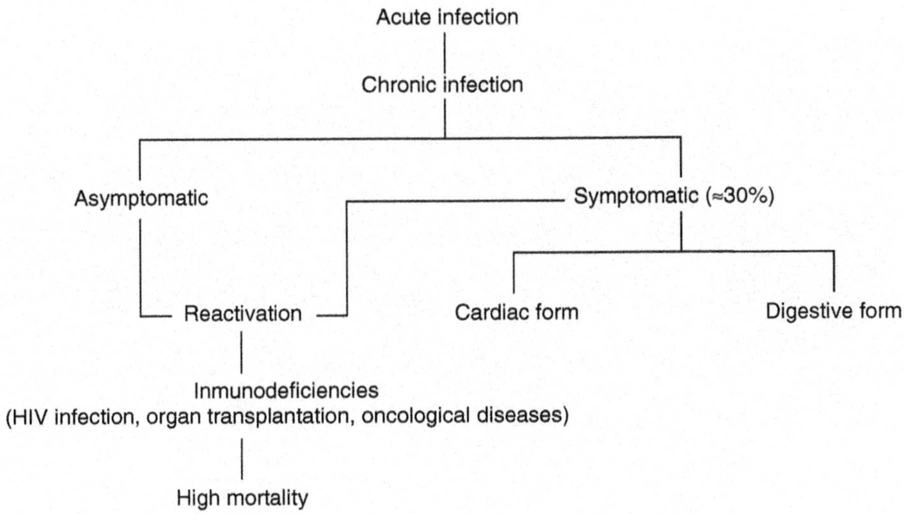

Fig. 1 Natural evolution of Chagas disease. *Source* Elaborated by the author

the United States of America. Besides, in an increasingly globalized world, patients live with this disease in the rest of the Americas, Europe, and other continents due to migrations [16–19]. Hence, knowledge on Chagas disease should have worldwide reach.

Chagas disease is suffered by the poorest populations and is strongly tied to political decisions regarding the allocation of economic resources.

An important demographic change has been taking place worldwide within the last decades; people abandon rural areas and migrate to cities seeking to improve their living conditions (Fig. 2). This sets the conditions for Chagas disease patients to inhabit in vector-free urban centers in different countries from several continents. An epidemiological change has occurred: Chagas disease has become urban. Besides, new cases are generated by transplacental infection, blood transfusion, or organ transplantation [20–22]. Currently, patients live in rural areas of endemic countries and in urban centers of endemic and non-endemic countries receiving Latin American migrants.

In endemic countries with little control of triatomines, most of the new cases are generated by contact with these vectors [23]. Conversely, in sites where vector control was established, new cases are due to transplacental transmission [24]. The oral route of infection occurs in both rural and urban areas, mainly by the uptake of food contaminated with infected vector feces [25, 26]. It can be observed within the Amazonia and countries with geographical links to this region: Colombia, Venezuela, and Brazil have reported a great amount of cases, many of them in the form of outbreaks.

The scenarios that Chagas disease patients face are extremely diverse: they live in cities with access to appropriate medical care or in rural areas with extreme poverty and limited resources. This implies a need for different strategies to approach their attention (Fig. 3).

Fig. 2 Rural dwelling in endemic area

Fig. 3 Vectors found inside a rural house in an endemic area

These are some of the most relevant sanitary features of this disease. However, in endemic countries and thanks to insecticide spraying actions in rural areas and specific controls at blood donation centers [27], indicators related to the abundance of vectors, the total number of patients, cardiopathy, and mortality have decreased. Nevertheless, there is still much work to do with the tools we currently count with.

Chagas disease exhibits a different panorama in non-endemic countries. There, public health systems should have precise policies regarding the detection of, and assistance to, infected patients, the study of blood and organ donors, and the systematic reach-out to the children of infected mothers [28–36].

Historically, the first patients to be diagnosed with acute *T. cruzi* infections were from rural areas and presented the Romaña's sign (unilateral bipalpebral lesion). Cardiac pathologies, on the other hand, were detected in healthcare centers from urban areas. In 1946 Taquini, in Buenos Aires [37], and Benchimol, in Rio de Janeiro [38], report cardiopathies caused by *T. cruzi* infection.

The association of the Romaña's sign [39] with acute Chagas disease diagnosis was, and is in part, an inconvenience: most of the physicians think of this malady only when the patient exhibits such lesion. It is estimated that only 5% of the acute Chagas patients present this sign. Other clinical manifestations in acute, vector-borne-infected Chagas disease patient can present prolonged febrile syndrome and liver and spleen enlargement.

In addition, we should keep in mind that most of the infected subjects, from newborn to adult, both acute and chronic, are usually asymptomatic [14]. Therefore, health practitioners should have a proactive attitude for their detection.

Since it affects the productivity of people of working age, and given that it produces disability and mortality, Chagas disease has a remarkable impact on Latin American economies, with losses estimated in 670,000 disability-adjusted life years (DALYs) per year. This ranks this infection as the parasitic disease with the highest impact on health and social systems in this region.

The disease was discovered in 1909 by Carlos Chagas in Brazil [40], the country where he was born and where he graduated as a physician in 1903. Among his many merits, one of the greatest is having been the only researcher to first describe the vector, the infectious agent, and the clinical manifestations of an infectious disease. All of this happened within the term of a few months.

This took place in Lassance, a municipality located in the North of the state of Minas Gerais, where he was sent by Oswaldo Cruz to conduct research on malaria during the construction of a railway. An engineer from the train company showed Carlos Chagas an insect, in whose feces parasites were detected. Simultaneously, he found the same parasite present in monkeys [41]. Carlos Chagas sent the insects to Oswaldo Cruz, in Rio de Janeiro, where a successful attempt to reproduce the infection in monkeys was conducted. Carlos Chagas would later find *T. cruzi* in the blood of a cat and of a young girl named Berenice, the first human acute case to be reported. The girl presented fever, hepatomegaly, and splenomegaly. However, Carlos Chagas was not able to find the protozoan in adults [40], which is not surprising, given that at the time diagnosis was performed by ocular inspection of blood smears under the microscope, a method that is only effective during the first few months after infection. In the following years, there were remarkable contributions to the understanding of this disease: the development of the complement fixation reaction (Machado Gerreiro), Brumpt's xenodiagnosis method, and the description, by Carlos Chagas, of the acute stage of the disease. All of this added up to Carlos Chagas being granted, in 1912, the Schaudinn Medal, awarded to the best research work in parasitology.

In 1912, Carlos Chagas claimed that "the discovered sanitary issue presented practical difficulties for its resolution, but since it was a matter linked to the duty of humanity and the pride of a nation, the energy required to tackle it in a definitive way would certainly not be lacking."

Carlos Chagas initially committed the mistake of attributing different thyroid lesions, goiter, and hypothyroidism to the parasite. This encouraged Dr. Kraus, between 1912 and 1916, to seek for this association in different places across Argentina. He would find numerous triatomines infected with *T. cruzi* in areas where thyroid pathology was inexistent. After the recognition of this mistake, Chagas disease was somewhat pushed into the background, in favor of research on other tropical diseases, like malaria.

Thanks to the presence, in Argentina, of Peter Mühlens, from the Institut für Tropenmedizin of Hamburg, in 1924, and of Charles Nicolle (Nobel Prize laurate in 1929) in 1925, from the Institut Pasteur of Tunisia, patients bearing *T. cruzi* in their blood were identified. As a consequence of these findings, the University of Buenos Aires created in 1926 the Mission for Argentine Regional Pathology Studies (*Misión de Estudios de Patología Regional Argentina*, MEPRA), a laboratory located in the north of this country dedicated to the exploration and study of regional diseases. The head of this laboratory was one of Nicolle's disciples, Dr. Salvador Mazza, a physician with a background in research and laboratory work. During his labor, Mazza detected 1400 patients with Chagas disease, which was important in order to bring it back into discussion [42].

The rest of the countries in the Americas reported their first cases of Chagas disease in different moments: El Salvador in 1913, Venezuela in 1919, Costa Rica in 1922, and Honduras in 1960. In Bolivia, a country which has historically been the one with the highest prevalence of infection, Dr. Torrico reported to have found the first patient in 1946 in the Cochabamba department. It was a girl who presented generalized edema and Romaña's sign. In this region, 84.9% of the *Triatoma infestans*, the main vector insect in the area, were infected with *T. cruzi*. In this context, Chagas disease stopped being regarded to as a Brazilian disease to become the American trypanosomiasis.

Historical writings can be found about the vectors. In the eighteenth century, a priest named Fray Reginaldo Lizarraga wrote about insects which the natives called "vinchucas" during a visit to an abbey in the Valley of Cochabamba (nowadays Bolivia): "They have clumsy feet and when they have filled their bellies with blood they have sucked, they cannot walk."

Triatomines caught Charles Darwin's attention during his travels through South America in 1835: "The night experienced an attack of the Benchuca (a specie of

Reduvius) the great black bug of the Pampas. It is most disgusting to feel soft wingless insects, about 1 in. long, crawling over one's body. Before suckling, they are quite thin, but afterwards become round and bloated with blood, and in this state, they are easily crushed" [43].

The first human presence in the Americas is estimated to have happened 20,000–25,000 years ago, but it took another 10,000–15,000 years for men and women to settle in South America. *T. cruzi* ancient DNA was detected by nested polymerase chain reaction in dry tissue samples from mummies of 9000 years of age. Besides, mummies found in Peru showed evidence of having clinical signs of Chagas disease [44]. According to researchers, an average of 40% of the prehistoric population of this area were infected with *T. cruzi* at the time of their death [45].

Being an enzootic disease at its early beginnings, it circulated among marsupials, mainly from the *Didelphis* genus. These eliminate *T. cruzi* through their odoriferous glands and can thereby transmit the parasite to other animals, perhaps even humans [46]. The advent of the hematophagous Hemiptera (*Triatonimae*) vectored the protozoan to other mammals. According to fossil registers, Hemiptera existed as an order since the Permian period, about 232 million years ago [47].

The domestic cycle of transmission finds its origin in the human invasion of the wild ecotope where the parasite, vectors, and reservoirs inhabited. The rural dwellings thus generated provided an appropriate environment for the reproduction of the vectors, where they could hide in the cracks and defects of walls and roofs of the houses, farmyards, and henhouses during the day and emerge at night to obtain a blood meal from humans or domestic animals. By ingesting blood from an infected animal, the insect enables *T. cruzi* reproduction in its gut, to later transmit it on its feces to another host.

Currently, we can consider the existence of a wild cycle and a domestic one, each with their own particular features regarding both the vector and the parasite genotypes.

More than 100 mammal species serve as reservoirs for *T. cruzi* in nature, including marsupials, bats, rodents, carnivores, and primates [48]. Given this massive diversity, we believe this parasitosis will never be wiped out completely from the planet. Even when we have achieved control of the domestic vector, other species within the wild cycle may eventually colonize human homes in the absence of a permanent control.

T. cruzi diverged from other trypanosomatids about 200 million years ago; it circulates in the blood of over 100 wild and domestic mammals and humans [49]. It is characterized by the presence of one flagellum and a single mitochondrion, where the kinetoplast is found. It reproduces by binary division within the tissues of the infected host. Over 80 researchers participated in the study of the genome of this parasite, which was published in 2005 [50]. Current classification, based on genetic features, comprises six discrete typing units (DTUs), named TcI to TcVI. These have a differential distribution over the Americas [51]. Studies on patients and vector samples revealed that a same individual can be infected with more than one DTU at a time.

1 Transmission Ways

The human host can acquire the parasite in different ways, the most frequent being vectorial, transplacental, and oral. Quite more uncommon are the transfusional way, organ transplantation, and laboratory accidents. The possibility of sexual transmission emerged as a new concern to be addressed, as parasite DNA was reported to have been found in the sperm of infected individuals [52].

1.1 Vectorial

Most people acquire the infection when their skin or mucosa comes in contact with the feces of infected bloodsucking insects, generally during the nighttime. These insects, mainly belonging to the genera *Panstrongylus*, *Rhodnius*, and *Triatoma*, inhabit inside houses with bahareque or adobe walls and thatched roofs, especially in endemic areas with scarce vector control actions. There, they nest in spaces like holes in the roof, under or behind furniture, and behind pictures. In some regions, the vector lives on trees in the surroundings of a house, entering it during the night.

Vector elimination through insecticide spraying at homes in rural areas is fundamental to avoid the generation of new cases of the disease. Organophosphorus compounds and carbamates have been employed, but since 1980 they were substituted by pyrethroids. Actions must be carried out in rural and peri-urban areas [53].

The Pan American Health Organization (PAHO) sets on several regional initiatives for the establishment and encouragement of actions against this disease. The first of them was the Southern Cone Initiative (*Iniciativa del Cono Sur*, INCOSUR), aiming at the elimination of domestic *Triatoma infestans* and the transfusional transmission of American trypanosomiasis. Created in 1991, it was integrated by Argentina, Bolivia, Brazil, Chile, Paraguay, and Uruguay. The success of this program led to the generation of another four initiatives comprising nearly all the countries across the Americas. As a consequence, the vector-borne transmission was halted in several countries, such as Uruguay, Chile, Paraguay and Brazil (except in the Amazonia region) [54, 55]. In many other countries, namely, Argentina, Bolivia, Colombia, and Peru, interruption of transmission was achieved in some areas. In several Central American countries, control of the main domestic vector, *Rhodnius prolixus was performed.* Several indicators are used to define vector-borne domestic transmission interruption in an endemic area: less than 1% of homes with presence of the insect and less than 0.5% of children under 5 years of age with *T. cruzi* infection. These criteria are evaluated by members of the PAHO together with local health officers.

The difficulties to establish appropriate actions against the vector are generally linked to budget and administrative issues and to the lack of interest from health authorities [56]. In some regions, this key task is often scarce, sporadic, and non-sustainable. Actions against the vector imply, firstly, insecticide spraying under the responsibility of qualified personnel and, secondly, permanent surveillance actions. This allows the assessment of the efficacy of these actions and commands to repeat the spraying in the sites of vector persistence. Community involvement is necessary for post-spraying surveillance, and therefore the public should be properly educated and made aware. Thanks to this type of program, the abundance of domestic vectors has diminished over time but is still persistent especially in the Great South American Chaco region, which comprises areas in Bolivia, Paraguay, and Argentina. This is attributable to geographic, demographic, and economic conditions, which hamper these actions. However, this aim has been accomplished in certain areas of this region [57, 58].

Despite vectors normally thrive in rural areas, there are urban centers in which they have been found in a permanent fashion, such as the cities of Arequipa in Peru and San Juan in Argentina.

1.2 *Transplacental*

Transplacental transmission occurs in endemic areas as much as in urban centers around the globe. Although Carlos Chagas suspected this mechanism existed, it was Dr. Dao in Venezuela who irrefutably demonstrated it in 1949 [59]. Different reports show diverse percentages of children being infected with *T. cruzi* during gestation, oscillating between 3% and 10%. These discrepancies may be regional but are probably conditioned by the exhaustiveness of the methodology utilized for their detection. Diagnosis and treatment of these patients is an overdue obligation of the health system, bearing in mind that this is the best moment in the life of a patient for the application of anti-parasitic treatment. Finally, we should stay aware of the fact that most of the children who are born with this infection are asymptomatic [3, 9].

1.3 *Oral Chagas Disease*

Since the first acute cases of orally transmitted Chagas disease reported by Shaw et al. [25, 60], hundreds of others have been described in the Brazilian Amazon region, which is nowadays considered to be endemic for oral transmission. Several outbreaks of acute Chagas disease transmitted by this way have been informed in different states of Brazil and other South American countries.

Oral transmission is probably the predominant way of infection among animals in the wild cycle, since several infection-susceptible mammals, such as small primates, usually eat triatomines that transmit *T. cruzi*. It is likely that this mechanism was first noticed at the time of the discovery of the disease (1909), as mentioned by Carlos Chagas.

Oral acute infection often presents worse clinical manifestation than the vector-borne infection. Edema is usually observed in the face, gums, and tongue, along with myocarditis.

Besides prolonged fever, present in most of the patients, splenomegaly and hepatomegaly are very common manifestations of the oral transmission cases, even overcoming myocarditis in frequency. Regarding the severity, cardiovascular compromise constitutes the highest percentage of death cause among victims of oral *T. cruzi* infection outbreaks [25, 61].

Most of the outbreaks have been associated to the consumption of beverages prepared from fruits or other vegetables, contaminated with triatomine feces or the secretions of infected mammals.

1.4 Transfusion

Since infected people may have parasitemia during the acute and chronic stages of Chagas disease, they are banned from donating blood, given the chance that they might transmit the infection to the recipient.

The first to suggest that Chagas disease might be transmitted by blood transfusion were Drs. Dias and Mazza in the 1940s, but it wasn't until 1949 that Dr. Pellegrino reported the first confirmed cases of infected donors in Brazil. Later, in 1952, Dr. Freitas described the first cases of patients acquiring the infection by this route [2, 16, 62].

Blood transfusion was regarded as the second most frequent way of infection with *T. cruzi* in endemic countries up to a few decades ago. Fortunately, screening programs were set up in any of these places in the last 20 years, which drastically reduced the risk of this type of transmission.

Whole blood, packed red blood cells, granulocytes, blood cryoprecipitate, and platelets are capable of carrying the parasite, while plasma derivatives are not [63].

The possibility of an individual acquiring the infection via blood transfusion depends on various factors: the amount of transfused blood, the infectivity of the parasites transfused with the blood, the parasite strain, the presence of parasitemia at the time of donation, the immune status of the donor, and the screening tests the transfused blood is subjected to [64].

Data collected during the 1960s and 1970s situates the infectivity rate for an infected whole blood unit in the range of 12–25% [63].

According to the WHO, a single high-sensitivity test is acceptable for determining the suitability of a blood unit for transfusion. ELISA is the most commonly applied method. However, each country has their own criteria for the evaluation of blood donor aptness. Thus, all blood donations must be analyzed for the presence of *T. cruzi* in endemic countries [65]. Meanwhile, this might not be required in non-endemic countries, where the number of at-risk donors is lower and blood supply protection is based on different approaches: the deferral of people at risk of having the infection and the acceptance of blood donation if specific laboratory assay results are negative. The latter is being introduced in countries where Latin American population is of a considerable size, such as the United States of America, Spain, and France [32, 66–68].

2 Immune Status of the Patient

The immune status of the patient and Chagas disease have several relation points to be considered. On one hand, there is the case of patients with chronic *T. cruzi* infection who develop an immunodeficiency, and on the other, there is the case of the

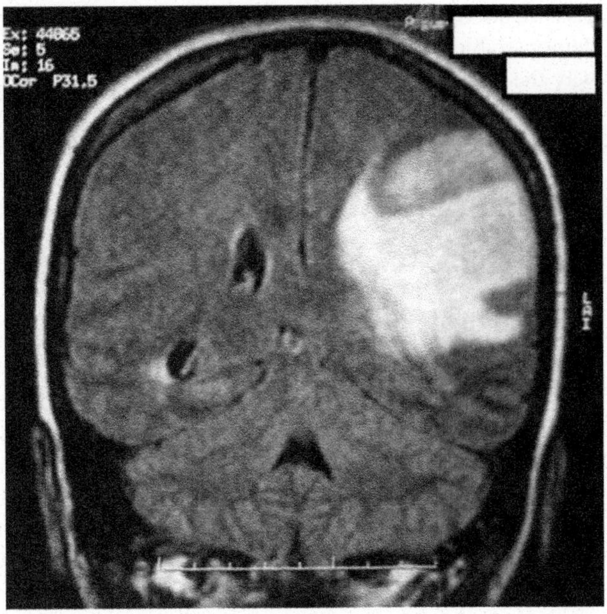

Fig. 4 Chagasic encephalitis in AIDS patient

patients with a certain degree of immunodeficiency who acquire the infection by the protozoan.

The first is the most frequently observed of these two situations; it comprises individuals who, while having a chronic *T. cruzi* infection, undergo a certain degree of immunodeficiency, which might be produced by HIV infection, an immunosuppressive treatment in the context of an organ transplantation, oncological treatment, or autoimmune disease (Fig. 4). Depending on the severity of the immunodeficiency, different situations may arise: very severe forms with high morbidity and mortality, a parasitological reactivation without clinical manifestation, or no signs of reactivation [69–71]. An usual problem is the delay on the diagnosis, as a consequence of not considering this possibility. The more precocious the diagnosis, the better is the therapeutic response.

3 Phases of the Disease

Traditionally, phases of the disease were defined as follows: (a) acute, elapsing approximately 2–4 months, is defined by high parasitaemia, detectable by direct parasitological methods (microhematocrit, Strout), and might present negative serological tests during the first weeks; (b) indeterminate, in which patients have positive serology, and normal ECG and thorax radiography; and (c) chronic, in which patients display reactive serology and some type of ECG abnormality or clinical cardiac or digestive manifestations. Many countries still use this classification. In 2010, after a meeting attended by professionals from several Latin American countries, a new classification was developed: (a) acute phase and (b) chronic phase, with or without demonstrable pathology [72, 73]. This classification reflects in a more appropriate way the biological reality of the disease and vanishes the ambiguous term "indeterminate," which led to confusion regarding its meaning. Switching terminology to chronic phase with or without demonstrable pathology enabled healthcare services with a developed infrastructure (echocardiography, Holter, stress test, etc.) to diagnose patients who do not exhibit clinical manifestations and have a normal EKG.

4 Diagnosis and Progression Markers

The gold standard of a diagnosis in infectology is the visualization of the infectious agent. In this disease, this is only possible in two circumstances: the acute phase and reactivation. Under these two conditions, laboratory diagnosis relies on the demonstrable visualization of *T. cruzi* by direct methods on blood or cerebrospinal fluid [74, 75]. Histology on biopsies of central nervous system, skin, or other organs is also often used, mainly in the frame of a reactivation [76].

During the chronic stage, diagnosis relies on the study of specific IgG antibodies: a patient is considered as infected if their serum shows positive results for at least two different serological techniques. Among these methods, we find ELISA based on total or recombinant parasite antigens, which is the most widespread method [77]. Specific IgM detection methods are not used during the acute phase.

Methods based on the detection of parasitic DNA, like PCR or LAMP [78, 79], are currently applied as "in-house" techniques, with the inconveniences that this entails. PCR is usually utilized as efficacy endpoint in trials of new anti-parasitic drugs [80]. In Latin America, given the infrastructural limitations, it is only used for research activities in a small number of medical services. The enormous asymmetry between health systems across the places where Chagas disease patients live should also be kept in mind. In urbanized centers, especially in Europe and the United States, it is relatively easy to have access to these high confidence level methods. This is not the case in rural endemic areas in Latin America, where direct parasitological methods and conventional serology are the essential tools.

New biomarkers which may help assessing the efficacy of treatment and the possibility of developing heart disease in infected asymptomatic patients are under development [81].

5 Cure in Chagas Disease

The real criterion of cure is the demonstration that a patient who received anti-parasitic treatment does not develop heart and/or digestive disease. This is hard to prove, given that a decades-long clinical follow-up is required [82]. Another possibility is to demonstrate that a patient treated with anti-parasitic drugs when they already have a certain degree of disease does not progress. We believe that children receiving anti-parasitic treatment will not develop cardiac disease during adulthood, but this needs to be demonstrated. Studying this cohort starting at age 20 years is troublesome.

The other criterion is demonstrating parasitological cure. This relies on two main tools: demonstrating serological negativization and/or negative PCR in numerous determinations after treatment ends [83]. Regarding the conventional serological negativization, it may take place months or decades later, depending on the age at which treatment is received. In adults it usually takes many years, while in children this time is shorter. The younger the patient, the faster negativization occurs. Another possibility is the use of parasite fragment-based serological methods [84]. Published data demonstrate that using these antigens, serological negativization is detected earlier than with the conventional serological method.

6 Anti-parasitic Drugs

Drugs currently available (benznidazole and nifurtimox) are effective against the parasite, especially in pediatric patients. There is still no consensus regarding the dose and times to be used [85], due to the lack of thorough clinical assays. Fortunately, within the last 10 years, pharmacokinetics studies have been set up which will allow dose adjustment for different ages and clinical situations [86]. These medicines usually have side effects of diverse severity, which requires close medical surveillance, especially during the first 20 days of treatment [87]. There are many clinical situations in which its usage needs to be resolved; there is an increasing accordance regarding its application during lactation. Keep in mind that the younger the patient, the better the efficacy of the treatment and the less adverse effects. Drugs used for other infectious diseases are under study, seeking to assess their anti-parasitic effect. The Drugs for Neglected Diseases Initiative plays a central role in these activities. An encouraging fact that secures the existence of the existing therapies is that the Argentinean laboratory Elea is currently producing benznidazole, and it has recently been approved by the FDA for pediatric application [88]. Recently, other Argentinean laboratory, Bayer is producing Nifurtimox. Another very important fact recently demonstrated is that young and teenage girls and adult women receiving anti-parasitic treatment avoid the possibility of giving birth to a child with transplacentally acquired Chagas disease.

7 Other Drugs?

Given that other factors are involved in cardiac pathogenesis, like immune system alterations and myocardial microcirculation disorders, it would be necessary to attempt the application of other medications, along with anti-parasitic drugs, which may act on these mechanisms in order to avoid the development of the cardiac lesion [89].

8 Endemic Rural Area

The comprehensive approach to this pathology in that region of Latin America is still a very complex issue. It is highly necessary to increase insecticide spraying in the dwellings in order to further decrease vector load. Ulterior surveillance actions are essential to detect resistant vectors and to direct respraying campaigns where vector presence is persistent. Involvement of the communities across all ages is required for this mission. These activities should be permanent and sustained; otherwise homes are infested anew [90].

Medical care of patients in local health systems is insufficient. Participation of several nongovernmental organizations has been crucial: Doctors Without Borders [91], Plan International, Jica, Bunge y Born foundation [92], and Mundo Sano foundation [93].

9 Patient Care

We are convinced that these patients should be taken care of by physicians in different levels of attention but mainly those in the primary level who are in narrowest contact with the patients. Specialized physicians must assist patients with complications: the infectologist should take up caring for those with Chagas disease and some immunodeficiency, while patients with severe cardiopathy should be assisted in cardiology services and those with digestive alterations, by gastroenterologists.

In order to increase detection of cases, the health system should adopt a proactive role on populational studies.

There are two extremely simple actions a physician can incorporate at their consulting room to enhance the detection of new patients. Thirty seconds should be dedicated to asking background questions regarding risk factors (permanence within endemic areas, having relatives diagnosed with Chagas disease, having received a blood transfusion). Serological studies should be solicited upon the existence of any of these factors. Another relevant point for this aim is the incorporation of a family approach to the task: every time a person is detected to have the infection, the rest of the family group should also be studied.

Anti-parasitic treatment must be supervised especially during the first 20 days, within the time when side effects usually appear [87]. Requiring a pregnancy test should be mandatory for teenage girls and adult women of fertile age before beginning anti-parasitic treatment.

Usually, when a physician has a Chagas disease patient under their care, the immediate thought is the risk of cardiopathy. This is not wrong, but neither is enough. The physician should pay attention to digestive pathology which, even when producing less morbidity, may be present with or without cardiac disorder [94].

The International Federation of People Affected by Chagas Disease (FINDECHAGAS) groups numerous associations from various countries and aims at enhancing the visibility of this endemic disease.

10 New Research

Several lines of research, comprising diverse aspects of biology, pharmacology, clinic, and social sciences, should be further explored to improve knowledge about this disease. Drugs used for other infectious agents are being evaluated. The fact that immunity factors are involved in the cardiac pathogenesis, led to the interest on developing vaccines aiming at modifying the immune response [95, 96]; the use of nanoparticles with anti-parasitic drugs for a better therapeutic response [97]; the continuation of the studies on the invasion mechanisms and metabolism of this complex and fascinating protozoan, the immune response; the evaluation of new insecticides; and the improvement of diagnostic tests (quick tests) for populational studies.

Chagas disease is very complex and generates morbidity, disability, high costs in health, and mortality. Up to today considerable success has been achieved, but much is yet to be done. It is fundamental to increase the acknowledgment of this disease among the members of the health system, in different stages of education, to generate a greater commitment. It is a responsibility of the governments and the health system to adopt a more emphatic approach; **the reality of Chagas disease would be remarkably different if everything already known would be translated into practice.**

References

1. Hotez PJ, Bottazzi ME, Franco-Paredes C, Ault SK, Rosrs Periago M. The neglected tropical diseases of Latin America and the Caribbean: a review of disease burden and distribution and a roadmap for control and elimination. PLoS Negl Trop Dis. 2008;2:e300.
2. World Health Organization. Sustaining the drive to overcome the global impact of neglected tropical diseases: second WHO report on neglected diseases. Geneva: WHO Press; 2013.
3. Lescure FX, Le Loup G, Freilij H, Develoux M, Paris L, Brutus L, et al. Chagas disease: changes in knowledge and management. Lancet Infect Dis. 2010;10(8):556–70.
4. Rassi A Jr, Rassi A, Marcondes de Rezende J. American trypanosomiasis (Chagas disease). Infect Dis Clin North Am. 2012;26:275–91.
5. Dias JSP, Machado EMM, Borges EC, et al. Doenca de Chagas em Lassance MG. Revaliacao clinic-epidemiologica 90 anos apos a descoberta de Carlos Chagas. Rev Soc Bras Med Trop. 2002;35:167–76.
6. Prata A. Clinical and epidemiological aspects of Chagas disease. Lancet Infect Dis. 2000;1:92–100.
7. Felman AM, Mac Namara D. Myocarditis. N Engl J Med. 2000;343:1388–98.
8. Carlier Y, Torrico F, Sosa-Estani S, Russomando G, Luquetti A, Freilij H, et al. Congenital Chagas disease: recommendations for diagnosis, treatment and control of newborns, siblings and pregnant women. PLoS Negl Trop Dis. 2011;5:e1250.
9. Freilij H, Altcheh J. Congenital Chagas' disease. Diagnostic and clinical aspects. Clin Infect Dis. 1995;21:551–5.
10. Segura EL, Sosa Estani S, Esquivel ML, G6mez A, Salomon OD. Control of the transmission of Trypanosoma cruzi in Argentina 1999. Medicina. 1999;59(Suppl 2):91–6.
11. Torrico F, Alonso-Vega C, Suarez E, Rodríguez P, Torrico MC, Dramaix M, et al. Endemic level of congenital Trypanosoma cruzi infection in the areas of maternal residence and the development of congenital Chagas disease in Bolivia. Rev Soc Bras Med Trop. 2005;38(Suppl 2):17–20.

12. Sánchez Negrette O, Mora MC, Basombrío MA. High prevalence of congenital Trypanosoma cruzi infection and family clustering in Salta, Argentina. Pediatrics. 2005;115(6):e668–72.
13. Biancardi MA, Torres N, Pepe C, Altcheh J, Freilij H. Seroprevalencia de la enfermedad de Chagas en 17 parajes del "Monte Impenetrable" de la Provincia del Chaco. Medicina. 2003;63:125. ISSN 1669-9106.
14. Chuit R, Subias E, Perez AC, Paulone I, Wisnivesky-Colli C, Segura EL. Usefulness of serology for the evaluation of Trypanosome cruzi transmission in endemic areas of Chagas' Disease. Rev Soc Brazil Trop Med. 1989;22(3):119–29. ISSN: 0037-8682-1989.
15. Viotti R, Vigliano C. Etiological treatment of chronic Chagas disease: neglected evidence-base medicine. Expert Rev Anti Infect Ther. 2007;5:717–26.
16. Strasen J, Williams T, Ertl G, et al. Epidemiology of Chagas disease in Europe: many calculations, little knowledge. Clin Res Cardiol. 2014;103:1–10.
17. Schmunis GA, Yadon ZE. Chagas disease: a Latin American health problem becoming a world health problem. Acta Trop. 2010;115:14–21.
18. Hotez PJ, Dumonteil E, Betancourt Cravioto M, Borrazi ME, Tapia-Conyer R, et al. An unfolding tragedy of Chagas disease in North America. PLoS Negl Trop Dis. 2013;7:e2300.
19. Guerri-Guttenberg RA, Ciannameo A, Di Girolamo C, Milei JJ. [Chagas disease: an emerging public health problem in Italy?]. Infez Med 2009;17:5–13. In Italian.
20. Chin-Hong PV, Schwartz BS, Bern C, et al. Screening and treatment of Chagas disease in organ transplant recipients in the United States: recommendations from the Chagas in Transplant Working Group. Am J Transplant. 2011;11:672–80.
21. Bisio M, Altcheh J, Lattner J, Moscatelli G, Fink V, Burgos JM, et al. Benznidazole treatment of chagasic encephalitis in pregnant woman with AIDS. Emerg Infect Dis. 2013;19(9):1490–2. https://doi.org/10.3201/eid1909.130667.
22. Soriano-Arandes A, Angheben A, Serre-Delcor N, Treviño-Maruri B, Gómez I, Prat J, Jackson Y. Control and management of congenital Chagas disease in Europe and other non-endemic countries: current policies and practices. Trop Med Int Health. 2016;21(5):590–6. https://doi.org/10.1111/tmi.12687.
23. Coura JR, Abad-Franch F, Aguillera X, Dias JCP, Gil H, Junqueira ACV, et al. The initiative for the control of Chagas disease in the Americas and in non-endemic countries. Rev Soc Bras Med Trop. 2009;42(Suppl. 2):106–10.
24. Bisio M, Seidenstein E, Burgos JM, Ballering B, Risso M, Marcelo Moreau M, et al. Urbanization of congenital transmission of Trypanosoma cruzi: prospective polymerase chain reaction study in pregnancy. Trans R Soc Trop Med Hyg. 2011;105(10):543–9. https://doi.org/10.1016/j.trstmh.2011.07.003.
25. Alarcón de Noya B, Díaz-Bello Z, Colmenares C, Ruiz-Guevara R, Mauriello L, Muñoz-Calderón A, Noya O. Update on oral Chagas disease outbreaks in Venezuela: epidemiological, clinical and diagnostic approaches. Mem Inst Oswaldo Cruz. 2015;110(3):377–86.
26. Shikanai-Yasuda MA, Carvalho NB. Oral transmission of Chagas disease. Clin Infect Dis. 2012;54:845–52.
27. Castro E. Chagas' disease: lessons from routine donation testing. Transfus Med. 2009;19(1):16–23. https://doi.org/10.1111/j.1365-3148.2009.00915.x.
28. Bern C, Montgomery SP, Herwaldt BL, Rassi A Jr, Marin-Neto JA, Dantas RO, et al. Evaluation and treatment of Chagas disease in the United States: a systematic review. JAMA. 2007;298:2171–81.
29. Carrilero B, Murcia L, Martinez-Lage L, Segovia M. Side effects of benznidazole treatment in a cohort of patients with Chagas disease in non-endemic country. Rev Esp Quimioter. 2011;24:123–6.
30. Perez-Ayala A, Perez-Molina JA, Norman F, et al. Chagas disease in Latin American migrants: a Spanish challenge. Clin Microbiol Infect. 2011;17:1108–13.
31. Leiby D, Nguyen ML, Proctor MC, Rebecca L, Townsend RL, Stramer SL. Frequency of Trypanosoma cruzi parasitemia among infected blood donors with a potential association between parasite lineage and transfusion transmission. Transfusion. 2017;57:1426–32. https://doi.org/10.1111/trf.14082.

32. Benjamin RJ, Stramer SL, Leiby DA, et al. Trypanosoma cruzi infection in North America and Spain: evidence in support of transfusion transmission. Transfusion. 2012;52:1913–21.
33. Guerri-Guttenberg RA, Grana DR, Ambrosio G, Milei J. Chagas cardiomyopathy: Europe is not spared. Eur Heart J. 2008;29:2587–91.
34. Jackson Y, Getaz L, Wolff H, Mauris A, Tardini A, Sztajzel J, et al. Prevalence, clinical staging and risk for blood-borne transmission of Chagas disease among Latin American migrants in Geneva, Switzerland. PLoS Negl Trop Dis. 2010;4(2):e592. https://doi.org/10.1371/journal.pntd.0000592.
35. Kessler DA, Shi PA, Avecilla ST, Shaz BH. Results of lookback for Chagas disease since the inception of donor screening at New York Blood Center. Transfusion. 2013;53:1083–7.
36. Basile L, Jansa JM, Carlier Y, Salamanca DD, Angheben A, Bartoloni A, et al. Chagas disease in European countries: the challenge of a surveillance system. Euro Surveill. 2011;16(37):pii: 19968.
37. Taquini AC, Lozada BB, Furman B. Cardiopatía chagásica crónica. Medicina. 1952;12:123–48.
38. Benchimol AB. Doenca de Chagas nos grandes centros urbanos. In: Annais do Congreso International sóbre a doenca de Chagas, Vol I, Rio de Janeiro, 1961, 189-203.
39. Romaña C. Acerca de un síntoma inicial de valor para el diagnóstico de forma aguda de la enfermedad de Chagas. MEPRA. 1935;22:16–28.
40. Chagas C. Nova Trypanosomiase humana. Estudos sobre a morfolojia e o ciclo evolutivodo Schizotrypanun cruzi, agente etiologicode nova entidade mórbida do homem. Mem Inst Oswaldo Cruz. 1909;1:159–218.
41. Sá MR. The history of tropical medicine in Brazil: the discovery of *Trypanosoma cruzi* by Chagas and the German school of protozoology. Parasitologia. 2005;47:309–17.
42. Mazza S. La enfermedad de Chagas en la República Argentina. Mem Inst Oswaldo Cruz. 1949;47:273–88.
43. Neiva A, Lent H. Triatomideos do Chile. Mem Inst Oswaldo Cruz. 1943;1:43–75.
44. Araujo A, Jansen AM, Reinhard K, Ferreira LF. Paleoparasitology of Chagas disease: a review. Mem Inst Oswaldo Cruz. 2009;104(Suppl I):9–16.
45. Aufderheide AC, Salo LW, Madden M, Streitz J, Ghul F, et al. A 9000 year record of Chagas disease. Proc Natl Acad Sci U S A. 2004;101(7):2034–9.
46. Deane MP, Lenzi HL, Jansen A. *Trypanosoma cruzi*: vertebrate and invertebrate cycles in the same mammal host, the opossum Didelphis marsupialis. Mem Inst Oswaldo Cruz. 1984;79:513–5.
47. Lenzi HL, Jansen AM, Deane MP. The recent discovery of what might be a primordial escape mechanism for Trypanosoma cruzi. Mem Inst Oswaldo Cruz. 1984;79:13–8.
48. Galvão C, Carvalho RU, Rocha DS, Juberg J. A check-list of the current valid species of the subfamily Triatominae Jeannel, 1919 (Hemiptera, Reduviidae) and their geographical distribution, with nomenclatural and taxonomic notes. Zootaxa. 2003;2002:1–36.
49. Stevens JR, Noyes HA, Dover GA, Gibson WG. The ancient and divergent origins of the human pathogenis trypanosomes, *Trypanosoma brucei* and *Trypanosoma. cruzi*. Parasitology. 1999;116:107–16.
50. El Sayed NM, Myler PJ, Bartholomeu DC, Nilsson D, Aggarwal G, Tran NA. The genome sequence of *Trypanosoma cruzi*, etiologic agent of Chagas disease. Science. 2005;309(5733):409–15.
51. Zingales B, Andrade S, Briones R, Campbell D, Chiari E, Fernandes O, et al. A new consensus for *Trypanosoma cruzi* intrespecific nomenclature: second revision meeting recommends TcI to TcVI. Mem Inst Oswaldo Cruz. 2009;1047:1051–4.
52. Araujo Perla F, Almeida Adriana B, Pimentel Carlos F, Silva Adriano R, Alessandro S, Valente Sebastião A, et al. Sexual transmission of American trypanosomiasis in humans: a new potential pandemic route for Chagas parasites. Mem Inst Oswaldo Cruz. 2017;112(6):437–46.
53. Gürtler RE. Sustainability of vector control strategies in the Gran Chaco Region: current challenges and possible approaches. Mem Inst Oswaldo Cruz. 2009;104(Suppl I):52–9.
54. Estimación cuantitativa de la enfermedad de Chagas en las Américas. OPS/HDM/CD/425-06. OPS/OMS.

55. Gurtler RE, Yadon ZE. Eco-bio-social research on community-based approaches for Chagas disease vector control in Latin America. Trans R Soc Trop Med Hyg. 2015;109(2):91–8. https://doi.org/10.1093/trstmh/tru203.
56. Henao-Martínez A, Colborn K, Parra-Hena G. Overcoming research barriers in Chagas disease—designing effective implementation science. Parasitol Res. 2017;116:35. https://doi.org/10.1007/s00436-016-5291-z.
57. Sartor P, Colaianni I, Cardinal MV, Bua J, Freilij H, Gurtler RE. Improving access to Chagas disease diagnosis and etiologic treatment in remote rural communities of the Argentine Chaco through strengthened primary health care and broad social participation. PLoS Negl Trop Dis. 2017;11:e0005336. https://doi.org/10.1371/journal.pntd.0005336.
58. Germano MD, Picollo MI. Demographic effects of deltamethrin resistance in the Chagas disease vector Triatoma infestans. Med Vet Entomol. 2016;30(4):416–25. https://doi.org/10.1111/mve.12196.
59. Dao L. Otros casos de enfermedad de Chagas en el Estado de Guarico (Venezuela) Observación sobre enfermedad de Chagas congénita. Rev Policlin Caracas. 1949;17:17–32.
60. Rueda K, Trujillo JE, Carranza JC, Vallejo GA. Transmisión oral de Trypanosoma cruzi: una nueva situación epidemiológica de la enfermedad de Chagas en Colombia y otros países suramericanos. Rev Biomédica. 2014;34(4):631.
61. Alarcón de Noya B, Díaz-Bello Z, Colmenares C, Ruiz R, Noya O, Zavala-Jaspe R, et al. Large urban outbreak of orally acquired acute Chagas disease at a school in Caracas, Venezuela. J Infect Dis. 2010;201:1308–15. https://doi.org/10.1086/651608.
62. Hernandez-Becerril N, Mejia AM, Ballinas-Verdugo MA, et al. Blood transfusion and iatrogenic risks in Mexico City. Anti-Trypanosoma cruzi seroprevalence in 43,048 blood donors, evaluation of parasitemia, and electrocardiogram findings in seropositive. Mem Inst Oswaldo Cruz. 2005;100:111–6.
63. Schmunis GA, Cruz JR. Safety of the blood supply in Latin America. Clin Microbiol Rev. 2005;18:12–29.
64. Schmunis GA. Trypanosoma cruzi, the etiologic agent of Chagas' disease: status in the blood supply in endemic and nonendemic countries. Transfusion. 1991;31:547–57.
65. Cerisola JA, Rabinovich A, Alvarez M, Corletto CA, Prumeda J. [Chagas' disease and blood transfusion]. Bol Oficina Sanit Panam 1972;73:203–221. In Spanish.
66. World Health Organization. WHO Consultation on International Biological Reference Preparations for Chagas Diagnostic Tests; Geneva. 2–3 July 2007. Accessed on 14 Feb 2015. Available at: http://www.who.int/bloodproducts/ref_materials/WHO_Report_1st_Chagas_BRP_consultation_7-2007_final.pdf.
67. Schmunis GA. Epidemiology of Chagas disease in non-endemic countries: the role of international migration. Mem Inst Oswaldo Cruz. 2007;102(Suppl 1):75–85.
68. Angheben A, Boix L, Buonfrate D, Gobbi F, Bisoffi Z, Pupella S, et al. Chagas disease and transfusion medicine: a perspective from non-endemic countries. Blood Transfus. 2015;13(4):540–50.
69. Cordova E, Boschi A, Ambrosioni J, Cudos C, Corti M. Reactivation of Chagas disease with central nervous system involvement in HIV-infected patients in Argentina, 1992-2007. Int J Infect Dis. 2008;12(6):587–92. https://doi.org/10.1016/j.ijid.2007.12.007.
70. Kun H, Moore A, Mascola L, Steurer F, Lawrence G, Kubak B, et al. Transmission of *Trypanosoma cruzi* by heart transplantation. Clin Infect Dis. 2009;48(11):1534–40.
71. Cura CI, Lattes R, Nagel C, Gimenez MJ, Blanes M, Calabuig E, et al. Early molecular diagnosis of acute Chagas disease after transplantation with organs from Trypanosoma cruzi-infected donors. Am J Transplant. 2013;13(12):3253–61.
72. http://www.sac.org.ar/wp-content/uploads/2014/04/Consenso-de-Enfermedad-de-Chagas-Mazza.pdf.
73. http://www.fac.org.ar/1/revista/11v40n3/consenso/chagas/mordini.php.
74. Freilij H, Muller Gonzalez Cappa DM. Direct micromethod for diagnosis of acute and congenital Chagas disease. J Clin Mocrobiol. 1983;18:327–30.
75. Strout RG. A method for concentrating hemoflagellates. J Parasitol. 1962;48:100.

76. Pitella JE. Central nervous system involvement in Chagas disease: a hundred-year-history. Trans R Soc Trop Med Hyg. 2009;103:973–8.
77. Moure Z, Angheben A, Molina I, Gobbi F, Espasa M, Anselmi M, et al. Serodiscordance in chronic Chagas disease diagnosis: a real problem in non-endemic countries. Clin Microbiol Infect. 2016;22(9):788–92. https://doi.org/10.1016/j.cmi.2016.06.001.
78. Rivero R, Bisio M, Velazquez EB, Esteva MI, Scollo K, Gonzalez NL, et al. Rapid detection of Trypanosoma cruzi by colorimetric loop-mediated isothermal amplification (LAMP): a potential novel tool for the detection of congenital Chagas infection. Diagn Microbiol Infect Dis. 2017;89(1):26–8.
79. Schijman AG, Altcheh J, Burgos JM, Biancardi M, Bisio M, Levin MJ, et al. Aetiological treatment of congenital Chagas' disease diagnosed and monitored by the polymerase chain reaction. J Antimicrob Chemother. 2003;52(3):441–9.
80. Morillo CA, Waskin H, Sosa-Estani S, Del Carmen Bangher M, Cuneo C. Benznidazole and posaconazole in eliminating parasites in asymptomatic T. cruzi carriers: the STOP-CHAGAS trial. J Am Coll Cardiol. 2017;69(8):939–47. https://doi.org/10.1016/j.jacc.2016.12.023.
81. Pinazo MJ, Thomas MC, Bua J, Perrone A, Schijman A, Viotti R, et al. Biological markers for evaluating therapeutic efficacy in Chagas disease, a systematic review. Expert Rev Anti Infect Ther. 2014;12(4):479–96.
82. Viotti R, Vigliano C, Lococo B, Bertocchi G, Petti M, Alvarez MG, et al. Long-term cardiac outcomes of treating chronic Chagas disease with benznidazol versus no treatment: a nonrandomized trial. Ann Intern Med. 2006;144:724–34.
83. Britto C, Cardoso MA, Vanni CM, Hasslocher A, Xaviert SS, Oelemann N, et al. Polymerase chain reaction detection of Trypanosoma cruzi in human blood samples as a tool for diagnosis and treatment evaluation. Parasitology. 1995;110(Pt 3):241–7.
84. Altcheh J, Corral R, Biancardi MA, Freilij H. Anti-F2/3 antibodies as cure marker in children with congenital Trypanosoma cruzi infection. Medicina. 2003;63(1):37–40.
85. Sales Junior PA, Molina I, Fonseca Murta SM, Sánchez-Montalvá A, Salvador F, Corrêa-Oliveira R, Carneiro CM. Experimental and clinical treatment of Chagas disease: a review. Am J Trop Med Hyg. 2017;97(5):1289–303. https://doi.org/10.4269/ajtmh.16-0761.
86. Altcheh J, Moscatelli G, Mastrantonio G, Moroni S, Giglio N, Marson ME, et al. Population pharmacokinetic study of benznidazole in pediatric Chagas disease suggests efficacy despite lower plasma concentrations than in adults. PLoS Negl Trop Dis. 2014;8(5):e2907.
87. Altcheh J, Moscatelli G, Moroni S, Garcia-Bournissen F, Freilij H. Adverse events after the use of benznidazole in infants and children with Chagas disease. Pediatrics. 2011;127(1):e212–8.
88. http://outbreaknewstoday.com/chagas-disease-benznidazole-first-treatment-approved-fda-63299/.
89. Rodrigues Silva R, Shrestha-Bajracharya D, Almeida-Leite CM, Leite R, Bahia MT, Talvani A. Short-term therapy with simvastatin reduces inflammatory mediators and heart inflammation during the acute phase of experimental Chagas disease. Mem Inst Oswaldo Cruz. 2012;107(4):513–21.
90. Bianchi F, Cucunubá Z, Guhl F, González NL, Freilij H, Nicholls R, et al. Follow-up of an asymptomatic Chagas disease population of children after treatment with nifurtimox (Lampit) in a sylvatic endemic transmission area of Colombia. PLoS Negl Trop Dis. 2015;9(2):e0003465. https://doi.org/10.1371/journal.pntd.0003465.
91. Yun O, Lima MA, Ellman T, Chambi W, Castillo S, Flevaud L, et al. Feasibility, drug safety, and effectiveness of etiological treatment programs for Chagas disease in Honduras, Guatemala, and Bolivia: 10-year experience of Médecins Sans Frontières. PLoS Negl Trop Dis. 2009;3(7):e488. https://doi.org/10.1371/journal.pntd.0000488.
92. Dreyer C, Armenti HA, Gurtler RE, Freilij H. Curso de educación a distancia, Chagas: del conocimiento a la acción. Fundación Bunge y Born. Available at: www.mundosano.org/files/web.mundosano.org/9513/5229/8630/Dreyer.pdf.
93. Lenardón M, Orsini P, Chopita M, Ramos P, Da Cruz AP, Suárez Crivaro F, et al. Chagas in a non-endemicarea: first level health care. Lights and shadows. PEAH – Policies for Equitable

Access to Health. Accessed on 29 Sep 2016. Available at: http://www.peah.it/2014/11/chagas-in-a-non-endemic-area-first-level-health-care-lights-and-shadows/.
94. Pinazo MJ, Lacima G, Elizalde JI, Posada EJ, Gimeno F, Aldasoro E, et al. Characterization of digestive involvement in patients with chronic T. cruzi infection in Barcelona, Spain. PLoS Negl Trop Dis. 2014;8(8):e3105. https://doi.org/10.1371/journal.pntd.0003105.
95. Biter AB, Weltje S, Hudspeth EM, Seid CA, McAtee CP, Chen WH, et al. Characterization and stability of Trypanosoma cruzi 24-C4 (Tc24-C4), a candidate antigen for a therapeutic vaccine against Chagas disease. J Pharm Sci. 2018;107:1468. https://doi.org/10.1016/j.xphs.2017.12.014. pii: S0022-3549(17)30884-5.
96. Sanchez Alberti A, Bivona AE, Cerny N, Schulze K, Weißmann S, Ebensen T, et al. Engineered trivalent immunogen adjuvanted with a STING agonist confers protection against Trypanosoma cruzi infection. Vaccine. 2017;2:9. https://doi.org/10.1038/s41541-017-0010-z.
97. Rial MS, Scalise ML, Arrúa EC, Esteva MI, Salomon CJ, Fichera LE. Elucidating the impact of low doses of nano-formulated benznidazole in acute experimental Chagasdisease. PLoS Negl Trop Dis. 2017;11(12):e0006119. https://doi.org/10.1371/journal.pntd.0006119.

Part II
The Agent

Trypanosoma cruzi Journey from the Insect Vector to the Host Cell

Catalina D. Alba Soto and Stella Maris González Cappa

Abstract *Trypanosoma cruzi*, etiological agent of Chagas disease, has evolved a complex interaction with mammalian cells and insect vector's intestine. During its journey between these environments, it is subject to stressful changes. To overcome them, parasites use numerous strategies. Different stages contact diverse compartments of hosts and vectors thus assuring survival and multiplication. Surface proteins, some identified in particular stages of the protozoan, are critical for interaction with the milieu although their relevance is not totally understood for many. Parasite molecules allow *T. cruzi* to progress in the vector intestine, to duplicate and differentiate in order to become the infective stage for mammals. Surface molecules also allow parasites to advance through intracellular matrix of the mammals to reach the cells and, after recognition, invade them and adapt to survive but also to multiply and differentiate to circulating trypomastigotes thus assuring life cycle continuity. In this chapter we summarize *T. cruzi* pathways of humans and other reservoirs of infection as well as the participation of different *T. cruzi* lineages in geographical distribution and human disease. Finally, we review some of the mechanisms used by the parasite to reach, enter, and survive inside the host cell described so far.

Trypanosoma cruzi is the etiological agent of American trypanosomiasis also known as Chagas disease. This is a complex health issue that should neither be merely approached from the understanding of parasite interaction with the insect vector and with the host cell, which is the matter of this chapter, nor from the point of view of patients and their clinical outcomes. Indeed, social and cultural determinants of people that suffer the disease must also be taken into consideration.

C. D. Alba Soto · S. M. González Cappa (✉)
Departamento de Microbiología, Parasitología e Inmunología. Facultad de Medicina, Instituto de Microbiología y Parasitología Médica (IMPaM-UBA/CONICET), Universidad de Buenos Aires, Buenos Aires, Argentina

J. M. Altcheh, H. Freilij (eds.), *Chagas Disease*, Birkhäuser Advances in Infectious Diseases, https://doi.org/10.1007/978-3-030-00054-7_2

1 The Etiological Agent

Trypanosoma cruzi is a protozoan belonging to the Trypanosomatidae family. Other digenetic parasites of medical importance as the genus *Leishmania* are included in this family which is placed in the order Kinetoplastida. Microorganisms of this order are characterized by the presence of a structure called kinetoplast, which contains about 20% of the total DNA distributed in maxi- and minicircles, and a flagellum that is released by the anterior end of the parasite. The flagellum, well developed in the extracellular stages, grows in the proximity of the kinetoplast and emerges through the flagellar pocket with a location in relation to the nucleus which differs according to the parasitic stage. In natural conditions this parasite accomplishes an indirect life cycle, requiring two hosts to complete it: mammals are definitive and triatominae insects intermediate hosts.

1.1 Parasitic Stages

- Trypomastigote: Nondividing elongated form (20–30 μm) with a vesicular nucleus and the kinetoplast located behind it. The flagellum grows near the kinetoplast and emerges from the side of parasite's body running under the cytoplasmic membrane to be released by the anterior end, thus giving the image of an extensive undulating membrane. This parasitic form is found in the insect vector and in the mammal.
    (a) Metacyclic trypomastigote: The result of epimastigote differentiation at the distal portion of the vector's intestine or rectum. Being the infective form for the mammalian host, it is deposited with the feces of the insect and penetrates mucous membranes or skin lesions. Inside the host cell, this stage differentiates to the amastigote.
    (b) Blood trypomastigote: The infective form for both the insect vector and the mammalian host. It differentiates from the intracellular amastigote and can disseminate through the bloodstream to invade new cells. It enters the vector with the ingested blood from an infected mammal and differentiates to epimastigote in the digestive tract to begin the cycle in the intermediate host.
- Epimastigote: Duplicative form that divides by binary fission, noninfective to the mammalian host and found in the midgut of the insect vector. It presents a fusiform aspect (20–30 μm in length), with a kinetoplast located between the nucleus and the flagellum. This last one grows next to the kinetoplast and emerges from the membrane body free in almost all its extension thus having a short undulating membrane.
- Spheromastigote: This is an extracellular stage that differentiates in the midgut of the insect vector and divides by binary fission. It measures between 2 and 4 μm being morphologically similar to the amastigote but with a free flagellum

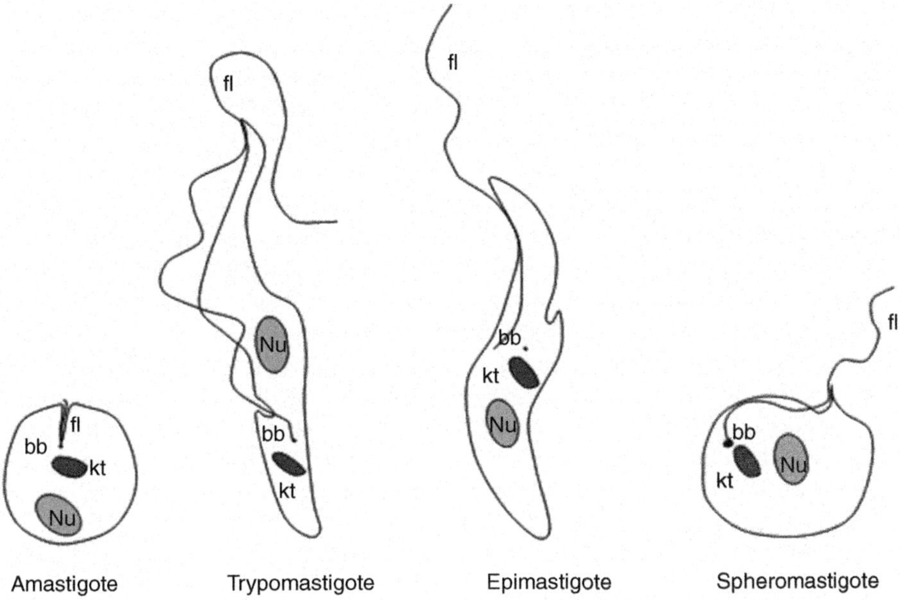

Fig. 1 *Trypanosoma cruzi* parasitic stage morphology. Nu, nucleus; kt, kinetoplast; bb, basal body; fl, flagellum

that emerges from, and surrounds, the parasite body. This form is usually found in reduced numbers but increases under stress conditions. Epimastigotes also differentiate from these spheromastigotes while progressing toward the distal portion of the triatomine intestine (reviewed by [1]; reviewed by [2]).

- Amastigote: The intracellular duplicative stage found in the mammalian host. It originates from the differentiation of metacyclic and blood trypomastigotes in the cytoplasm of infected host cells. It has a rounded shape, a large nucleus, a kinetoplast, and a short flagellum sequestered inside a flagellar pocket. Measuring 2–2.5 μm, it multiplies by binary fission.
- Scheme of the most relevant *T. cruzi* stages can be seen in Fig. 1.

1.2 Parasite Body

The parasite body is delimited by a plasma membrane formed by a lipid bilayer. Most of the surface proteins are inserted through a glycosylphosphatidylinositol anchor which is a distinctive feature of the trypanosome parasites [3]. Under the inner leaflet, a network of subpellicular microtubules constituted by filaments of actin and tubulin organize the cytoskeleton. The membrane participates in complex functions such as cell differentiation, motility, and tissue migration. These protozoa possess well-developed endoplasmic reticulum, ribosomes, and Golgi apparatus.

The kinetoplast is within the only mitochondria that runs through the cytoplasm of the protozoa. They also possess other specialized organelles: glycosomes that concentrate and compartmentalize the enzymes of the glycolytic pathway [4] and acidocalcisomes, with acidic pH and high content of calcium ions, that participate in the maintenance of intracellular pH and osmoregulation [5, 6].

The flagellar pocket from which the flagellum emerges is formed by a plasma membrane invagination. The flagellum has a typical structure of nine pairs of peripheral microtubules and a central pair and is associated with the basal corpuscle which in turn is composed of nine triplets of peripheral microtubules. The plasma membrane which constitutes the flagellar pocket does not present the subpellicular microtubule network. In this region and in the cytostome, another invagination of the plasma membrane found in some stages of the parasite, pinocytic vesicles can be observed that allow the incorporation of macromolecules and other substances into the cellular cytoplasm.

Considering that the parasite must adapt to diverse habitats such as insect vector's intestine and mammalian host's circulation and tissues, it is reasonable that each stage has its own antigenic, chemical, physiological, and morphological characteristics.

1.3 Taxonomic Classification

Protist kingdom
Subkingdom: Protozoa
Phylum: Sarcomastigophora
Class: Zoomastigophora
Order: Kinetoplastida
Family: Trypanosomatidae
Genus: *Trypanosoma*
Section: Stercoraria
Subgenera: *Schizotrypanum*
Species: *Trypanosoma cruzi*

According to Hoare [7], the genus *Trypanosoma* is divided into two sections: Salivaria, which comprises pathogenic species whose infective form is transmitted in saliva of the insect vectors (African trypanosomes), and Stercoraria, whose species complete their evolutionary cycle in the hindgut and are transmitted with the feces of the insect vectors. The species of the Stercoraria section are nonpathogenic except for *T. cruzi*.

1.4 Intraspecific Variation

Intraspecific variation in *T. cruzi* was first observed by Carlos Chagas [8] that mentioned, among others, differences in the virulence of this parasite recovered from a patient when injected into different experimental hosts as well as in the morphology of the circulating trypomastigotes. Differences in morphology among strains [9–11] as well as those related to virulence and lethal capacity [12, 13] have been confirmed in mouse experimental model of infection. Others, such as tissue tropism [14, 15], modulation of immune response [16, 17], pathological features [18], and placental pattern infection and interference in fertility in a strain-dependent manner [19, 20], had been reported, too.

First attempts to group *T. cruzi* strains were related to characterization of biodemes, zymodemes, and schizodemes. The biodemes were described by Andrade [21, 22], zymodemes by Miles et al. [23], and schizodemes by Morel et al. [24]. In [25], Souto et al., based on DNA markers, defined two major phylogenetic lineages: TcI and TcII. The latter, due to its heterogenicity, was subdivided into five groups (TcIIa–e). An experts meeting revised the *T. cruzi* nomenclature to reach a consensus, and *T. cruzi* lineages were classified into six "discrete typing units" (DTU) [26]. The term DTU was proposed by Tibayrenc in [27] to underline that these discrete genetic entities "can be characterized by given genetic markers or given sets genetic markers." Still, all DTUs are able to infect humans. The geographical distribution of each DTU has been already described [28, 29] in a nice review recently published.

TcI has a great geographic range dispersion in the three American continents and is associated with sylvatic and domestic cycles. In the sylvatic cycle, *Rhodnius* species with arboreal niches are its primary vectors, and mammals with sylvatic habits are the primary hosts. This DTU has been reported to have important genetic heterogenicity. Hernández et al. [30] referred two subdivisions of TcI (TcIDom and TcI sylvatic). In Colombia, they reported as the main DTU TcI and when referred to human infection the predominant genotype was TcIDom during the chronic phase. In México, Central America, and the northern of South America, TcI predominates in the human infection being associated in these regions with cardiomyopathy [31].

TcII has been isolated in the southern and central regions in South America. In humans it has been associated with cardiac and digestive manifestations. In the Southern Cone, this DTU is rarely found except in some areas of central and east of Brazil [28] and is usually related to the domestic rather than with the sylvatic cycle [32].

Regarding DTU TcIII, human infection is extremely rare with only few cases reported in Colombia [28, 31]. It has been isolated from domestic dogs in the northeastern of Argentina, and *Triatoma infestans* is the most feasible vector ([28];· [33]). In the Paraguayan Chaco, Acosta et al. [34] reported that armadillo species are the principal reservoir of this DTU in the sylvatic area.

TcIV is present in North and South America, showing distinct lineages in both geographic areas [35]. It is the secondary agent for Chagas disease in Venezuela and associated with human outbreaks in the Western Brazilian Amazon [31, 36].

In the southern and central areas of the southern South American countries, TcV and TcVI are the most prevalent DTU infecting humans. Both are natural hybrids of lineages II and III and are responsible for chronic Chagas disease as well as the infections by vertical transmission [28, 29]. In the northern of Argentina, Diosque et al. [37] reported five genotypes circulating in the area under study. One of these genotypes was TcV—associated with humans—and the other TcVI, associated with dogs. Authors did not find apparent association between these *T. cruzi* genotypes and the domiciliated vector *T. infestans*. Also in the north of Argentina (Santiago del Estero Province), TcVI was the most frequent DTU in the same vector in rural areas, while TcV predominated in vectors recovered from communities where house infestation was higher [38]. The same group reported, in the Argentinean Chaco, TcVI as the first DTU infecting *T. infestans* and TcV the second. Bugs infected with TcV were captured in domicilies, while those infected with TcVI, both in domicilies and peridomicilies [39]. The initially described geographic restriction of these DTUs to the Southern Cone of South America may not be so strict. In Colombia, even when the most prevalent DTU was TcI, all remaining DTUs have been reported in domestic and sylvatic foci, including TcV and TcVI [40].

Research has been conducted by several authors trying to correlate particular DTUs with the pathogenesis and clinical outcome of the disease. Studies on the field regarding tissue tropisms [41, 42], cardiomyopathy [15], immune response [43], immune suppression [44, 45], cytokines profile [46], and sensitivity/resistance to parasiticide drugs [47–49] have been reported.

However, the complexity of *T. cruzi* population added to the diversity of the genetic background of human hosts makes it difficult to confirm any association [50, 51]. On the other hand, both vector and mammals can be infected with more than one DTU. This might result in misinterpretations due to differences in duplication speed or in tropisms for the diverse parasite populations [28]. Up to date, major progresses have been related to the ecoepidemiology of *T. cruzi* DTUs. Other correlations require additional studies.

1.5 **T. cruzi** ***Surface Molecules***

Several surface proteins have been identified in the different stages of this protozoan, but for a large number of them, their relevance is not yet understood. Alves et al. [52] analyzed *T. cruzi* glycoprotein profile and reported 334 different glycoproteins exclusive of tissue culture-derived trypomastigotes (the in vitro equivalent to the circulating trypomastigote stage) and 170 exclusive of the epimastigote stage. Other authors, studying molecules released by the trypomastigote stage to the culture medium, identified 540 different proteins [53]. Many of these proteins are constitutively or transiently present at the parasite surface, whereas others are secreted to the media. For some, relevance to survival or the biological cycle of the parasite is known, either because of their role in adhesion, host cell recognition, and invasion or in mechanisms of the parasite differentiation. Diverse stage-specific proteins have been identified whose functions are being studied and disclosed.

In Table 1 some relevant molecules of the infective parasitic stages as well as their main features and assigned functions are mentioned.

Table 1 Surface molecules in the different infective parasitic stages of *T. cruzi*

Stage	Molecule	Receptor	Main feature/ function	Reference
Amastigote	SA85 (glycoprotein)	C-type lectin	Uptake by phagocytic cells	Kahn et al. [54]
	ASP-1 (amastigote surface protein-1)	Macrophage mannose receptor	Major surface antigen	Kahn et al. [54], Santos et al. [55]
Trypomastigote	Mucin p45	Galectin 3	Adherence to myoblasts	Turner et al. [56]
	Gp85/TS (trans-sialidases with/ without enzymatic activity)		Cellular invasion Destabilization of the parasitophorous vacuole	Alves and Colli [57]
	Gp85-11 (member of the Gp85 family)	Cytokeratin 18	Adhesion to the extracellular matrix, basal lamina, and host cell membranes	Magdesian et al. [58]
	T. cruzi-TS (trans-sialidase with enzymatic activity)		Virulence factor Cellular invasion Destabilization of the parasitophorous vacuole Immune modulation	Freire-de-Lima et al. [59], Schenkman et al. [60], Rubin-de-Celis et al. [61], Nardy et al. [62]
	TSSA (trypomastigote small surface antigen)		Adhere to non-phagocytic cells The different isoforms might be related with differences in infectivity	Cánepa et al. [63]
	TcTASV (Trypomastigote Alanine Serine Valine-rich protein)		Trypomastigote surface antigen Major cargo of parasite shed vesicles	García et al. [64], Bernabó et al. [65]
Metacyclic trypomastigote	Gp82 (equivalent to Gp85)	Peptide 7 of the gastric mucin	Cell adhesion and invasion Adhesion to extracellular fibronectin	Staquicini et al. [66], Cortez et al. [67], Correa et al. [68], Yoshida [69]
	Gp90		Negative regulation of the invasion process	Yoshida [69], Rodrigues et al. [70]

(continued)

Table 1 (continued)

Stage	Molecule	Receptor	Main feature/ function	Reference
	Gp35/40 (mucin-like molecule)		Secondary role in the process of cellular invasion Protection from the host gastric environment	Yoshida [69]
Present in all parasitic stages	Cruzipain (the most abundant protease; diverse isoforms)		Virulence factor Cellular invasion	Cazzulo [71], San Francisco et al. [72], Stoka et al. [73]

2 Intermediate Host

The first record of triatominaes in America dates from the year 1590, when the priest Ronaldo de Lizarraga described them and their hematophagous habits on its expedition to Argentina [74]. On the other hand, Charles Darwin made a clear account of these bloodsucking insects in his journey through South America. He noted in the diary he kept during his voyage at the Beagle, on the 25th of March 1835: "At night I experienced an attack (for it deserves no less a name) of the *Benchuca* (a species of Reduvius) the great black bug of the Pampas. It is most disgusting to feel soft wingless insects, about an inch long, crawling over one's body. Before sucking they are quite thin, but afterwards become round and bloated with blood, and in this state they are easily crushed" [75].

2.1 *Taxonomic Classification*

Animalia Kingdom
Phylum: Arthropoda
Class: Insecta
Order: Hemiptera
Family: Reduviidae
Subfamily: Triatominae
Genus: *Panstrongylus*, *Triatoma*, *Rhodnius*, etc.

2.2 *Main Species of Vectors and Geographic Distribution*

There are more than 130 bloodsucking Hemiptera of the family Reduviidae that can potentially participate as intermediate hosts of *T. cruzi*, although only a few species are competent vectors [76]. Among the most important for human transmission are those that acquired domiciliary habits as *Triatoma infestans*, *Rhodnius prolixus*, *Panstrongylus megistus*, and *Triatoma dimidiata*.

T. infestans is the main vector in endemic sub-Amazonian areas and countries of the Southern Cone. In Brazil, at the sub-Amazonian Atlantic coast, it coexists with *P. megistus*. *R. prolixus* is distributed in Central America and Colombia, and *T. dimidiata* extends from northern Mexico and Central America to the Pacific coast in Colombia and Ecuador [77]. In Argentina and Chile, the main responsible species for the transmission of *T. cruzi* is *T. infestans*, an anthropophilic vector that extends to the parallel 45° South.

2.3 Biological Cycle of the Insect Vector

The life cycle of triatomines includes egg stages, five nymphal or immature stages that molt into the male and female adults (imagoes) (Figs. 2 and 3). This hematophagous insect feeds through a proboscis (sucking organ). Its digestive system comprehends an initial portion of the intestine (pharynx and esophagus), followed by the middle intestine which is divided into the anterior midgut (stomach where the concentration and lysis of red blood cells begin) and the posterior midgut (where blood is digested and nutrients absorbed) (Fig. 4). The middle intestine is of endodermal origin, ending in the rectum, which, like the pharynx and esophagus, is of ectodermal origin. Having an incomplete metamorphosis, the nymphs share the same habitat and eating habits of the adult stages thus being hematophagous and able to become infected and transmit *T. cruzi*. As in other bloodsucking insects, the salivary glands secrete anticoagulant substances that prevent ingested blood from clotting. Blood feeding is necessary for molting (ecdysis). A full blood meal distends the abdomen, and the mechanical effect acts on neurosecretory cells that stimulate the secretion of ecdysone to the hemolymph, thus regulating the process of ecdysis as well as the levels of the juvenile hormone involved in the growth of nymphal stages [1, 2, 78]. After each meal insects can spend prolonged periods without feeding. Usually only one feed is required between each molting.

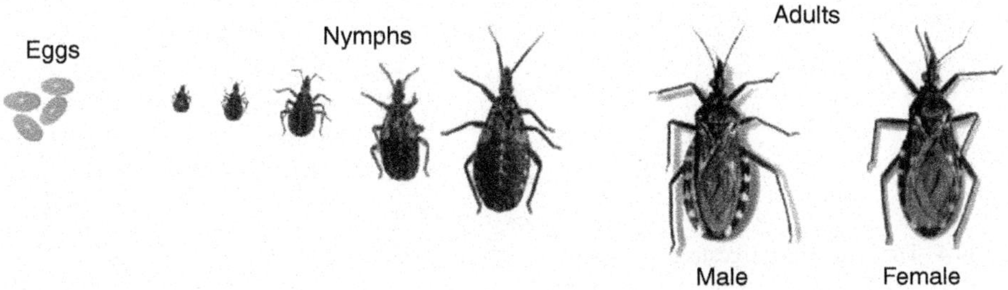

Fig. 2 Evolutive stages of triatomines. From: Ministerio de Salud de la Nación (www.msal.gov.ar)

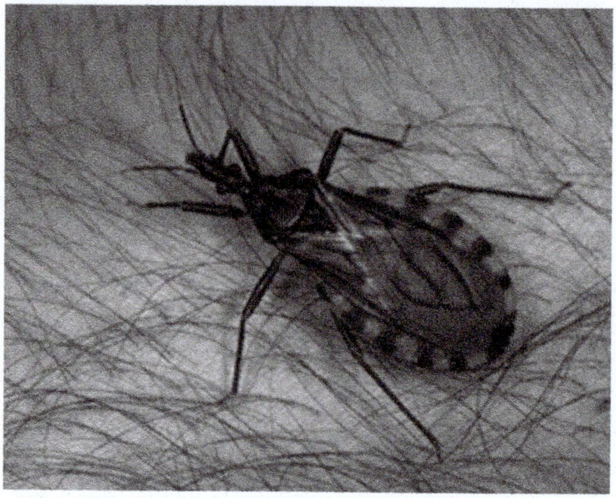

Fig. 3 Adult of *Triatoma infestans*. From: Ministerio de Salud de la Nación (www.msal.gov.ar)

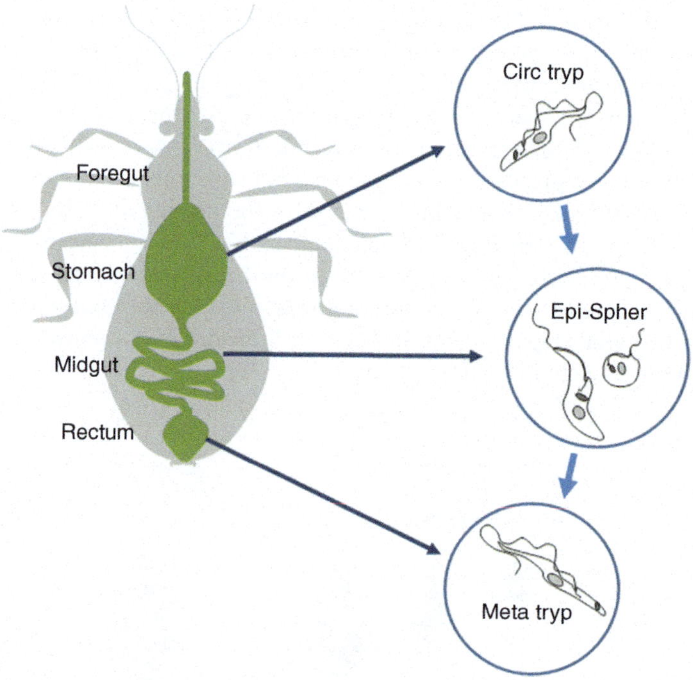

Fig. 4 The triatomine digestive tract and *T. cruzi* differentiation

3 *Trypanosoma cruzi* Life Cycle

The vector can ingest circulating trypomastigotes when it feds on an infected host. Infection depends not only on the number of parasites present in the blood meal but also on the virulence of the ingested parasite population [79, 80]. Blood trypomastigotes differentiate into epimastigotes in the anterior midgut of the insect vector (stomach). Some spheromastigotes may also be found which, like epimastigotes, are duplicative but not infective stages (reviewed by [1]). Epimastigotes progress to the posterior intestine, and, at the level of the rectum, they differentiate into metacyclic trypomastigotes. This process is known as metacyclogenesis (Fig. 5). Metacyclic trypomastigotes are eliminated with the feces of the vector when it defecates while feeding from the definitive host. Infective to mammals, metacyclic trypomastigotes enter the definitive host through the bite, skin lesions, or mucous membranes. Metacyclic trypomastigotes infect phagocytic and non-phagocytic cells near the site of entry. Inside the cells they differentiate to amastigotes which divide actively and subsequently differentiate into trypomastigotes. This stage can enter new cells in the proximity or can reach circulation to disseminate through other tissues or to be ingested by a new vector. Following penetration into the cells, trypomastigotes are surrounded by a parasitophorous vacuole in the host cell cytoplasm. Immediately, the process of transformation to the amastigote stage begins, together with the disruption of the membrane vacuole. Subsequently, the amastigote, free in the cytoplasm, begins the process of binary cell division that continues for several days, depending of the *T. cruzi* strain [81]. Then the process of differentiation of these forms into trypomastigotes begins. In this transition the parasite acquires a free and long flagellum being the activity of motile parasitic form apparently responsible for the cell rupture and its release to the intercellular space [82]. The cycle restarts when a triatomine is fed on the infected mammal.

3.1 Domestic Cycle

T. cruzi household life cycle occurs between man or domestic animals (such as dogs, cats, rodents) and insects with domiciliary habits (such as *T. infestans*). This cycle takes place in endemic areas in the context of precarious dwellings. In these houses all vector stages raise in holes, wall irregularities, and ceilings depending on the material used (straw, palm leaves, wood) or are located behind pictures and furniture. Adults and nymphal stages leave these places at night to feed. Dogs are ideal reservoirs for the maintenance of the cycle since, in general, they present high parasitemia and cohabit with man and insects inside the house [83–85].

Vector control to interrupt the domiciliary cycle has been addressed in the American region in collaboration between the countries of the region and PAHO. With the implementation of several initiatives (Southern Cone, Central

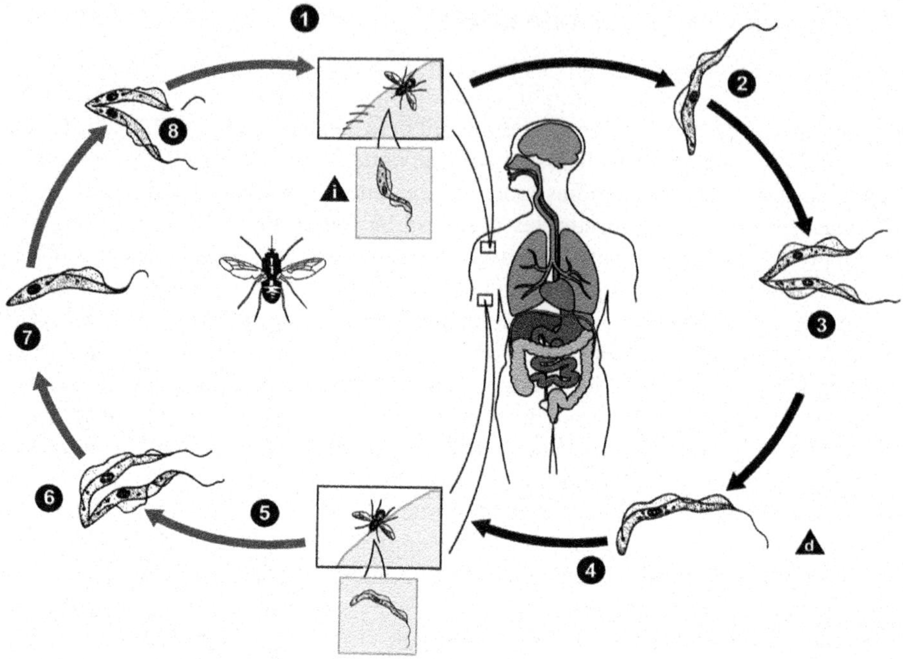

Fig. 5 Biological cycle of *T. cruzi*. (1) Triatomine bug takes a meal and passes metacyclics in feces; trypomastigotes enter through a wound or through mucosal membrane such as conjunctiva. (2) Metacyclic trypomastigotes penetrate various cells at bite wound site. Inside these cells they differentiate into amastigotes. (3) Amastigotes multiply by binary fission inside the cell. (4) Intracellular amastigotes differentiate into trypomastigotes, burst out the cells, and enter the bloodstream. Circulating trypomastigotes can invade other cells differentiating into amastigotes in new cell sites. (5) Triatomine bugs take a blood meal containing circulating trypomastigotes. (6) In the midgut they differentiate to epimastigotes. (7) Epimastigotes duplicate in the midgut. (8) Epimastigotes differentiate to metacycle in the hindgut. *i* infective stage, *d* diagnostic stage. Adapted from PHIL.CDC/Alexander J. da Silva/Melanie Moser

America and Belize, Andean, Amazon basin), interrupting *T. cruzi* transmission by *T. infestans* has been certified in countries such as Chile, Uruguay, and Brazil—excluding the area of the Amazon. In other regions this has been partially achieved as in some provinces of Argentina [86, 87]. However, we must bear in mind that dwellings free of domiciliary vectors can be reinfested if a strict and sustained epidemiological surveillance is not carried out [88].

The peridomicile is the space near the house used by the man to carry out its activities, including the nocturnal rest, and to maintain domestic animals as in corrals, chicken coops, dovecotes, etc. Being an extension of the house, domestic animals as well as wild ones can enter and leave, and light and blood supply can attract wild triatominae to this area. Therefore, this space can function as a bridge between domestic and wild cycle. In a study carried out in northwestern Argentina, in the Province of Misiones, an area reported free of vector transmission by *T. infestans*, the relevance of this bridge has been evidenced by the high infection rates found in marsupials of rural areas near the houses and by the documented infection of two vampire bats caught inside a house [89].

3.2 Sylvatic Life Cycle

The primary cycle of *T. cruzi* is precisely the wild one, being the human infection very recent in an evolutionary scale. Being essentially enzootic, the protozoon circulates in extensive areas of America among numerous mammals and vectors that vary according to environmental factors and food availability. It is considered that the infection causes very mild or does not cause damage to natural hosts. In this cycle definitive hosts comprise wild animals such as marsupials, rodents, foxes, bats, or monkeys, among others. Vectors are adapted to live in nests of birds, weasels, and rodents, caves of bats or other animals, tree holes, etc. Palms are colonized in extense areas of Central America, Ecuador, and Venezuela mainly, but not exclusively, by species of the *Rhodnius* genus. Palms are also a common habitat of sylvatic vectors in Brazil [90, 91].

4 Non-vectorial Pathways of *T. cruzi* Infection

This protozoan parasite can infect humans by routes not requiring the participation of the insect vector. Such routes, mainly blood transfusion and mother to child transmission, are responsible for Chagas disease transmission in vector-free regions and for disease emergence in Europe, the United States of America, Canada, and Asian developed countries [92, 93].

4.1 Blood Transfusions

Most chronically infected people which are asymptomatic but with fluctuating parasitemia are unaware of their infection status. This situation makes blood donation a risk for *T. cruzi* transmission that is proportional to Chagas disease prevalence among blood donors. To avoid this risk, screening coverage in blood banks must be adopted and blood units from reactive donors discarded. In areas where the infection rate of potential donors is very high and/or blood screening unfeasible, some strategies of parasite reduction—mainly against bloodstream trypomastigotes—have been proposed. The only strategy currently used is the treatment with gentian violet 1: 4000 for 24 h at 4 °C. This treatment, although highly effective, does not guarantee 100% elimination of the parasites, depending on the parasite population [94, 95].

In most endemic countries, the coverage has reached almost 100% in the last years. In these countries, serological control is mandatory and has been legislated, which has significantly reduced the risk of transmission [92].

4.2 Congenital or Connatal Route

Any women with Chagas disease can transmit *T. cruzi* to the fetus both transplacentally during pregnancy and at the time of delivery. Child infection status should be studied at birth and during the first months of life using direct diagnostic methods and serologically after 8–10 months. When positive, infants should be treated with parasiticide drugs [96]. The transmission rate is usually low (2–10%) with variations between endemic regions. An infected mother who has transmitted the infection to a child will not necessarily transmit the infection in succeeding pregnancies. In a study conducted by Juiz et al. [97], an association between polymorphisms in genes expressed in the placenta and susceptibility to congenital infection was demonstrated. The authors suggest that polymorphisms of proteins participating in extracellular matrix remodeling can mediate susceptibility to vertical transmission of the parasite.

4.3 Organ or Tissue Transplantation

The ideal conditions to carry out a solid organ transplant are having both donor and recipient free of *T. cruzi* infection. When this aim is not feasible mainly due to the high prevalence of Chagas disease in some endemic areas, there are rules strictly regulated by the competent institutions of affected countries. With few exceptions, when the patient is infected and the organ to be transplanted comes from a healthy individual, the transplant can be performed. The transplant receptor must be periodically studied to determine whether the immunosuppressive treatment administered to avoid graft rejection lead to reactivation of the infection to promptly initiate parasiticide treatment. When the donor is infected, the recommendation to transplant depends on the organ to be transplanted and on the infection status of the recipient. It is feasible to perform transplants of various solid organs with varying degrees of success [98–101]. In Argentina the institution which reviews and updates the procedures as established by transplants Law No. 24,193 is INCUCAI. Its Resolution 269/99 indicates for Chagas disease specifically the conditions to perform transplants (*https://www.incucai.gov.ar/files/docs-incucai/Legislacion/03-ResIncucai/Procuracion-y-trasplante-de-organos/07-res_incucai_269_99.pdf*).

4.4 Oral Route

This mode of infection occurs upon ingestion of food or beverages contaminated with feces from *T. cruzi* infected triatomine bugs. It is more common in regions with sylvatic cycles and less frequent outside. In recent years outbreaks of acute infection have been reported in Brazil, Venezuela, and Colombia. Metacyclic trypomastigotes express at their surface the glycoprotein Gp82, which has an adhesion site for gastric mucins. The parasite adheres to them and penetrates the epithelial cells of the stomach multiplying then spreading through the bloodstream [66, 67].

4.5 Laboratory Accidents

Infection can occur among health workers and scientists through the manipulation of blood, tissues, or culture media containing trypomastigotes or amastigotes of this parasite. The risk of having an accident can be minimized by following biosecurity recommendations. The routes of infection can be percutaneous, inoculation, or oral.

5 The Parasite in the Insect Vector

The digestive tract of triatominaes is the anatomical site where the parasite develops part of its biological cycle. Insects feed on blood, both from mammals and birds. While birds are resistant to infection by *T. cruzi*, mammals are not. Accordingly, circulating trypomastigotes enter vectors that feed on infected mammals including man. The digestive tract of the vector is a hostile medium for the parasites in which they are subject to changes in temperature or the presence of molecules and intestinal microbiota particular of insects, among others. During the first 24 h post-meal, the number of ingested trypomastigotes decreases significantly at the stomach level. With the purpose of adapting to the new microenvironment and to avoid the deleterious action of insects on the trypomastigotes, the protozoa differentiate into intermediate forms to finally transform into the duplicative epimastigotes. When they pass to the posterior intestine, epimastigotes adhere to the cells of the outer membrane of the microvilli and pursue multiplication. After a couple of months, numerous epimastigotes reach the rectum, adhere to the hydrophobic wax layer of its cuticle, and differentiate into metacyclic trypomastigotes ([102]; review by [2]; review by [103]).

5.1 Metacyclogenesis

Numerous stimuli have been described that play a role in inducting differentiation to metacyclic forms. The activity of vector intestine extracts in metacyclogenesis has been reported [104–106], and isolated factors of insect's hindgut inducing in vitro differentiation have been characterized [107]. Among these factors are a fragment of αD-globin and a peptide that activates adenylyl cyclase in the membrane of epimastigotes [108], and others are free fatty acids, mainly oleic acid [109, 110]. The differentiation mechanism is promoted by the physicochemical characteristics of vector's intestine [111]. Other authors have demonstrated that insect's long fasting periods that cause metabolic stress trigger metacyclogenesis. This effect can be reproduced in vitro when culturing epimastigotes in poor culture media that induce nutritional stress such as TAU and M16 media [1, 112, 113]. Hamedi et al. [114] have reported that during in vitro starvation-induced metacyclogenesis, there is an increase in the expression of adenylyl cyclase, coincident with the observations made by Fraidenraich et al. [108] using insect extract stimulation.

5.2 *Selectivity Among the Parasitic Populations and the Vectors*

Each triatominae species has a different affinity for the diverse parasitic populations. So, the infection rate and the success of metacyclogenesis will correlate to the greater or lesser affinity between vector and parasite. In this sense, it has been reported that, in a single triatominae species, the rate of metacyclogenesis can vary depending on the *T. cruzi* strain used. On the other hand, different numbers of metacyclic trypomastigotes can be recovered among different triatominae species infected with the same parasitic population [79, 115–117].

6 The Mammalian Host

T. cruzi has a low specificity for definitive hosts thus having the capacity to infect virtually any mammal. Humans, and domestic animals like dogs and cats, usually become infected inside human dwellings or in the peridomicilia. Domestic animals as goats, guinea pigs, and domiciliated rodents can also become infected in the peridomicilia. All of them are reservoirs of the parasite (among which the dog stands out as already mentioned), and they are of the utmost importance as they can be responsible for the transmission of *T. cruzi* to man. In wild areas, several mammals, including the opossum, the only marsupial species in South American, are naturally infected, and some of them, having high parasitemia, act as very efficient reservoirs. Monkeys, armadillos, wild rodents, and bats can become infected at variable rates.

During the vectorial cycle of *T. cruzi*, metacyclic trypomastigotes are responsible for the initiation of infection in the definitive host. This stage is unable to penetrate intact skin but takes advantage of skin abrasions or the mucous membranes to enter the host. Once inside host tissues, trypomastigotes transit through the extracellular matrix to contact the target cell [118–120]. This stage invades cells near the site of entry first and differentiates into intracellular diving amastigotes. Depending on the parasite population, the number of successive divisions varies [81]. Then, amastigotes differentiate into circulating trypomastigotes, which break down the cell, disseminate through blood, and reach various tissues where they initiate new cycles of intracellular multiplication.

In the different periods of infection, parasitemia levels vary, being during the chronic phase extremely low and fluctuating but enough to infect a vector insect while ingesting blood from an infected mammal.

6.1 *Any Cell Can Be Host of* T. cruzi

As already mentioned, *T. cruzi* can invade a wide range of vertebrate hosts being capable of dwelling inside virtually any nucleated cell including diverse laboratory cell lines. The process of invasion, infection, and differentiation occurs even in cells whose nuclei have been removed, thus indicating the independence of this process on host transcriptional machinery [121, 122]. In natural mammalian infection, metacyclic trypomastigotes locally invade epithelial cells, macrophages, fibroblasts, and adipocytes, and then circulating trypomastigotes disseminate to diverse tissues infecting muscle cells, cardiomyocytes, and nerve cells [123]. This high degree of versatility can only be achieved through the development of multiple invasion strategies by this ubiquitous protozoan. Controversies also arise considering the myriad of host cell-entry mechanisms described so far being difficult to conceal a uniform model. Different host cell types, parasite isolates, or strains and even different parasite stages take part in in vitro models of host-parasite interaction giving sometimes opposite results. These features can also reflect *T. cruzi* ability to use tailored mechanisms of entry according to the characteristics of host cell contacted.

6.2 Navigating Across the Extracellular Matrix

Invasive stages of *T. cruzi* (e.g., bloodstream trypomastigotes, metacyclic trypomastigotes, and extracellular amastigotes) are known to interact with different components of the extracellular matrix (ECM) as the proteins laminin, thrombospondin, collagen, and fibronectin and proteoglycans (heparin, heparan sulfate). *T. cruzi* exploits them to reach host cells and initiate the process of cell invasion. Infective forms of the parasite upregulate the expression of ECM components laminin gamma 1 or thrombospondin 1, but their silencing by RNAi reveals a dramatic reduction of cell infection [120]. To infect muscle cells, *T. cruzi* must cross the basal lamina that surrounds them. For this, the trypomastigote stage-specific surface molecule Gp83 mediates the upregulation of laminin gamma 1, thus participating in parasite adhesion to the host cell. This observation correlates with the finding of laminin deposited in hearts of *T. cruzi*-infected patients [124]. Calreticulin is a well-conserved intracellular calcium-binding chaperone; it has been described in diverse parasite species including *T. cruzi* [125]. Parasite calreticulin (TcCRT/Tc45) is also expressed at the surface of infective trypomastigotes and may act as a receptor for the collagen tail of C1q and for mannan-binding lectin to promote phagocytosis through C1qR. Surface-expressed TcCRT is also known to interact with host thrombospondin 1 enhancing cellular infection [126]. The ECM human lectin, galectin-3, binds to a 45 kDa trypomastigote surface mucin in a lectin manner. This lectin simultaneously binds to laminin of basement membranes via carbohydrate recognition domains, thus providing a bridge between the parasite and muscle cells [127, 128]. Considering that galectin-3 concentrations in fluids can increase during microbial infection [129], *T. cruzi* may have adapted this trapping mechanism to recruit parasites at the ECM enabling cellular invasion. The prolyl oligopeptidase Tc80 secreted by infective trypomastigotes is involved in collagen and fibronectin hydrolysis. Selective inhibitors of its proteinase activity can block parasite entry to the host cell [130]. Some reports suggest that parasite proteases as Tc85 a glycoprotein from the Gp85/trans-sialidase (Gp85/TS) multigene family interact with diverse components of the ECM as laminin, the intermediate filament proteins cytokeratin and vimentin, fibronectin, mucin, and the prokineticin-2 receptor [118, 131]. Regarding the interaction of the metacyclic trypomastigotes, the infective stage from the insect vector, with components of the extracellular matrix, Maeda and coworkers described that this parasitic stage binds and degrades fibronectin by means of the stage-specific glycoprotein Gp82 [119]. Cruzipain, the major *T. cruzi* cysteine proteinase which also digests fibronectin, acts synergistically to increased metacyclic trypomastigote internalization. Considering the diverse nature and composition of the extracellular matrix between different tissues and the dynamics of this extracellular mesh, it is tempting to suggest that *T. cruzi*-ECM interactions are relevant for parasite virulence and tissue tropism. Furthermore, changes in the physiological condition of ECM driven by chronic infection may have implications for the pathogenesis of the disease in target tissues.

6.3 *Attaching to the Host Cell Surface*

Most of the knowledge on the interaction between parasites and cells has been obtained from in vitro studies. Trypomastigotes can initiate the process of cell invasion once they have recognized cell surface receptors at the host cell. These receptors have not been clearly identified, but they interact with a large number of well-characterized parasite's surface-membrane protein ligands. Parasite ligands are grouped into different families as mucins, trans-sialidases (TS), TS-like proteins, and membrane proteins. The best characterized are members of the Gp85/trans-sialidase family [132]. However, recently some underrepresented families as the TcTASV members have attracted attention [64, 65]

6.3.1 Role of Gp82 and Gp90 in Oral Infection

Mucin-like surface glycoproteins are expressed in the outer membrane of metacyclic trypomastigotes which makes this parasite stage resistant to gastric environment. Owing to this, this stage is able to invade intestinal epithelial cells of mammals.

At least two main glycoproteins, Gp82 and Gp90, which belong to the Gp85/TS superfamily, are known to participate in host cell invasion. While Gp82 has a central role in host cell recognition and invasion, Gp90 negatively regulates parasite entry [68].

Gp82 is highly conserved among *T. cruzi* strains and binds gastric mucins [67] being the peptide p7 the chief mucin-binding site, followed by p10 [66]. Gp82 is attached to the outer membrane of the parasite by a glycosylphosphatidylinositol anchor and confers resistance to the low pH of the stomach. Gp82-mucin signaling though results in an increase of intracellular Ca^{2+}, thus launching the invasion process. Those *T. cruzi* strains that are deficient in Gp82 are poorly invasive.

Parasite strains that express high levels of Gp90 are likewise poorly invasive. This molecule binds the target cell but fails to induce or induces low Ca^{2+} mobilization [133]. This glycoprotein possesses various isoforms, some of them susceptible to gastric pepsin digestion. Thus, infectivity of a *T. cruzi* strain may be exacerbated according to Gp90 sensitivity to these gastric juices [69].

6.3.2 Role of Gp85/TS in Adhesion and Survival into Mammalian Host Cells

First identified as a 85 kDa surface protein of *T. cruzi*, it is now known to belong to a multigene glycoprotein superfamily named Gp 85/TS [134]. Every member of this superfamily shares a sialidase domain with a peptide motive called FLY. Due to its complexity, the superfamily was subdivided into groups [135]. Proteins from group I are those which possess TS with enzymatic activity, while those of group II lack from this activity.

Gp85 without enzymatic activity have been associated with mammalian cell invasion. Mattos et al. [136] reported that the N-terminus of Gp85 binds to laminin and induces dephosphorylation of cytoskeletal proteins. Otherwise, at the C-terminus, the FLY sequence binds to cytokeratin, major constituents of intermediate filaments in the host cells, which results in increased cell invasion. A small peptide motif (TS9) reported by Teixeira et al. [131] shared by all Tc85 proteins is also capable to bind cytokeratin, as well as vimentin. Both FLY and TS9 are located inside of a laminin-G-like domain at the C-terminus. The authors propose that TS9 and FLY, which are far separated in a linear sequence, could be located next to each other in a tridimensional conformation, thus displaying a nonlinear keratin-binding site. This plasticity of parasite surface protein helps to overcome the barriers of ECM and those of host cell membrane [58]. FLY conserved sequence also binds to endothelium with a significant avidity for heart vasculature reinforcing its role in cell invasion and implications in homing [137].

Zingales et al. [138] described how trypomastigotes incorporate to their surface sialic acid from compounds as fetuin but not free sialic acid directly. Parasites with TS enzymatic activity can transfer sialic acid from glycoconjugates of the hosts to acceptor mucin molecules exposed in their outer membrane. Sialidation of parasite surface proteins is required for invasion of mammalian cells [60]. TS molecule comprises two domains, the C-terminus which is an immunodominant antigen made of amino acid repeats (shed acute-phase antigen—SAPA) and the N-terminus which possesses the enzymatic activity [139, 140]. Even though both are antigenic, SAPA is immunodominant producing an early strong antibody response during acute infection. These antibodies react specifically against the nonenzymatic portion of the whole molecule [141]. It has been speculated that this is a parasite strategy to delay an antibody response against the enzymatic activity leading to its neutralization.

TS is the major virulence factor of *T. cruzi*. It is engaged in numerous steps that allow host cell invasion and adaptation to the intracellular lifestyle (review by [59]). The first obstacles that parasites must avoid are innate mammal defense mechanisms as complement lytic activity and macrophage activation. Trypomastigotes, but not epimastigotes, are resistant to complement lysis due to the presence of a surface molecule similar to human decay-accelerating factor which interferes with the C3 convertase, thus inhibiting complement cascade [142]. After skewing this obstacle and getting across the ECM, the parasite proceeds to invade the cell using diverse mechanisms described in Sects. 7.1–7.3. Inside of the parasitophorous vacuole (PV; see Sect. 9), trypomastigotes and amastigotes secrete lytic factors which are active at low pHs and specially on desialidated membranes. So, when lysosomes fused and acidified inside the PV, TS enhances parasite resistance to this environment and contributes to the disruption of the PV membranes, setting parasites free in the cytoplasm (see Sect. 9.2). This is one of the most relevant TS activities [61, 143].

TS also plays an important role in the modulation of host immune response [144, 145]. Ruiz Díaz et al. [146] reported how TS deflects the Th1 phenotype, which is protective against *T. cruzi* but also promotes host tissue damage, while stimulating a Th2 phenotype response. This strategy is useful for *T. cruzi* establishment and to reach an equilibrium between host and parasite, with survival of the former and persistence of the latter.

7 One Parasite with Multiple Mechanisms to Enter the Host Cell

In a remarkable fashion, trypomastigotes can enter both phagocytic and nonprofessional phagocytic cells to finally access host cell lysosomes. Establishment in a lysosomal-based vacuole is required for a successful infection. For this, the parasite has evolved to exploit assorted cellular pathways to reach this host cell compartment. Excellent reviews have been published on this subject; thus we offer here a brief description of some of the proposed mechanisms [122, 147, 148].

7.1 Lysosomal-Dependent Exocytic Pathway

In 1991, Schenkman et al. [149] demonstrated that *T. cruzi* entered nonprofessional phagocytic cells by an actin-independent mechanism distinct from phagocytosis. Later, it was verified that lysosomes are recruited to the plasma membrane at parasite attachment sites and that a vacuole is formed around the parasite by the fusion of lysosomes with the plasma membrane [150, 151]. The vacuole formation involves lipid remodeling by the delivery of acid sphingomyelinase that enriches in ceramides the outer leaflet of the plasma membrane [152]. By this route, parasites are readily directed to an acidic lysosomal vacuole. Reduction of peripheral lysosomes or inhibition of lysosomal fusion with plasma membrane results in reduced *T. cruzi* infection. The exocytosis of lysosomes ensues rapidly following parasite contact in a unidirectional movement toward the plasma membrane-parasite attachment site [153]. This process appears to depend on the Ca^{2+}-binding protein synaptotagmin VI and kinesin motors on microtubules that require Ca^{2+}-calmodulin [154, 155]. The parasite has the ability to trigger a mobilization of Ca^{2+} stores near the attachment site in a phospholipase C and inositol 1,4,5-triphosphate-dependent manner [156, 157]. Different works have suggested a participation of host cell plasma membrane lipid rafts and their associated proteins in *T. cruzi* invasion process. These microdomains are involved in *T. cruzi* entry in phagocytic and non-phagocytic cells as membrane cholesterol depletion by methyl-beta cyclodextrin, which hampers lysosomal recruitment to the plasma membrane, decreased in vitro *T. cruzi* infectivity [158]. Interestingly, this process of lysosome fusion with plasma membrane, first described for *T. cruzi* invasion, has been recognized as a plasma membrane wound repair mechanism regulated by intracellular calcium levels [159, 160]. These findings offer a conceivable explanation for the tissue tropism exhibited by *T. cruzi* within its vertebrate host and for a part of Chagas disease pathogenesis, as muscle cells exhibit increased in vivo plasma membrane injury with respect to other cell types [152, 161].

7.2 Endocytic Pathway

Early studies showed that, in conditions where parasite internalization and lysosomal association were inhibited, a fraction of vacuoles were initially devoid of lysosomal markers but gradually acquired lysosome-associated membrane protein 1 (LAMP-1) and a fluid-phase endocytic tracer from the lysosomal compartment [149, 162, 163]. These experiments revealed that the parasite can also enter a vacuole composed of invaginated plasma membrane that subsequently fuses with endosomes and lysosomes. These lysosome-independent pathways have been observed in diverse mammalian cell lines and primary cardiomyocytes [163]. The membrane trafficking that leads to the final lysosomal location of the parasite is not completely understood. The GTPase dynamin, which plays an important role in clathrin-mediated endocytosis, and the GTPases Rab5 from early endosomes as well as Rab7 from late endosomes are required for host cell infection [164]. The findings revealed for the first time that the sources of donor membranes are diverse and include early and late endosomes prior to lysosome fusion of the endocytic vacuole [164]. Quantitative analysis has shown that this process represents 50% of parasite internalization and in 20% of them, the participation of early endosomes is detected by early endosome antigen 1 labeling of the vacuoles 10 min after infection [162].

7.3 The Autophagic Mode of Entry

The work from Romano and coworkers has recently established a connection between *T. cruzi* and host cell autophagic pathway. LC3, the most relevant autophagosome marker, colocalized with the *T. cruzi* parasitophorous vacuole (PV) following in vitro infection of Chinese hamster ovary cells with trypomastigotes. Real-time video microscopy revealed that, less than 1 h after infection, green fluorescent protein-tagged LC3 molecules already decorate autophagosomes which concentrated at the cytosolic face of the plasma membrane in the vicinity of *T. cruzi* contact sites. This interaction with LC3-positive compartments occurred early from the moment of PV formation and persisted until parasites located freely in the cytosol. Indeed, amastigotes located free in the cytoplasm (48–72 h postinfection) no longer colocalized with LC3 molecules. The relevance of autophagy mechanisms for *T. cruzi* invasion was demonstrated in studies where autophagy was induced by starvation or treatment with the mTOR inhibitor rapamycin [165]. Autophagy induction increased infection, and the absence of genes necessary for the early steps of autophagic pathway beclin-1 or Atg5 decreased this rate [165, 166]. Under autophagic conditions as starvation, host cells produce new sets of autolysosomal vesicles. These vesicles can be clustered to the cell membrane by infective trypomastigotes that exploit lysosomal exocytosis machinery to generate a PV to shelter them. Under full nutrient media in vitro conditions, cells produce a limited number of acidic lysosomes, thus reducing *T. cruzi* colonization. In experimental murine infection, it has been demonstrated that under starvation conditions tissues like heart and skeletal muscles are prone to establish sustained autophagy [167]. Romano et al. [168] suggest that, in human Chagas disease, life conditions of *T. cruzi*-infected individuals as nutritional deficit could induce autophagy, thus promoting tissue infection and chronic disease

7.4 *Phosphatidylinositol 3-Kinase Signaling and the Parasitophorous Vacuole*

The mechanisms of *T. cruzi* cell invasion described so far show that in the early steps of PV assembly, donor membranes can be from plasma membrane, lysosomes, endosomes, and even autophagosomes. However, the host cell signaling pathways that must be activated during *T. cruzi* invasion are less understood. Wilkowsky and coworkers demonstrated for the first time that non-phagocytic pathways of cell invasion decreased by wortmannin a phosphatidylinositol 3-kinase inhibitor [169]. They showed that molecules in the plasma membrane of trypomastigotes, but not of epimastigotes, were able to activate the host PI3K/PKB/Akt signal transduction cascade, relevant for cell proliferation, cytoskeleton reorganization, and membrane trafficking. In parallel, Chuenkova et al. [170] demonstrated that *T. cruzi* trans-sialidase was able to activate PI3/AKT signaling. First, it was acknowledged that PI3 kinase was relevant for lysosomal-independent *T. cruzi* invasion, but further experiments demonstrated the participation of this cascade in lysosomal-dependent pathway, thus revealing the universality of this signaling [163]. As seen with other signal transduction molecules like Ca^{2+}, the PI3K/PKB cascade could be a common pathway used by several pathogens to invade and survive within host cells. Further work from Andrade and Andrews [171] described an alternative pathway of invasion which is lysosome-independent and wortmannin insensitive.

8 How to Survive Inside a Professional Phagocytic Cell?

Macrophages are one of the major targets of *T. cruzi* particularly following vectorial infection where metacyclic trypomastigotes deposited by the insect host can be spotted and captured by tissue-resident macrophages. Later, amastigotes released by disrupted cells are also engulfed by these professional phagocytic cells. Parasites are directed to phagolysosomes where they can be efficiently destroyed by oxidative species, a major effector mechanism of the antiparasitic innate immune response [172, 173]. The capacity of invading parasites to modulate macrophage activation before they can reach their cytoplasm is central to parasite survival. To cope with oxidative burst, parasites have evolved an antioxidant network shaped by enzymes and nonenzymatic redox-active molecules with varied subcellular distribution (e.g., endoplasmic reticulum, glycosomes, mitochondrion, and cytosol) [174]. It has been proposed that the potency of the antioxidant "shield" may vary between *T. cruzi* isolates thus dictating their final fate within the macrophage. Accordingly, the antioxidant complex can be considered as a *T. cruzi* virulence factor ([174, 175].; [176]). Alternatively, some parasites may survive inside professional phagocytic cells by invading them in non-phagocytic/active pathways as revealed by experiments where inhibition of actin polymerization failed to completely abrogate macrophage infection [177].

9 The Parasitophorous Vacuole (PV)

9.1 PV Always Ends Up as a Lysosomal Compartment

Regardless of the endocytic mechanism that takes place for parasite entry to the host cell, PVs always result in an acidic lysosomal compartment. Fusion with lysosomes occurs at the site of parasite entry or after entry with an already preformed PVs. Microtubules appear to be necessary for targeting lysosomes to these places. Upon entry, PVs also recruit endocytic vesicles (initial and late endosomes) in a process that together with the fusion of lysosomes leads to PV maturation through acidification. Lysosomal fusion is essential to the retention of parasite inside the cell as its blockade by wortmannin results in their exit to the extracellular milieu [171]. In addition to this finding, Woolsey and Burleigh demonstrated that the actin cytoskeleton-dependent processes of PV fusion with endosomal and lysosomal vesicles are necessary for parasite retention. In fact, the process of reversible invasion increased notably in cytochalasin D-treated host cells [162].

9.2 Degradation of the PV and Cytosolic Settlement

The establishment of productive intracellular infection depends on parasite location inside PV transformed into lysosomal compartments prior to their multiplication free in the host cell cytosol. In this continuous process, the parasite breaks down the vacuolar membrane in a lysosome- and pH-dependent mechanism [178, 179]. Early studies demonstrated that parasite trypomastigotes and amastigotes secreted a pore-forming factor related to complement component C9, the hemolysin TcTox [180], and another C9 cross-reactive membrane-targeted protein with hemolytic activity, LYT1 [181], whose precise role in PV disruption has not been demonstrated. It is proposed that both lytic molecules which are expressed and optimally active at the acidic pH brought by lysosomal fusion of the PV would promote membrane breakdown and cytosolic localization of the parasite. Surface TS enhances parasite resistance to this acidic environment while facilitating membrane destabilization. In fact, it was demonstrated that residence inside the vacuolar space increases parasite secretion of TS. Desyalidation of vacuolar membrane molecules, like LAMP-1, is a process suggested to foster the insertion of TcTox into the lipid bilayer [144].

10 Parasite Differentiation

The differentiation of infective trypomastigotes to intracellular amastigotes in the host cell cytoplasm is poorly understood, but in vitro evidence shows that it can take from 2 to 8 h depending on the infective strain. Launching trypomastigote-amastigote differentiation appears to require, among other signaling factors, the acidic pH provided by the lysosomal compartment [182] in a process also dependent on L-proline [183] and on phosphorylation/dephosphorylation events [184]. Quiescent amastigotes reenter the cell cycle and replicate in the cytosol a limited number of times until they occupy the host cell volume to further differentiate into trypomastigotes. This motile *T. cruzi* stage destroys the host cell plasma membrane [185] to initiate another infection cycle. The mechanism of host cell death remains controversial as infection leads to fibroblast death in a non-apoptotic way [186], whereas other cell types as cardiomyocytes and macrophages appear to die from apoptosis [187].

11 Extracellular Amastigote Entry to the Host Cells

Amastigotes which are prematurely released by ruptured host cells have the chance to return to the replication cycle when taken up by neighboring cells. In contrast to the active mechanisms used by trypomastigotes, extracellular amastigotes enter the cell by elicitation of actin-dependent mechanisms even in non-phagocytic cells [188]. Remarkably, the capacity of amastigotes to induce phagocytosis in non-phagocytic cells is parasite strain-dependent being extracellular amastigotes from the less infective strains mainly engulfed by macrophages [189, 190]. Once inside the host cells, amastigotes have the same ability than trypomastigotes to disrupt the PV to replicate freely in the cytosol and differentiated to infective trypomastigotes.

12 Concluding Remarks

In this chapter we intended to highlight the complexity of this parasite and its biological cycle. Different parasitic stages with differences in duplicative and/or infective capacity are accompanied by a significant intraspecific genotypic diversity. This parasite has evolved the capacity to invade a vast number of insect vector species and a wide range of mammalians in domestic and sylvatic cycles. This has made virtually impracticable all attempts of parasite eradication. A large number of molecules interact with the various cell types that can become host for the intracellular form of the parasite which can escape from immune surveillance and drugs in hidden niches. Such complexity has been demonstrated by the limited success in controlling this parasite. So far, much progress has been made, but we are still far from understanding how to successfully hamper with the progress of the infection and thus prevent Chagas disease.

References

1. Kollien AH, Schaub GA. The development of *Trypanosoma cruzi* in Triatominae. Parasitol Today. 2000;16(9):381–7.
2. Garcia ES, Genta FA, de Azambuja P, Schaub GA. Interactions between intestinal compounds of triatomines and *T. cruzi*. Trends Parasitol. 2010a;26(10):499–505.
3. Ferguson MA. The estructure, biosintesis and functions of glycosylphospatidyilinositol anchors, and the contributions of trypanosome research. J Cell Sci. 1999;112:2799–809.
4. Opperdoes RF, Michels PA. The glicosome of the kinetoplastida. Biochimie. 1993;75(3-4):231–4.
5. Docampo R, Moreno SN. Acidocalcisome: a novel Ca2+ storage compartment in trypanosomatids and apiocomplexan parasites. Parasitol Today. 1999;15(11):443–8.
6. Docampo R, de Souza W, Miranda K, Rohloff P, Moreno SN. Acidocalcisomes – conserved from bacteria to man. Nat Rev Microbiol. 2005;3(3):251–61.
7. Hoare C. The trypanosomes of mammals. Oxford, England: Blackwell Scientific Publications; 1972. p. 60. Chapter 5.
8. Chagas C. Nova tripanozomiase humana: estudos sobre a morfolojia e o ciclo evolutivo do *Schizotrypanum cruzi* n. gen., n. sp., ajente etiolojico de nova entidade morbida do homem. Mem Inst Oswaldo Cruz. 1909;1:159–218. Cited in Cruz RE, Macedo AM, Barnabe C, Freitas JM, Chiari E, Veloso CM, et al. Further genetic characterization of the two *Trypanosoma cruzi* Berenice strains (Be-62 and Be-78) isolated from the first human case of Chagas disease (Chagas, 1909) Acta Trop. 2006; 97: 239–46.
9. Bertelli M, Brener Z. Infection of Tissue Culture Cells with Bloodstream Trypomastigotes of *Trypanosoma cruzi*. J Parasitol. 1981;66(6):992–7.
10. Brener Z. The behavior of slender and stout forms of *Trypanosoma cruzi* in the blood-stream of normal and immune mice. Ann Trop Med Parasitol. 1969;63(2):215–20.
11. Schmatz DM, Boltz RC, Murray PK. *Trypanosoma cruzi*: separation of broad and slender trypomastigotes using a continuous hypaque gradient. Parasitology. 1983;87(Pt 2):219–27.
12. Borges MM, De Andrade SG, Pilatti CG, do Prado Júnior JC, Kloetzel JK. Macrophage activation and histopathological findings in Calomys callosus and Swiss mice infected with several strains of *Trypanosoma cruzi*. Mem Inst Oswaldo Cruz. 1992;87(4):493–502.
13. Risso MG, Garbarino GB, Mocetti E, Campetella O, Gonzalez Cappa SM, Buscaglia CA, et al. Differential expression of a virulence factor, the trans-sialidase, by the main *Trypanosoma cruzi* phylogenetic lineages. J Infect Dis. 2004;189(12):2250–9.
14. Botero LA, Mejía AM, Triana O. Caracterización biológica y genética de dos clones pertenecientes a los grupos I y II de *Trypanosoma cruzi* de Colombia. Biomedica. 2007;27(Suppl 1):64–74.
15. Melo RC, Brener Z. Tissue tropism of different *Trypanosoma cruzi* strains. *J Parasitol.* 1978;64(3):475–82.
16. Alba Soto CD, Mirkin GA, Solana ME, González Cappa SM. *Trypanosoma cruzi* infection modulates in vivo expression of major histocompatibility complex class II molecules on anti-

gen-presenting cells and T-cell stimulatory activity of dendritic cells in a strain-dependent manner. Infect Immun. 2003;71(3):1194–9.
17. Mirkin GA, Celentano AM, Malchiodi EL, Jones M, González Cappa SM. Different *Trypanosoma cruzi* strains promote neuromyopathic damage mediated by distinct T lymphocyte subsets. Clin Exp Immunol. 1997;107(2):328–34.
18. Mirkin GA, Jones M, Sanz OP, Rey R, Sica RE, González Cappa SM. Experimental Chagas' disease: electrophysiology and cell composition of the neuromyopathic inflammatory lesions in mice infected with a myotropic and a pantropic strain of *Trypanosoma cruzi*. Clin Immunol Immunopathol. 1994;73(1):69–79.
19. Andrade SG. The influence of the strain of *Trypanosoma cruzi* in placental infections in mice. Trans R Soc Trop Med Hyg. 1982;76(1):123–8.
20. Solana ME, Celentano AM, Tekiel V, Jones M, González Cappa SM. *Trypanosoma cruzi*: effect of parasite subpopulation on murine pregnancy outcome. J Parasitol. 2002;88(1):102–6.
21. Andrade SG. Caracterizaçao de cepas de *Trypanosoma cruzi* isoladas no Recôncavo Baiano. Rev Patol Trop. 1974;3:65–121.
22. Andrade SG. Influence of *Trypanosoma cruzi* strain on the pathogenesis of chronic myocardiopathy in mice. Mem Inst Oswaldo Cruz. 1990;85(1):17–27.
23. Miles MA, Toye PJ, Oswald SC, Godfrey DG. The identification by isoenzyme patterns of two distinct strain-groups of *Trypanosoma cruzi*, circulating independently in a rural area of Brazil. Trans R Soc Trop Med Hyg. 1977;71(3):217–25.
24. Morel C, Chiari E, Camargo EP, Mattei DM, Romanha AJ, Simpson L. Strains and clones of *Trypanosoma cruzi* can be characterized by pattern of restriction endonuclease products of kinetoplast DNA minicircles. Proc Natl Acad Sci U S A. 1980;77(11):6810–4.
25. Souto RP, Fernandes O, Macedo AM, Campbell DA, Zingales B. DNA markers define two major phylogenetic lineages of *Trypanosoma cruzi*. Mol Biochem Parasitol. 1996;83(2):141–52.
26. Zingales B, Andrade SG, Briones MR, Campbell DA, Chiari E, Fernandes O, et al. Second Satellite Meeting. A new consensus for *Trypanosoma cruzi* intraspecific nomenclature: second revision meeting recommends TcI to TcVI. Mem Inst Oswaldo Cruz. 2009;104(7):1051–4.
27. Tibayrenc M. Genetic epidemiology of parasitic protozoa and other infectious agents: the need for an integrated approach. Int J Parasitol. 1998;28(1):85–104.
28. Zingales B. *Trypanosoma cruzi* genetic diversity: something new for something known about Chagas disease manifestations, serodiagnosis and drug sensitivity. Acta Trop. 2018;184:38. https://doi.org/10.1016/j.actatropica.2017.09.017. Review. pii: S0001-706X(17)30426-6.
29. Zingales B, Miles MA, Campbell DA, Tibayrenc M, Macedo AM, Teixeira MM, et al. The revised *Trypanosoma cruzi* subspecific nomenclature: rationale, epidemiological relevance and research applications. Infect Genet Evol. 2012;12(2):240–53.
30. Hernández C, Cucunubá Z, Flórez C, Olivera M, Valencia C, Zambrano P, et al. Molecular diagnosis of Chagas disease in colombia: parasitic loads and discrete typing units in patients from acute and chronic phases. PLoS Negl Trop Dis. 2016;10(9):e0004997.
31. Ramírez JD, Guhl F, Rendón LM, Rosas F, Marin-Neto JA, Morillo CA. Chagas cardiomyopathy manifestations and *Trypanosoma cruzi* genotypes circulating in chronic Chagasic patients. PLoS Negl Trop Dis. 2010;4(11):e899.
32. Brenière SF, Waleckx E, Barnabé C. Over six thousand *Trypanosoma cruzi* strains classified into discrete typing units (DTUs): attempt at an inventory. PLoS Negl Trop Dis. 2016;10(8):e0004792.
33. Cardinal MV, Lauricella MA, Ceballos LA, Lanati L, Marcet PL, Levin MJ, et al. Molecular epidemiology of domestic and sylvatic *Trypanosoma cruzi* infection in rural northwestern Argentina. Int J Parasitol. 2008;38(13):1533–43.
34. Acosta N, López E, Lewis MD, Llewellyn MS, Gómez A, Román F, et al. Hosts and vectors of *Trypanosoma cruzi* discrete typing units in the Chagas disease endemic region of the Paraguayan Chaco. Parasitology. 2017;144(7):884–98.

35. Marcili A, Lima L, Valente VC, Valente SA, Batista JS, Junqueira AC, et al. Comparative phylogeography of *Trypanosoma cruzi* TCIIc: new hosts, association with terrestrial ecotopes, and spatial clustering. Infect Genet Evol. 2009;9(6):1265–74.
36. Monteiro WM, Magalhães LK, de Sá AR, Gomes ML, Toledo MJ, Borges L, et al. *Trypanosoma cruzi* IV causing outbreaks of acute Chagas disease and infections by different haplotypes in the Western Brazilian Amazonia. PLoS One. 2012;7(7):e41284. https://doi.org/10.1371/journal.pone.0041284.
37. Diosque P, Barnabé C, Padilla AM, Marco JD, Cardozo RM, Cimino RO, et al. Multilocus enzyme electrophoresis analysis of *Trypanosoma cruzi* isolates from a geographically restricted endemic area for Chagas' disease in Argentina. Int J Parasitol. 2003;33(10):997–1003.
38. Fernández MP, Cecere MC, Lanati LA, Lauricella MA, Schijman AG, Gürtler RE, et al. Geographic variation of *Trypanosoma cruzi* discrete typing units from *Triatoma infestans* at different spatial scales. Acta Trop. 2014;140:10–8.
39. Maffey L, Cardinal MV, Ordóñez-Krasnowski PC, Lanati LA, Lauricella MA, Schijman AG, et al. Direct molecular identification of *Trypanosoma cruzi* discrete typing units in domestic and peridomestic *Triatoma infestans* and *Triatoma sordida* from the Argentine Chaco. Parasitology. 2012;139(12):1570–9.
40. Guhl F, Ramírez JD. Retrospective molecular integrated epidemiology of Chagas disease in Colombia. Infect Genet Evol. 2013;20:148–54.
41. Cura CI, Lucero RH, Bisio M, Oshiro E, Formichelli LB, Burgos JM, et al. *Trypanosoma cruzi* discrete typing units in Chagas disease patients from endemic and non-endemic regions of Argentina. Parasitology. 2012;139(4):516–21.
42. Duz AL, Vieira PM, Roatt BM, Aguiar-Soares RD, Cardoso JM, Oliveira FC, et al. The TcI and TcII *Trypanosoma cruzi* experimental infections induce distinct immunoresponses and cardiac fibrosis in dogs. Mem Inst Oswaldo Cruz. 2014;109(8):1005–13.
43. Marin-Neto JA, Cunha-Neto E, Maciel BC, Simões MV. Pathogenesis of chronic Chagas heart disease. Circulation. 2007;115:1109–23.
44. Burgos JM, Begher S, Silva HM, Bisio M, Duffy T, Levin MJ, et al. Molecular identification of *Trypanosoma cruzi* I tropism for central nervous system in Chagas reactivation due to AIDS. Am J Trop Med Hyg. 2008;78(2):294–7.
45. Burgos JM, Diez M, Vigliano C, Bisio M, Risso M, Duffy T, et al. Molecular identification of *Trypanosoma cruzi* discrete typing units in end-stage chronic Chagas heart disease and reactivation after heart transplantation. Clin Infect Dis. 2010;51(5):485–95.
46. Poveda C, Fresno M, Gironès N, Martins-Filho OA, Ramírez JD, Santi-Rocca J. Cytokine profiling in Chagas disease: towards understanding the association with infecting *Trypanosoma cruzi* discrete typing units (a BENEFIT TRIAL sub-study). PLoS One. 2014;9(3):e91154.
47. Mejía-Jaramillo AM, Fernández GJ, Montilla M, Nicholls RS, Triana-Chávez O. *Trypanosoma cruzi* strains resistant to benznidazole occurring in Colombia. Biomedica. 2012;32(2):196–205.
48. Rumi MM, Pérez Brandán C, Gil JF, D'Amato AM, Ragone PG, Lauthier JJ, et al. Benznidazole treatment in chronic children infected with *Trypanosoma cruzi*: serological and molecular follow-up of patients and identification of Discrete Typing Units. Acta Trop. 2013;128(1):130–6.
49. Zingales B, Araujo RG, Moreno M, Franco J, Aguiar PH, Nunes SL, et al. A novel ABCG-like transporter of *Trypanosoma cruzi* is involved in natural resistance to benznidazole. Mem Inst Oswaldo Cruz. 2015;110(3):433–44.
50. Andrade LO, Machado CR, Chiari E, Pena SD, Macedo AM. Differential tissue distribution of diverse clones of *Trypanosoma cruzi* in infected mice. Mol Biochem Parasitol. 1999;100(2):163–72.
51. Freitas JM, Andrade LO, Pires SF, Lima R, Chiari E, Santos RR, et al. The MHC Gene Region of Murine Hosts Influences the Differential Tissue Tropism of Infecting *Trypanosoma cruzi* Strains. PLoS One. 2009;4(4):e5113.
52. Alves MJ, Kawahara R, Viner R, Colli W, Mattos EC, Thaysen-Andersen M, et al. Comprehensive glycoprofiling of the epimastigote and trypomastigote stages of *Trypanosoma cruzi*. J Proteomics. 2017;151:182–92.

53. Queiroz RM, Ricart CA, Machado MO, Bastos IM, de Santana JM, de Sousa MV, et al. Insight into the exoproteome of the tissue-derived trypomastigote form of *Trypanosoma cruzi*. Front Chem. 2016;4:42.
54. Kahn SJ, Wleklinski M, Ezekowitz RA, Coder D, Aruffo A, Farr A. The major surface glycoprotein of *Trypanosoma cruzi* amastigotes are ligands of the human serum mannose-binding protein. Infect Immun. 1996;64(7):2649–56.
55. Santos MA, Garg N, Tarleton RL. The identification and molecular characterization of *Trypanosoma cruzi* amastigote surface protein-1, a member of the trans-sialidase gene superfamily. Mol Biochem Parasitol. 1997;86(1):1–11.
56. Turner CW, Lima MF, Villalta F. *Trypanosoma cruzi* uses a 45-kDa mucin for adhesion to mammalian cells. Biochem Biophys Res Commun. 2002;290(1):29–34.
57. Alves MJM, Colli W. Adhesion to the host cell and intracellular survival. Critical review. IUBMB Life. 2007;59(4-5):274–9.
58. Magdesian MH, Giordano R, Ulrich H, Juliano MA, Juliano L, Schumacher RI, et al. Infection by *Trypanosoma cruzi*. Identification of a parasite ligand and its host cell receptor. J Biol Chem. 2001;276(22):19382–9.
59. Freire-de-Lima L, Fonseca LM, Oeltmann T, Mendonça-Previato L, Previato JO. The trans-sialidase, the major *Trypanosoma cruzi* virulence factor: three decades of studies. Glycobiology. 2015;25(11):1142–9.
60. Schenkman S, Jiang MS, Hart GW, Nussenzweig V. A novel cell surface trans-sialidase of *Trypanosoma cruzi* generates a stage-specific epitope required for invasion of mammalian cells. Cell. 1991a;65(7):1117–25.
61. Rubin-de-Celis SS, Uemura H, Yoshida N, Schenkman S. Expression of trypomastigote trans-sialidase in metacyclic forms of *Trypanosoma cruzi* increases parasite escape from its parasitophorous vacuole. Cell Microbiol. 2006;8(12):1888–90.
62. Nardy AF, Freire-de-Lima CG, Pérez AR, Morrot A. Role of *Trypanosoma cruzi* Trans-sialidase on the Escape from Host Immune Surveillance. Front Microbiol. 2016;7:348. https://doi.org/10.3389/fmicb.2016.00348. Mini Review.
63. Cánepa GE, Degese MS, Budu A, Garcia CR, Buscaglia CA. Involvement of TSSA (trypomastigote small surface antigen) in *Trypanosoma cruzi* invasion of mammalian cells. Biochem J. 2012;444(2):211–8.
64. García EA, Ziliani M, Agüero F, Bernabó G, Sánchez DO, Tekiel V. TcTASV: a novel protein family in *Trypanosoma cruzi* identified from a subtractive trypomastigote cDNA library. PLoS Negl Trop Dis. 2010b;4(10):e841. https://doi.org/10.1371/journal.pntd.0000841.
65. Bernabó G, Levy G, Ziliani M, Caeiro LD, Sánchez DO, Tekiel V. TcTASV-C, a protein family in *Trypanosoma cruzi* that is predominantly trypomastigote-stage specific and secreted to the medium. PLoS One. 2013;8(7):e71192.
66. Staquicini DI, Martins RM, Macedo S, Sasso GR, Atayde VD, Juliano MA, et al. Role of GP82 in the selective binding to gastric mucin during oral infection with *Trypanosoma cruzi*. PLoS Negl Trop Dis. 2010;4(3):e613.
67. Cortez C, Sobreira TJ, Maeda FY, Yoshida N. The gp82 surface molecule of *Trypanosoma cruzi* metacyclic forms. Subcell Biochem. 2014;74:137–50.
68. Correa PR, Cordero EM, Gentil LG, Bayer-Santos E, da Silveira JF. Genetic structure and expresión of the surface glycoprotein GP82, the main adhesin of *Trypanosoma cruzi* metacyclic trypomastigotes. Scientific World Journal. 2013;2013:156734.
69. Yoshida N. Molecular mechanisms of *Trypanosoma cruzi* infection by oral route. Mem Inst Oswaldo Cruz. 2009;104(Suppl 1):101–7.
70. Rodrigues JPF, Santana GHT, Juliano MA, Yoshida N. Inhibition of host cell lysosome spreading by *Trypanosoma cruzi* metacyclic stage-specific surface molecule gp90 downregulates parasite invasion. Infect Immun. 2017;85(9):e00302–17. https://doi.org/10.1128/IAI.00302-17.
71. Cazzulo JJ. Proteinases of *Trypanosoma cruzi*: patential targets for the chemotherapy of Changas desease. Curr Top Med Chem. 2002;2(11):1261–71.

72. San Francisco J, Barría I, Gutiérrez B, Neira I, Muñoz C, Sagua H, et al. Decreased cruzipain and gp85/trans-sialidase family protein expression contributes to loss of *Trypanosoma cruzi* trypomastigote virulence. Microbes Infect. 2017;19(1):55–61.
73. Stoka V, Nycander M, Lenarcic B, Labriola C, Cazzulo JJ, Björk I, et al. Inhibition of cruzipain, the major cysteine proteinase of the protozoan parasite, *Trypanosoma cruzi*, by proteinase inhibitors of the cystatin superfamily. FEBS Lett. 1995;370(1-2):101–4.
74. Pinto Dìas JC, Schofield CJ. History of Chagas Disease as a public health problem in Latin America. In: Teixeira A, Vinaud M, Castro AM, editors. Emerging Chagas Disease. Sharjah: Bentham; 2009. p. 1–9. Chapter 1.
75. Darwin C. Journal of researches into the geology and natural history of the various countries visited by H.M.S. Beagle, under the command of captain Fitzroy, R.N. from 1832 to 1836. London: Henry Colburn; 1839.
76. Zeledón R, Rabinovich JE. Chagas Disease: an ecological appraisal with special enphasis on its insect vector. Annu Rev Entomol. 1981;26:101–33.
77. Shelock IA. Vectores. In: Brener Z, Andrade ZA, Barral-Netto M, editors. *Trypanosoma cruzi* e doença de Chagas, vol. 3. 2nd ed. Río de Janeiro: Wanabara Koogan SA; 2000. p. 21–40.
78. Garcia ES, de Azambuja P. Fisiología de Triatomíneos: desenvolvimento, reproduçao e interaçao com *Trypanosoma cruzi*. In: Brener Z, Andrade Z, Barral-Netto M, editors. *Trypanosoma cruzi* e doença de Chagas, vol. 4. 2nd ed. Río de Janeiro: Wanabara Koogan SA; 2000. p. 41–7.
79. García DS, Dvorak JA. Growth and development of two *Trypanosoma cruzi* clones in the arthropod *Dipetalogaster maximun*. Am J Trop Med Hyg. 1982;31(2):259–62.
80. Lammel EL, de Isola EL, Korn C, Gonzalez Cappa SM. *Trypanosoma cruzi*: comparative Studies of infectivity of parasite ingested by *Triatoma infestans* and those present in their feces. Acta Trop. 1981;38(2):107–14.
81. Dvorak JA. A new *in vitro* approach to quantitation of *Trypanosoma cruzi*-vertebrate cell interaction. In: Proc. Symposium on New approaches in American Trypanosomiasis research. Washington, DC: PAHO/WHO; 1976. p. 109–20. Scientific Publication No 318.
82. de Souza W. O parasito e sua interaçao com os hospedeiros. In: Brener Z, Andrade Z, Barral-Netto M, editors. *Trypanosoma cruzi* e doença de Chagas, vol. 7. 2nd ed. Río de Janeiro: Wanabara Koogan SA; 2000. p. 88–126.
83. Enriquez GF, Bua J, Orozco MM, Wirth S, Schijman AG, Gürtler RE, et al. High levels of *Trypanosoma cruzi* DNA determined by qPCR and infectiousness to *Triatoma infestans* support dogs and cats are major sources of parasites for domestic transmission. Infect Genet Evol. 2014;25:36–43.
84. Gürtler RE, Cécere MC, Rubel DN, Petersen RM, Schweigmann NJ, Lauricella MA, et al. Chagas disease in north-west Argentina: infected dogs as a risk factor for the domestic transmission of *Trypanosoma cruzi*. Trans R Soc Trop Med Hyg. 1991;85(6):741–5.
85. Ramírez JD, Turriago B, Tapia-Calle G, Guhl F. Understanding the role of dogs (*Canis lupus familiaris*) in the transmission dynamics of *Trypanosoma cruzi* genotypes in Colombia. Vet Parasitol. 2013;196(1-2):216–9.
86. Moncayo A. Current epidemiological trends after the interruption of vectorial and transfusional transmission in the southern Cone Countries. Mem Inst Oswaldo Cruz. 2003;98(5): 577–91.
87. Moncayo A, Silveira AC. Current epidemiological trends for Chagas disease in Latin America and future challenges in epidemiology, surveillance and health policy. Mem I Oswaldo Cruz. 2009;104(Suppl 1):17–309.
88. Reyes M, Torres A, Esteban L, Flórez M, Angulo VM. Riesgo de transmisión de la enfermedad de Chagas por intrusión de triatominos y mamíferos silvestres en Bucaramanga, Santander, Colombia. Biomédica. 2017;37:68–78.
89. Argibay HD, Orozco MM, Cardinal MV, Rinas MA, Arnaiz M, Mena Segura C, et al. First finding of *Trypanosoma cruzi II* in vampire bats from a district free o domestic vector-born transmssion in Northeastern Argentina. Parasitology. 2016;143(11):1358–68.

90. Gurgel-Gonçalves R, Cura C, Schijman AG, Cuba CA. Infestation of Mauritia flexuosa palms by triatomines (Hemiptera: Reduviidae), vectors of *Trypanosoma cruzi* and *Trypanosoma rangeli* in the Brazilian savanna. Acta Trop. 2012;121(2):105–11.
91. Miles MA, Arias JR, de Souza AA. Chagas' disease in the Amazon basin: V. Periurban palms as habitats of *Rhodnius robustus* and *Rhodnius pictipes*-triatomine vectors of Chagas' disease. Mem Inst Oswaldo Cruz. 1983;78(4):391–8.
92. Schmunis GA. Prevention of transfusional *Trypanosoma cruzi* infection in Latin America. Mem Inst Oswaldo Cruz. 1999;94(Suppl 1):93–101.
93. Yadon ZE, Schmunis GA. Congenital Chagas disease: estimating the potential risk in the United States. Am J Trop Med Hyg. 2009;81(6):927–33.
94. Celentano AM, González Cappa SM. Chagas' disease and blood transfusion: trypanocidal activity of maprotiline hydrochloride and gentian violet. Medicina. 1988;48(3):265–8.
95. Hammond DJ, Croft SL, Hogg J, Gutteridge WE. A strategy for the prevention of the transmission of Chagas' disease during blood transfusion. Acta Trop. 1986;43:367–78.
96. Carlier Y, Torrico F, Sosa-Estani S, Russomando G, Luquetti A, Freilij H, et al. Congenital Chagas disease: recommendations for diagnosis, treatment and control of newborns, siblings and pregnant women. PLoS Negl Trop Dis. 2011;5(10):e1250.
97. Juiz NA, Cayo NM, Burgos M, Salvo ME, Nasser JR, Búa J, et al. Human polymorphisms in placentally expressed genes and their association with susceptibility to congenital *Trypanosoma cruzi* infection. Infect Dis. 2016;213(8):1299–306.
98. Barcán L, Luna C, Clara L, Sinagra A, Valledor A, De Rissio AM, et al. Transmission of *T. cruzi* infection via liver transplantation to a nonreactive recipient for Chagas' disease. Liver Transpl. 2005;11(9):1112–6.
99. Bocchi EA, Fiorelli A. The paradox of survival results after heart transplantation for cardiomyopathy caused by *Trypanosoma cruzi.* First Guidelines Group for Heart Transplantation of the Brazilian Society of Cardiology. Ann Thorac Surg. 2001;71(6):1833–8.
100. Casadei D, Chagas' Disease Argentine Collaborative Transplant Consortium. Chagas' disease and solid organ transplantation. Transplant Proc. 2010;42(9):3354–9.
101. Kransdorf EP, Zakowski PC, Kobashigawa JA. Chagas disease in solid organ and heart transplantation. Curr Opin Infect Dis. 2014;27(5):418–24.
102. Kleffmann T, Schmidt J, Schaub GA. Attachment to *Trypanosoma cruzi* epimastigotes to hydrophobic substrates and use of this property to separate stages and promote metacyclogenesis. J Eukaryot Microbiol. 1998;45(5):548–55.
103. Guarneri AA, Lorenzo MG. Triatomine physiology in the context of trypanosome infection. J Insect Physiol. 2017;97:66–76.
104. de Isola EL, Lammel EM, Katzin VJ, GonzalezCappa SM. Influence of organ extracts of *Triatoma infestans* on differentiation of *Trypanosoma cruzi.* J Parasitol. 1981;67(1):53–8.
105. Isola EL, Lammel EM, Giovanniello O, Katzin AM, González Cappa SM. *Trypanosoma cruzi* morphogenesis: preliminary purification of an active fraction from hemolymph and intestinal homogenate of *Triatoma infestans.* J Parasitol. 1986b;72(3):467–9.
106. Lammel EM, Barbieri MA, Wilkowsky SE, Bertini F, Isola EL. *Trypanosoma cruzi*: involvemente of intracellular calciun in multiplication and differentiation. Exp Parasitol. 1996;83(2):240–9.
107. Isola EL, Lammel EM, González Cappa SM. *Trypanosoma cruzi* diferenciation after interaction of epimastigote and *Triatoma infestans* intestine homogenate. Exp Parasitol. 1986a;62(3):329–35.
108. Fraidenraich D, Peña C, Isola EL, Lammel EM, Coso O, Añel AD, et al. Stimulation of *Trypanosoma cruzi* adenyl cyclase by an alpha D-globin fragment from Triatoma hindgat: effect on differentiation of epimastigote to trypomastigote forms. Proc Natl Acad Sci U S A. 1993;90(21):10140–4.
109. Belauzaran ML, Lammel EM, Giménez G, Wainszelbaum MJ, de Isola EL. Envolvement of protein kinase C isoenzymes in *Trypanosoma cruzi* metaciclogenesis induced by oleica cid. Parasitol Res. 2009;105(1):47–55.

110. Wainszelbaum MJ, Belaunzarán ML, Lammel EM, Florin-Christensen M, Florin-Christensen J, Isola EL. Free fatty acids induce cell differenciatin to infective forms in *Trypanosoma cruzi*. Biochem J. 2003;375(3):705–12.
111. Isola EL, Lammel EM, González Cappa SM. *Trypanosoma cruzi*: differentiation to metacyclic trypomastigotes in the presence of ADP-ribosyltransferase inhibitors. Exp Parasitol. 1987;64(3):424–9.
112. Contreras VT, Araujo-Jorge TC, Bonaldo MC, Thomaz N, Barbosa HS, Meirelles Mde N, et al. Biological aspects of the D28c clone of *Trypanosoma cruzi* after metacyclogenesis in chemical defined media. Mem Inst Oswaldo Cruz. 1988;83(1):123–33.
113. Contreras VT, Salles JM, Thomas N, Morel CM, Goldenberg S. In vitro differentiation of *Trypanosoma cruzi* under chemically defined conditions. Mol Biochem Parasitol. 1985;16(3):315–27.
114. Hamedi A, Botelho L, Britto C, Fragoso SP, Umaki AC, Goldenberg S, et al. In vitro metacyclogenesis of *Trypanosoma cruzi* induced by starvation correlates with a transient adenilyl cyclase stimulation as well as with a constitutive upregulation of adenylyn cyclase expression. Mol Biochem Parasitol. 2015;200(1-2):9–18.
115. Dworak ES, Araújo SM, Gomes M, Massago M, Ferreira ÉC, Toledo MJO. Sympatry influence in the interaction of *Trypanosoma cruzi* with triatomine. Rev Soc Bras Med Trop. 2017;50(5):629–37.
116. Lammel EM, Müller LA, Isola EL, González Cappa SM. Effect of vector on infectivity of *Trypanosoma cruzi*. Acta Trop. 1985;42(2):149–55.
117. Szumlewicz AP, Muller CA. Studies in search of a suitable experimental insect model for xenodiagnosis of hosts with Chagas' disease. 2 Attempts to upgrade the reliability and the efficacy of xenodiagnosis in chronic Chagas' disease. Mem Inst Oswaldo Cruz. 1987;82(2):259–72.
118. Alves MJ, Colli W. Role of the gp85/trans-sialidase superfamily of glycoproteins in the interaction of *Trypanosoma cruzi* with host structures. Subcell Biochem. 2008;47:58–69.
119. Maeda FY, Cortez C, Izidoro MA, Juliano L, Yoshida N. Fibronectin-degrading activity of *Trypanosoma cruzi* cysteine proteinase plays a role in host cell invasion. Infect Immun. 2014;82(12):5166–74.
120. Nde PN, Simmons KJ, Kleshchenko YY, Pratap S, Lima MF, Villalta F. Silencing of the laminin gamma-1 gene blocks *Trypanosoma cruzi* infection. Infect Immun. 2006;74(3):1643–8.
121. Coimbra VC, Yamamoto D, Khusal KG, Atayde VD, Fernandes MC, Mortara RA, et al. Enucleated L929 cells support invasion, differentiation, and multiplication of *Trypanosoma cruzi* parasites. Infect Immun. 2007;75(8):3700–6.
122. Epting CL, Coates BM, Engman DM. Molecular mechanisms of host cell invasion by *Trypanosoma cruzi*. Exp Parasitol. 2010;126(3):283–91.
123. Burleigh BA, Andrews NW. The mechanisms of *Trypanosoma cruzi* invasion of mammalian cells. Annu Rev Microbiol. 1995;49:175–200.
124. Milei J, Sánchez J, Storino R, Yu ZX, Denduchis B, Ferrans VJ. Antibodies to laminin and immunohistochemical localization of laminin in chronic chagasic cardiomyopathy: a review. Mol Cell Biochem. 1993;129(2):161–70.
125. Ferreira V, Molina MC, Valck C, Rojas A, Aguilar L, Ramírez G, et al. Role of calreticulin from parasites in its interaction with vertebrate hosts. Mol Immunol. 2004;40(17):1279–91.
126. Johnson CA, Kleshchenko YY, Ikejiani AO, Udoko AN, Cardenas TC, Pratap S, et al. Thrombospondin-1 interacts with *Trypanosoma cruzi* surface calreticulin to enhance cellular infection. PLoS One. 2012;7(7):e40614. https://doi.org/10.1371/journal.pone.0040614.
127. Kleshchenko YY, Moody TN, Furtak VA, Ochieng J, Lima MF, Villalta F. Human galectin-3 promotes *Trypanosoma cruzi* adhesion to human coronary artery smooth muscle cells. Infect Immun. 2004;72(11):6717–21.
128. Moody TN, Ochieng J, Villalta F. Novel mechanism that *Trypanosoma cruzi* uses to adhere to the extracellular matrix mediated by human galectin-3. FEBS Lett. 2000;470(3):305–8.
129. Sato S, Hughes RC. Control of Mac-2 surface expression on murine macrophage cell lines. Eur J Immunol. 1994;24(1):216–21.

130. Grellier P, Vendeville S, Joyeau R, Bastos IM, Drobecq H, Frappier F, et al. *Trypanosoma cruzi* prolyl oligopeptidase Tc80 is involved in nonphagocytic mammalian cell invasion by trypomastigotes. J Biol Chem. 2001;276(50):47078–86.
131. Teixeira AA, de Vasconcelos Vde C, Colli W, Alves MJ, Giordano RJ. *Trypanosoma cruzi* binds to cytokeratin through conserved peptide motifs found in the laminin-G-like domain of the gp85/trans-sialidase proteins. PLoS Negl Trop Dis. 2015;9(9):e0004099.
132. Weston D, Patel B, Van Voorhis WC. Virulence in *Trypanosoma cruzi* infection correlates with the expression of a distinct family of sialidase superfamily genes. Mol Biochem Parasitol. 1999;98(1):105–16.
133. Ruiz RC, Favoreto S, Dorta ML, Oshiro MEM, Ferreira AT, Manque PM, et al. Infectivity of *Trypanosoma cruzi* strains is associated with differential expresión of surface glycoproteins with differential CA^2 signaling activity. Biochem J. 1998;330:505–11.
134. Alves MJ, Abuin G, Kuwajima VY, Colli W. Partial inhibition of trypomastigote entry into cultured mammalian cells by monoclonal antibodies against a surface glycoprotein of *Trypanosoma cruzi*. Mol Biochem Parasitol. 1986;21(1):75–82.
135. Freitas LM, dos Santos SL, Rodrigues-Luiz GF, Mendes TAO, Rodrigues TS, Gazzinelli RT, et al. Genomic analices, gene expresison and antigenic profile of the trans-sialidase superfamily of *Trypanosoma cruzi* reveal an undetected level of complexity. PLoS One. 2011;6: e25914.
136. Mattos EC, Tonelli RR, Colli W, Alves MJ. The Gp85 surface glycoproteins from *Trypanosoma cruzi*. Subcell Biochem. 2014;74:151–80.
137. Tonelli RR, Giordano RJ, Barbu EM, Torrecilhas AC, Kobayashi GS, Langley RR, et al. Role of the gp85/trans-sialidases in *Trypanosoma cruzi* tissue tropism: preferential binding of a conserved peptide motif to the vasculature in vivo. PLoS Negl Trop Dis. 2010;4(11): e864.
138. Zingales B, Carniol C, de Lederkremer RM, Colli W. Direct sialic acid transfer from a protein donor to glycolipids of trypomastigote forms of *Trypanosoma cruzi*. Mol Biochem Parasitol. 1987;26(1-2):135–44.
139. Campetella O, Sánchez D, Cazzulo JJ, Frasch AC. A superfamily of *Trypanosoma cruzi* surface antigens. Parasitol Today. 1992;8(11):378–81.
140. Frasch AC. Trans-sialidase, SAPA amino acid repeats and the relationship between *Trypanosoma cruzi* and the mammalian host. Parasitology. 1994;108(Suppl):S37–44.
141. Affranchino JL, Ibañez CF, Luquetti AO, Rassi A, Reyes MB, Macina RA, et al. Identification of a *Trypanosoma cruzi* antigen that is shed during the acute phase of Chagas' disease. Mol Biochem Parasitol. 1989;34(3):221–8.
142. Tambourgi DV, Kipnis TL, da Silva WD, Joiner KA, Sher A, Heath S, et al. A partial cDNA clone of trypomastigote decay-accelerating factor (T-DAF), a developmentally regulated complement inhibitor of *Trypanosoma cruzi*, has genetic and functional similarities to the human complement inhibitor DAF. Infect Immun. 1993;61(9):3656–63.
143. Hall BF, Webster P, Ma AK, Joiner KA, Andrews NW. Desialylation of lysosomal membrane glycoproteins by *Trypanosoma cruzi*: a role for the surface neuraminidase in facilitating parasite entry into the host cell cytoplasm. J Exp Med. 1992;176(2):313–25.
144. Risso MG, Pitcovsky TA, Caccuri RL, Campetella O, Leguizamón MS. Immune system pathogenesis is prevented by the neutralization of the systemic trans-sialidase from *Trypanosoma cruzi* during severe infections. Parasitology. 2007;134(Pt 4):503–10.
145. Vercelli CA, Hidalgo AM, Hyon SH, Argibay PF. *Trypanosoma cruzi* trans-sialidase inhibits human lymphocyte proliferation by nonapoptotic mechanisms: implications in pathogenesis and transplant immunology. Transplant Proc. 2005;37(10):4594–7.
146. Ruiz Díaz P, Mucci J, Meira MA, Bogliotti Y, Musikant D, Leguizamón MS, et al. *Trypanosoma cruzi* trans-sialidase prevents elicitation of Th1 cell response via interleukin 10 and downregulates Th1 effector cells. Infect Immun. 2015;83(5):2099–108.
147. Barrias ES, de Carvalho TM, De Souza W. *Trypanosoma cruzi*: entry into mammalian host cells and parasitophorous vacuole formation. Front Immunol. 2013;4:186.
148. Caradonna KL, Burleigh BA. Mechanisms of host cell invasion by *Trypanosoma cruzi*. Review. Adv Parasitol. 2011;76:33–61.

149. Schenkman S, Robbins ES, Nussenzweig V. Attachment of *Trypanosoma cruzi* to mammalian cells requires parasite energy, and invasion can be independent of the target cell cytoskeleton. Infect Immun. 1991b;59(2):645–54.
150. Andrews NW. Lysosome recruitment during host cell invasion by *Trypanosoma cruzi*. Trends Cell Biol. 1995;5(3):133–7.
151. Tardieux I, Webster P, Ravesloot J, Boron W, Lunn JA, Heuser JE, et al. Lysosome recruitment and fusion are early events required for trypanosome invasion of mammalian cells. Cell. 1992;71(7):1117–30.
152. Fernandes MC, Cortez M, Flannery AR, Tam C, Mortara RA, Andrews NW. *Trypanosoma cruzi* subverts the sphingomyelinase-mediated plasma membrane repair pathway for cell invasion. J Exp Med. 2011;208(5):909–21.
153. Rodríguez A, Samoff E, Rioult MG, Chung A, Andrews NW. Host cell invasion by trypanosomes requires lysosomes and microtubule/kinesin-mediated transport. J Cell Biol. 1996;134(2):349–62.
154. Caler EV, Chakrabarti S, Fowler KT, Rao S, Andrews NW. The exocytosis-regulatory protein synaptotagmin VII mediates cell invasion by *Trypanosoma cruzi*. J Exp Med. 2001;193(9):1097–104.
155. Chakrabarti S, Kobayashi KS, Flavell RA, Marks CB, Miyake K, Liston DR, et al. Impaired membrane resealing and autoimmune myositis in synaptotagmin VII-deficient mice. J Cell Biol. 2003;162(4):543–9.
156. Rodríguez A, Rioult MG, Ora A, Andrews NW. A trypanosome-soluble factor induces IP3 formation, intracellular Ca2+ mobilization and microfilament rearrangement in host cells. J Cell Biol. 1995;129(5):1263–73.
157. Tardieux I, Nathanson MH, Andrews NW. Role in host cell invasion of *Trypanosoma cruzi*-induced cytosolic-free Ca2+ transients. J Exp Med. 1994;179(3):1017–22.
158. Fernandes MC, Cortez M, Geraldo Yoneyama KA, Straus AH, Yoshida N, Mortara RA. Novel strategy in Trypanosoma cruzi cell invasion: implication of cholesterol and host cell microdomains. Int J Parasitol. 2007;37(13):1431–41.
159. Andrews NW. Lysosomes and the plasma membrane: trypanosomes reveal a secret relationship. Review. J Cell Biol. 2002;158(3):389–94.
160. Reddy A, Caler EV, Andrews NW. Plasma membrane repair is mediated by Ca(2+)-regulated exocytosis of lysosomes. Cell. 2001;106(2):157–69.
161. McNeil PL, Steinhardt RA. Plasma membrane disruption: repair, prevention, adaptation. Review. Annu Rev Cell Dev Biol. 2003;19:697–731.
162. Woolsey AM, Burleigh BA. Host cell actin polymerization is required for cellular retention of *Trypanosoma cruzi* and early association with endosomal/lysosomal compartments. Cell Microbiol. 2004;6(9):829–38.
163. Woolsey AM, Sunwoo L, Petersen CA, Brachmann SM, Cantley LC, Burleigh BA. Novel PI 3-kinase-dependent mechanisms of trypanosome invasion and vacuole maturation. J Cell Sci. 2003;116(17):3611–22.
164. Wilkowsky SE, Barbieri MA, Stahl PD, Isola EL. Regulation of *Trypanosoma cruzi* invasion of nonphagocytic cells by the endocytically active GTPases dynamin, Rab5, and Rab7. Biochem Biophys Res Commun. 2002;291(3):516–21.
165. Romano PS, Arboit MA, Vázquez CL, Colombo MI. The autophagic pathway is a key component in the lysosomal dependent entry of *Trypanosoma cruzi* into the host cell. Autophagy. 2009;5(1):6–18.
166. Martins RM, Alves RM, Macedo S, Yoshida N. Starvation and rapamycin differentially regulate host cell lysosome exocytosis and invasion by *Trypanosoma cruzi* metacyclic forms. Cell Microbiol. 2011;13(7):943–54.
167. Mizushima N, Yamamoto A, Matsui M, Yoshimori T, Ohsumi Y. In vivo analysis of autophagy in response to nutrient starvation using transgenic mice expressing a fluorescent autophagosome marker. Mol Biol Cell. 2004;15(3):1101–11.
168. Romano PS, Cueto JA, Casassa AF, Vanrell MC, Gottlieb RA, Colombo MI. Molecular and cellular mechanisms involved in the *Trypanosoma cruzi*/host cell interplay. IUBMB Life. 2012;64(5):387–96.

169. Wilkowsky SE, Barbieri MA, Stahl P, Isola EL. *Trypanosoma cruzi*: phosphatidylinositol 3-kinase and protein kinase B activation is associated with parasite invasion. Exp Cell Res. 2001;264(2):211–8.
170. Chuenkova MV, Furnari FB, Cavenee WK, Pereira MA. *Trypanosoma cruzi* trans-sialidase: a potent and specific survival factor for human Schwann cells by means of phosphatidylinositol 3-kinase/Akt signaling. Proc Natl Acad Sci U S A. 2001;98(17):9936–41.
171. Andrade LO, Andrews NW. Lysosomal fusion is essential for the retention of *Trypanosoma cruzi* inside host cells. J Exp Med. 2004;200(9):1135–43.
172. Kierszenbaum F, Knecht E, Budzko DB, Pizzimenti MC. Phagocytosis: a defense mechanism against infection with *Trypanosoma cruzi*. J Immunol. 1974;112(5):1839–44.
173. Muñoz-Fernández MA, Fernández MA, Fresno M. Activation of human macrophages for the killing of intracellular *Trypanosoma cruzi* by TNF-alpha and IFN-gamma through a nitric oxide-dependent mechanism. Immunol Lett. 1992;33(1):35–40.
174. Piacenza L, Alvarez MN, Peluffo G, Radi R. Fighting the oxidative assault: the *Trypanosoma cruzi* journey to infection. Curr Opin Microbiol. 2009;12(4):415–21. https://doi.org/10.1016/j.mib.2009.06.011.
175. Celentano AM, González Cappa SM. In vivo macrophage function in experimental infection with *Trypanosoma cruzi* subpopulations. Acta Trop. 1993;55(3):171–80.
176. Zago MP, Hosakote YM, Koo SJ, Dhiman M, Piñeyro MD, Parodi-Talice A, et al. TcI Isolates of *Trypanosoma cruzi* Exploit the Antioxidant Network for Enhanced Intracellular Survival in Macrophages and Virulence in Mice. Infect Immun. 2016;84(6):1842–56.
177. Kipnis TL, Calich VL, da Silva WD. Active entry of bloodstream forms of *Trypanosoma cruzi* into macrophages. Parasitology. 1979;78(1):89–98.
178. Andrews NW. From lysosomes into the cytosol: the intracellular pathway of *Trypanosoma cruzi*. Braz J Med Biol Res. 1994;27(2):471–5.
179. Ley V, Robbins ES, Nussenzweig V, Andrews NW. The exit of *Trypanosoma cruzi* from the phagosome is inhibited by raising the pH of acidic compartments. J Exp Med. 1990;171(2):401–13.
180. Andrews NW, Abrams CK, Slatin SL, Griffiths G. A *T. cruzi*-secreted protein immunologically related to the complement component C9: evidence for membrane pore-forming activity at low pH. Cell. 1990;61(7):1277–87.
181. Manning-Cela R, Cortés A, González-Rey E, Van Voorhis WC, Swindle J, González A. LYT1 protein is required for efficient in vitro infection by *Trypanosoma cruzi*. Infect Immun. 2001;69(6):3916–23.
182. Tomlinson S, Vandekerckhove F, Frevert U, Nussenzweig V. The induction of *Trypanosoma cruzi* trypomastigote to amastigote transformation by low pH. Parasitology. 1995;110(Pt 5):547–54.
183. Tonelli RR, Silber AM, Almeida-de-Faria M, Hirata IY, Colli W, Alves MJ. L-proline is essential for the intracellular differentiation of *Trypanosoma cruzi*. Cell Microbiol. 2004;6(8):733–41.
184. Grellier P, Blum J, Santana J, Bylèn E, Mouray E, Sinou V, et al. Involvement of calyculin A-sensitive phosphatase(s) in the differentiation of *Trypanosoma cruzi* trypomastigotes to amastigotes. Mol Biochem Parasitol. 1999;98(2):239–52.
185. Costales J, Rowland EC. A role for protease activity and host-cell permeability during the process of *Trypanosoma cruzi* egress from infected cells. J Parasitol. 2007;93(6):1350–9.
186. Clark RK, Kuhn RE. *Trypanosoma cruzi* does not induce apoptosis in murine fibroblasts. Parasitology. 1999;118(2):167–75.
187. de Souza EM, Araújo-Jorge TC, Bailly C, Lansiaux A, Batista MM, Oliveira GM, et al. Host and parasite apoptosis following *Trypanosoma cruzi* infection in in vitro and in vivo models. Cell Tissue Res. 2003;314(2):223–35.

188. Mortara RA. *Trypanosoma cruzi*: amastigotes and trypomastigotes interact with different structures on the surface of HeLa cells. Exp Parasitol. 1991;73(1):1–14.
189. Ley V, Andrews NW, Robbins ES, Nussenzweig V. Amastigotes of *Trypanosoma cruzi* sustain an infective cycle in mammalian cells. J Exp Med. 1988;168(2):649–59.
190. Mortara RA, Andreoli WK, Taniwaki NN, Fernandes AB, Silva CV, Fernandes MC, et al. Mammalian cell invasion and intracellular trafficking by *Trypanosoma cruzi* infective forms. An Acad Bras Cienc. 2005;77(1):77–94.
191. Perlowagora-Szumlewics A, Moreira CJ. In vivo differentiation of *Trypanosoma cruzi*--1. Experimental evidence of the influence of vector species on metacyclogenesis. Mem Inst Oswaldo Cruz. 1994;89(4):603–18.
192. Previato JO, Andrade AF, Pessolani MC, Mendonça-Previato L. Incorporation of sialic acid into *Trypanosoma cruzi* macromolecules. A proposal for a new metabolic route. Mol Biochem Parasitol. 1985;16(1):85–96.

A Panoramic View of the Immune Response to *Trypanosoma cruzi* Infection

Gonzalo R. Acevedo, Magali C. Girard, and Karina A. Gómez

Abstract Chagas disease is a complex disorder in which the immunological response developed by the host plays a fundamental role, not only in the clearance of the parasite but also in the inflammatory status observed in specific affected tissues. Chagas disease has two phases, acute and chronic, the latter being established in those cases where treatment with currently available anti-parasitic drugs (nifurtimox and benznidazole) is either not applied or not effective. During the chronic phase, the disease may remain without any detectable symptoms for several decades or progress toward cardiac, digestive, neurological forms, or even a combination of these alterations. The immune response developed in all of these conditions is flowery and comprises humoral and cellular components; however the clearance of the parasite is incomplete due to the multiple mechanisms that *T. cruzi* deploys in order to perpetuate itself within the host.

Here, we make an extensive review of *T. cruzi*-host immune response interactions with special attention on human models, also referring to the particular clinical scenario of etiological treatment in Chagas disease.

1 Introduction

Many chronically infecting pathogens, like *T. cruzi*, share a common property that is crucial for their survival: the capacity of evading and/or modulating the mechanisms of the immune system. It is highly probable that this feature was acquired to

G. R. Acevedo · M. C. Girard · K. A. Gómez (✉)
Instituto de Investigaciones en Ingeniería Genética y Biología Molecular
(INGEBI-CONICET), Buenos Aires, Argentina
e-mail: gomez@dna.uba.ar

J. M. Altcheh, H. Freilij (eds.), *Chagas Disease*, Birkhäuser Advances in Infectious Diseases, https://doi.org/10.1007/978-3-030-00054-7_3

an extensive coevolutionary history of the pathogen and its host's immune processes. Hence, *T. cruzi* has evolved sophisticated means to escape, inactivate, or subvert different components of the human immune system.

Despite the different actors of this system work in tight interdependence, subdividing its study is often useful for ease of comprehension. Thus, innate immunity refers to the cellular and molecular mechanisms active even before an infection or injury occurs and which respond essentially in the same way upon repeated exposure to the same pathogen. On the other hand, adaptive immunity alludes to the mechanisms that enable the immune system to respond, ideally, with increasing magnitude and efficacy to a pathogen upon reiterative exposure. It is defined by two main features: an extremely refined specificity for particular molecules or parts of molecules and the capacity to "remember" such molecules or parts to enable a response that increases its efficacy upon re-exposure [1].

The mechanisms of each immunity "type" are given by their molecular and cellular components. In the case of innate immunity, these are:

- Physical and chemical barriers at the entrances to the organism (epithelia and antimicrobial molecules, such as lysozymes in epithelial secretions)
- Blood molecules, in particular the complement system factors and other mediators of inflammation
- Phagocytes (neutrophils and macrophages), dendritic cells, and natural killer (NK) lymphocytes
- Cytokines and chemokines, molecules that signal and coordinate innate immunity mechanisms

In the case of adaptive immunity, these components are:

- B cells and B cell activation products, mainly antibodies
- T cells

In this chapter, we intend to compile the current knowledge of the different aspects of the immune response covering from the acute to the chronic phase of the disease, with an emphasis on those phenomena that have been studied and confirmed in the human host (Fig. 1). One special scenario of the innate and adaptive immunity against the parasite is discussed: the anti-parasitic treatment of patients in the chronic phase of Chagas disease.

It is important to clarify that the majority of the data reported about the acute phase has been obtained from animal models of infection using different strains of *T. cruzi*. The lack of information on human acute infections is due mainly to the non-specific nature of the signs and symptoms of infection and, even worse, to the difficulty access to healthcare often poses to segregated populations in highly endemic areas. On the contrary, an extended number of reports describe the immunology of human *T. cruzi* infection during the chronic phase. Thus, the authors of this chapter invite you to keep in mind the limitations of the investigation done so far.

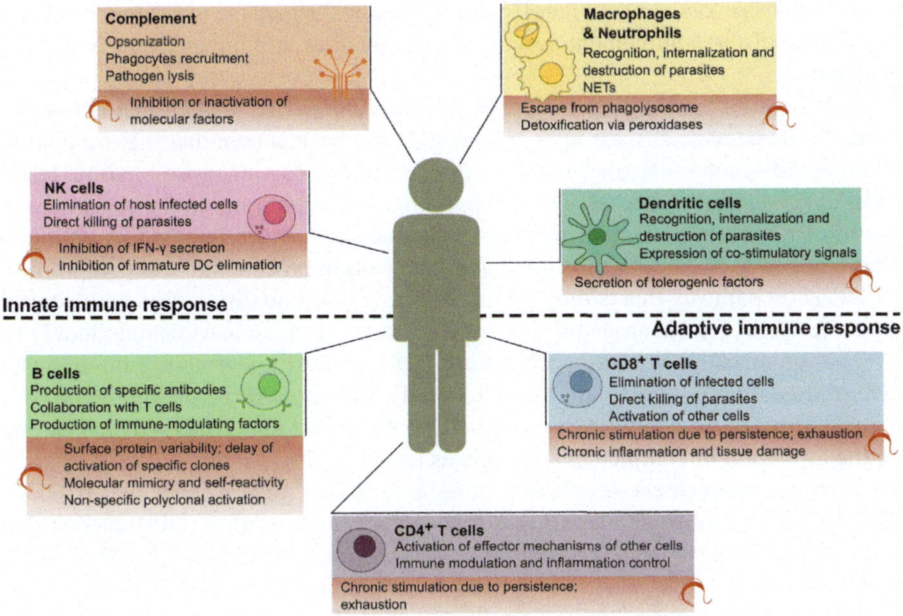

Fig. 1 Schematic representation of the interplay between *T. cruzi* and human immune response, made by G.A: Acevedo

2 Innate Immunity

2.1 Complement System

The complement system consists of several plasma circulating proteins, which opsonize pathogens and recruit phagocytes to the infection site and, in some cases, destroy the pathogen in a direct fashion. It functions as a cascade of proteolytic events in which an enzymatic precursor (known as zymogen) is activated and therefore becomes an active protease, capable of cleaving the following component, which in term acquires proteolytic activity and activates the next factor in the complement cascade. As a whole, this mechanism amplifies the signal generated by the presence of a pathogen to favor its efficient clearance. The first activation step can occur by three different ways, known as the classical, alternative, and lectin pathways. All of them converge in the formation of C3 convertase complex (although the composition of this complex differs between the classical pathway and the other two) and the consequent separation of C3 in C3a and C3b. The effector mechanisms of the complement depend on this step: the complement-promoted phagocytosis is mediated by C3b receptors expressed by neutrophils and macrophages, and C3a is a pro-inflammatory agent. Furthermore, C3b is necessary for the generation of the C5 convertase, which gives rise to C5a, a strong pro-inflammatory signal, and for pore formation on the pathogen's membrane.

T. cruzi has evolved a set of molecules that allow the parasite to escape or subvert the effects of the complement system, by inactivating or dampening its activation pathways.

Calreticulin (TcCRT) is an endoplasmic reticulum (ER) protein that is translocated to the parasite surface after infection. It has been shown that it is capable of binding different molecular pattern sensors, like C1q, mannose-binding lectin (MBL) [2], and L-ficolin [3]. Therefore, it affects the first step on the classical and lectin complement activation pathways. Nevertheless, mice deficient in a protein that binds to MBL during the complement activation process suffer a compromise of the lectin pathway that is only partial, and as a result no differences with control mice were observed in susceptibility upon infection in this experimental model [4].

The trypomastigote decay-accelerating factor (T-DAF), a protein from the inactive trans-sialidase family, and its analogue the human DAF modulate the physiological decay of the alternative and classical pathway C3 convertases, by interfering in the formation of complement complexes [5–7].

T. cruzi complement regulatory protein (TcCRP), also known as gp160, is another inactive trans-sialidase that is glycophosphatidylinositol (GPI)-anchored to the cell membrane of the trypomastigote. TcCRP is capable of binding C3b and C4b and interferes in the classical and alternative complement pathways [8]. Furthermore, even though it has not been empirically tested, its C4 binding capacity could also affect the lectin pathway [7]. Normally, it is not expressed on the epimastigote and amastigote forms surface, but their exogenous expression on epimastigotes renders them less susceptible to complement-mediated lysis [8, 9]. Some authors have proposed a correlation between the TcCRP expression level and the virulence in different *T. cruzi* strains [10]. However, evidence on that regards is scarce and inconclusive.

The *T. cruzi* complement C2 receptor inhibitor trispanning (TcCRIT) is a transmembrane protein homologous to *Schistosoma haematobium* ShCRIT (previously known as trispanning orphan receptor, TOR) and hCRIT in humans. TcCRIT (like its analogues) and C4 compete for the binding of C2, therefore modulating the C3 convertase formation in the classical pathway [7, 11].

T. cruzi 58/68 glycoprotein (gp58/68) functions as an analogue of endogenous complement regulatory molecules, diminishing the alternative pathway C3 convertase formation by inhibiting the union of the complement factor B protein with parasite-bound C3b. Its release to the environment by trypomastigotes has also been reported, although the biological relevance of this mechanism is still to be clarified [7].

2.2 Macrophages and Neutrophils

Phagocytes, especially macrophages, neutrophils, and dendritic cells (DCs), are the first line of defense against pathogens on their way through the epithelial barriers. Two major features are key to their function: (1) the recognition of the pathogen-associated molecular patterns (PAMPs) and damage-associated molecular patterns

(DAMPs) through membrane receptors (like Toll like-receptors, TLR), enabling the detection and phagocytosis of microorganisms and cellular debris and internalization and destruction of microorganisms and (2) the secretion of cytokines that promote inflammation and enhance the activation of other cells at the site of infection [1].

The mechanism by which *T. cruzi* enters the so-called professional phagocytes has been a matter of controversy, since discordant results have been reported regarding whether it takes place by means of parasite-extrinsic mechanisms or the parasite participates actively in the process. Current consensus includes both hypotheses and assumes that at least two different internalization mechanisms take place during infection: a phagocytic, actin-dependent one and another one that relies on microdomain-like cell membrane structures on the macrophage [12].

Tissue-resident macrophages are considered the first host cells to be invaded by *T. cruzi* upon infection. Although this cell type can internalize both trypomastigotes and epimastigotes, only the former can escape the phagolysosome [5, 12]. To detoxify the oxidant agents that the activated macrophage produces for its elimination, the parasite counts on an antioxidant metabolic network composed of several enzymes and nonenzymatic molecules distributed across different subcellular compartments. Among them, five peroxidases have been described: the cytosolic and mitochondrial tryparedoxin peroxidase (TcCPX and TcMPX, respectively), the ascorbate-dependent hemoperoxidase (TcAPX), and the glutathione peroxidases I and II [5]. Additionally, in the murine infection model, the enzyme cruzipain, a cysteine protease from the parasite, diminishes the macrophage's trypanocide activity by increasing the arginase activity, which competes with iNOS for its substrate L-arginine [13]. It is worth mentioning that in Chagas disease pediatric patients, despite an observed shrinkage of the morphologically macrophage-like population in flow cytometry assessments, the surface molecule expression analysis demonstrated that monocytes with a pro-inflammatory phenotype have an increased frequency in these patients, as compared with non-infected children of the same age [14].

The interaction between *T. cruzi* and neutrophils has been studied in murine model of susceptibility (BALB/c mice) and resistance (C57BL/6 mice) to infection with Tulahuen strain parasites. In the susceptible model, animals display increased parasitemia, number of tisular pseudocysts, and mortality when monocytes and neutrophils are depleted, as compared to non-depleted control mice. In contrast, the same depletion on the resistant model leads to decreased parasitemia without changes neither in the number of pseudocysts nor in mortality. A comparative analysis of the secreted cytokine profile upon monocytes and neutrophils depletion suggested a protective role against infection for IL-2, IFN-γ, and TNF-α from macrophages, while IL-10 secretion seems not to be affected [15]. Differences in the neutrophil induction of nitric oxide (NO) production on macrophages, which in time depends on TNF-α and elastase on one hand (a greater production of these molecules in C57BL/6 mice would have a protective effect) and prostaglandin E2 and TGF-β on the other (a greater production of these would lead to susceptibility in BALB/c mice, as a result of an increased parasite replication), might explain the differential susceptibility to infection in these mice strains [16]

In humans, the expression of metalloproteinases (MMPs, enzymes involved in the remodeling of the extracellular matrix) and different cytokines in monocytes and neutrophils from asymptomatic and cardiac Chagas disease patients and in non-infected individuals has been studied. In vitro observations showed that these two cellular types are involved in the coordination of the adaptive immune response, the regulation of inflammation through cytokines secretion, and extracellular matrix remodeling in the heart, which could be related to Chagas heart disease-associated fibrosis. Regarding the latter, the authors suggest that MMP-2 may have a preventive action on cardiac tissue damage, in correlation with IL-10 and TNF-α secretion, while MMP-9 would have a detrimental effect as an enhancer of inflammation [17].

In addition, Sousa-Rocha et al. demonstrated that trypomastigotes and soluble *T. cruzi* antigens are capable of inducing the release of neutrophil fibrous structures (also known as NETs), by triggering TLR-2 and TLR-4 receptors, and that this response depends on the activation of the respiratory burst and the production of ROS [18]. Additionally, it was demonstrated that these NETs do not have trypanocide capacity, as they do not affect parasite viability, but they are able to reduce their infectivity by inducing trypomastigotes to differentiate into amastigotes in the extracellular environment [18].

2.3 Dendritic Cells

Like macrophages and neutrophils, DCs are also activated in the presence of PAMPs and DAMPs. Once activated, a DC responds by expressing co-stimulatory molecules and cytokines which are necessary, besides the antigen itself, for the activation of T cells. The profile of produced cytokines depends on the nature of the activating pathogen and will direct the differentiation of naïve T cells toward different functional profiles [1, 19].

In the context of *T. cruzi* infection, it has been demonstrated that the parasite is internalized by DCs, although murine model experiments show there are different degrees of infectivity which depend on the parasite strain. This dependence did not show any relation to the strain classification in discrete typing units (DTU), nor to biological parameters of the DC [20].

Experiments with human DCs showed their function is affected by factors secreted by the parasite, which induce a tolerogenic profile by decreasing IL-12 and TNF-α production [21]. It should be mentioned that under this model, cytoplasm invasion by the pathogen was observed. Also, a decrease of class I and II major histocompatibility complex (MHC) molecules and CD40 co-receptor expression was shown to be induced on DCs by *T. cruzi* soluble factors, tampering with this cell's antigen presentation capacity [21, 22]. These effects have been attributed, at least partially, to parasite-produced glycosylinositolphospholipids (GIPL) [23]. These phenomena have a direct effect on T cell activation [22]. Experiments with mouse bone marrow-derived DCs showed that inhibitory receptor SIGLEC-E activated by the sialylated ligands on the parasite surface downregulates IL-12 secretion and upregulates that of IL-10 [24] and may therefore be involved in immunomodulatory mechanisms associated with *T. cruzi* infection.

2.4 Natural Killer Lymphocytes

Natural killer (NK) cells play a central role in the innate immune response, especially in infections by intracellular pathogens. In contrast to T and B lymphocytes, NK cells do not require a clonal expansion and differentiation process to fulfill their function.

NK cells bias T cell differentiation toward an inflammatory profile and activate macrophages via IFN-γ secretion, which was demonstrated to take place in the murine *T. cruzi* infection model. It has been shown that NK cells produce a peak of IFN-γ secretion shortly after infection, in a process that is dependent of adherent cells in the thymus, but independent of T cells, and that also requires the presence of viable parasites (heat- or radiation-killed parasites failed to activate this cytokine secretion). This IFN-γ blast may be crucial for the early control of parasitemia during the acute phase of the infection. Depleting NK cells from these mice results in IFN-γ secretion abrogation and IL-10 production increase, probably causing the immune system to tolerate the presence of the parasite [25].

In addition, NK cells can eliminate extracellular parasites directly, by means of the quick formation of intercellular contacts that result in the immediate loss of motility and cell membrane damage for the parasite. This mechanism depends on the NK cell activation by IL-12 and involves the exocytosis of cytotoxic granules [26, 27]. These observations lead to the conclusion that the role of NK cells in controlling the parasite burden has more to do with direct extracellular parasite killing than with the elimination of *T. cruzi*-infected cells.

The observation of a decrease in the infective capacity of the parasite on cultured mouse fibroblasts in the presence of total splenocytes that is reverted when NK cells are depleted has been pointed out as evidence for a regulatory role for this cell subset on nonimmune cells in the context of *T. cruzi* infection. Looking into this phenomenon, Lieke et al. corroborated that it is mediated by IFN-γ secretion from NK cells, which induces an increase of iNOS expression in the fibroblasts. Also, type I interferons were identified as messengers in this cross talk between NK cells and fibroblasts, being produced by both cell types in response to the parasite. It was demonstrated, however, that the effect of these interferons and IL-12 on the trypanocide mechanism induced by NK cells on fibroblast was neglectable [28].

In the context of human infection, Ferreira et al. conducted a microarray-based transcriptomic signature study on peripheral blood cells from Chagas disease patients, in which ex vivo transcriptional profiles were compared between patients classified in different categories: severe cardiopathy, mild cardiopathy, asymptomatic with negative PCR for *T. cruzi* DNA, asymptomatic with positive PCR, and control non-infected subjects [29]. Despite this grouping being questionable, due to the usage of different criteria for the definition of the categories, it is worth mentioning that the expression of genes related to NK cell activity was found to be upregulated in positive PCR asymptomatic patients and mild cardiopathy patients and downregulated in patients with severe cardiopathy [29].

A phenotypic study of the cell subsets circulating in peripheral blood from pediatric and adult Chagas disease patients, with and without cardiac symptoms, suggested a role for pre-NK cells in the activation of macrophage effector mechanisms, during early stages of the asymptomatic phase. These pre-NK cells are predominantly cytokine secretors, in contrast to mature NK cells which are predominantly cytotoxic. An augmented frequency of mature NK cells was also observed in asymptomatic patients, as compared to cardiac patients, suggesting a contribution of this cell subset to the establishment and/or maintenance of the lack of symptoms during chronic infection [14]. Additionally, peripheral and cord blood experimental infection assays highlighted NK cells as the most potent IFN-γ-producing subset in response, besides also secreting IL-15 [30], suggesting a relevance of this subset in the primary response to infection.

3 Adaptive Immunity

3.1 B Lymphocytes

B cells are the only cells capable of producing antibodies. In mice and humans, there are three principal classes of B cells, classified on the basis of their ontogeny and anatomic localization: B1 and B2 B cells, consisting of the marginal zone (MZ) and follicular (FO) B cells [31]. B1 and MZ B cells respond like innate cells in mediating rapid IgM antibody responses, while FO B cells, which reside in spleen and lymph nodes, are the conventional B lymphocytes of the adaptive immune system and are the most numerous of all B cell lineages.

It is known that *T. cruzi* infection generates effects on B cell maturation and differentiation, which conduct to an antibody response that is not capable of effectively eliminating the parasite or protecting against re-exposition [32]. It has been described that patients with chronic Chagas disease evidence a higher frequency of B ($CD19^+$) cells with increased expression of co-stimulatory molecules CD80 and CD86, as compared to non-infected individuals [33]. This is a property of activated B cells, and therefore this observation might be attributed to the permanent presence of activating antigens, or to a pathological chronic activation phenomenon.

An increase in the frequency of circulating B cell seems to occur during the late acute phase of *T. cruzi* infection but only becoming significant at the beginning of the chronic phase [34]. Nonetheless, it has been proposed that, in patients with established chronic Chagas disease, the number of $CD19^+$ cells is not different from that of non-infected individuals [33, 35], suggesting a subsequent contraction of this population. In spite of this apparent unaltered number of circulating B cells, the same is not true about the representation of the different B cell subsets: Fernández et al. reported a selective decrease in B cells from the memory subset with different degrees of differentiation, and in terminally differentiated plasma cells, along with an increase in the unconventional double-negative B cells [35]. This implies an increase in B cells incapable of generating a substantial anti-*T. cruzi* IgG antibody response. In addition, Fares et al. found an augmented CD21-expressing B cell population [33], being CD21 a CD19-associated co-receptor which enhances the activation signals generated by membrane Ig in contact with complement-bound antigen and T cell-dependent response antigens [36].

The specificity of antibodies generated upon infection has been studied since the early days of Chagas disease immunology research. Up to now, the immune epitopes database (IEDB, http://www.iedb.org) has 88 *T. cruzi* molecule entries, which contain in total over 2×10^3 epitopes bound by specific antibodies in human infection and in animal models. Some of these antibodies are often referred to as "lytic," meaning they enable complement-mediated parasite lysis. Among target antigens to these antibodies are gp190, T-DAF, the 90 kDa surface protein, and several GPI-anchored mucin-like glycoproteins. The mechanism by which these antibodies allow direct lysis of the parasites is thought to be related to the blockade of the complement evasion pathways discussed previously [37]. Among other proteins recognized by—non-necessarily lytic—antibodies, produced in response to *T. cruzi* infection, there are mucins (TcMUC), mucin-associated surface proteins (MASPs), trans-sialidases, amastigote surface proteins (ASPs), paraflagellar rod protein (TcPRP), kinetoplastid membrane protein 11 (KMP-11), glycoprotein gp82, the enzyme neuraminidase, heat shock protein hsp70, ribosomal proteins, and several others (IEDB, accessed December 2017), besides carbohydrates like the so-called gal epitope (gal α1-3 gal) [38]. It is worth mentioning that the recent application of high-throughput technologies and informatics prediction methods have expanded the possibilities of exploring the repertoire of antibodies produced against infection by this complex parasite and constitute promising tools for the study of the humoral adaptive immune response in Chagas disease [39, 40].

The importance of anti-*T. cruzi* antibodies to the control of infection has been demonstrated in the mouse infection model: in spite of living significantly longer than T cell-deficient animals, mutant mice incapable of producing antibodies ($b2m^{-/-}$) are unable to control parasitic growth and succumb to infection during the acute phase [41]. This suggests that T cell response rather than B cell response is crucial for the initial control of the infection, but B cells are keys to long-term survival.

Despite this, the antibodies produced against *T. cruzi* do not effectively eliminate the parasite, providing an opportunity to establish a persistent infection. This deficient antibody response is possibly due to three main factors. The first is antigenic variability: the parasite exposes a variety of antigens on its surface, such as mucins, trans-sialidases, and MASPs, encoded by highly polymorphic multigenic families. This high diversity of molecules expressed at the same time delays the activation of specific B cell clones and therefore also the production and maturation of high-affinity antibodies with neutralizing capacity [5, 42, 43].

The second factor is the reduction of immature B cells in the bone marrow (BM). It was proposed that, by affecting the BM, *T. cruzi* could compromise the whole humoral immune response, limiting the generation of mature B cells in the periphery. Zuniga et al. have shown that during acute *T. cruzi* murine infection, there is a loss of immature B cells in the BM and of transitional (Tr) B cells in the periphery, probably as a consequence of an increased apoptosis rate in immature B cells [44]. Of note, Tr B cells are developmentally intermediate between immature BM B cells and fully mature näive B cells found in the peripheral blood and secondary lymphoid tissues.

The third factor is unspecific polyclonal B cell activation. It was demonstrated in murine models of infection that some parasite molecules cause non-specific, T-independent activation and proliferation of B cells, generating splenomegaly and hypergammaglobulinemia associated with the production of non-*T. cruzi*-specific antibodies [5, 45–47]. This phenomenon has been linked to susceptibility to infection in BALB/c mice in contrast with resistance in C57BL/6 mice. Monitoring total and anti-*T. cruzi* (measured as TcCRP-specific antibodies) IgM and IgG production, it was shown that while BALB/c mice do not produce IgM in response to the parasite, they suffer from hypergammaglobulinemia and low parasite-specific response and that C57BL/6 exhibit an initial increase in total IgG and IgM followed by a rise in anti-CRP antibody titer of both isotypes [48]. A case of accidental infection permitted the observation of this phenomenon in the context of a human acute infection case, exhibiting expansion of total antibodies, initially IgM and IgA, followed by IgG with specificities not related to *T. cruzi* antigens [49]. Glutamate dehydrogenase (TcGDH) [50], proline racemase [51], and TcTS [52] are among the parasitic proteins identified as polyclonal B cell mitogens. Two different effects were proposed for this polyclonal B cell activation. On one hand, an early host defense with non-specific antibodies against a big spectrum of conserved structures present in *T. cruzi* and other pathogens may allow the clearance of the parasite during the acute phase. In this scenario, the concomitant reduced specific humoral response could lead to the establishment of a persistent infection. On the other hand, polyclonal B cell activation could be a mechanism triggered by the parasite to avoid the host-specific immune response, by lowering the frequency of specific antibody-producing B cells [32].

At this point, it is worth mentioning that one of the most widespread hypotheses explaining the pathogenesis of Chagas disease is the existence of self-reactive immune mechanisms developed as a consequence of the persistent infection [53]. In particular, evidence of molecular mimicry has been found between host and parasite molecules, and it has been suggested that this may lead to the production of self-reactive antibodies with deleterious effects. In particular, molecular mimicry has been demonstrated between *T. cruzi* peptide B13 and human myosin heavy chain, ribosomal P proteins and muscarinic and adrenergic receptors, and protein FL-160 and a neuronal 47 kDa protein, among others summarized in Table 1 [54–57].

Given that self-reactivity has been observed most frequently in the chronic phase of Chagas disease in humans and experimental models, some authors propose that it is a result of a low-level stimulation of self-reactive cells during a long period of time [58]. However, cellular and humoral self-reactivity against myosin and antibodies against actin and tubulin have been found in acutely infected mice [58, 59], and antibodies specific for laminin were detected in acutely infected humans [49], suggesting that this phenomenon may begin during the acute phase of infection.

Several mechanisms have been suggested to participate in triggering self-reactivity after *T. cruzi* infection including (1) exposure to large amounts of self-antigens released by the mechanical tissue damage caused by the parasite, together with an appropriate inflammatory environment (bystander activation), (2) polyclonal activation of B cells generating autoantibody production, and (3) molecular mimicry (as mentioned above), in which T and B cells recognize parasite antigens

Table 3.1 Host and parasite components involved in molecular mimicry and self-reactive immune response

Host component	*T. cruzi* antigen	Type of response
Neurons, liver, kidney, testis	Unknown	Humoral
Neurons	Sulfated glycolipids	Humoral
47 kDa neuron protein	FL-160	Humoral
Heart and skeletal muscle	Microsomal fraction	Humoral
Smooth and striated muscle	150 kDa protein	Humoral
Human cardiac myosin heavy chain	B13 protein	Humoral and T cells
Human cardiac myosin heavy chain	Cruzipain	Humoral
95 kDa myosin tail	*T. cruzi* cytoskeleton	Humoral
Skeletal muscle Ca^{2+}-dependent SRA	SRA	Humoral
Glycosphingolipids	Glycosphingolipids	Humoral
MAP (brain)	MAP	Humoral
Myelin basic protein	*T. cruzi* soluble extract	Humoral and T cells
28 kDa lymphocyte membrane protein	55 kDa membrane protein	Humoral
23 kDa ribosomal protein	23 kDa ribosomal protein	Humoral
Ribosomal P protein	Ribosomal P protein	Humoral
38-kDa heart antigen	R13 peptide from ribosomal protein P1, P2	Humoral
β1-adrenoreceptor, M2 muscarinic receptor	Ribosomal P0 and P2β proteins	Humoral
β1-adrenoreceptor, M2 cholinergic receptor	150 kDa protein	Humoral
Cardiac muscarinic acetylcholine receptors (mAChR)	Unknown	Humoral
Cardiac muscarinic acetylcholine receptors (mAChR)	Cruzipain	Humoral
Cha antigen	SAPA, 36 kDa TENU2845	Humoral and T cells

Adapted from Cunha-Neto et al. [57]

that share structurally similar epitopes in host antigens, leading to autoreactive responses.

Besides their antibody-secreting function, B cells play an immune-modulatory role, important to the establishment and profiling of T cell response. Several reports highlight the role of B cells in the establishment of memory T cell populations upon infection with *T. cruzi* [60], which in time, as it is discussed below, are essential for controlling the infection [61].

In order to evaluate B1 cell role in *T. cruzi* infection, some studies were performed on infected BALB Xid mice carrying an X-linked mutation. This mutation prevents B1 cell development. These mice showed poor B cell response to infection and low levels of specific and non-specific antibodies in serum. However, they were able to control parasitemia and develop almost no pathology in chronic phase.

The resistance of these mice to experimental *T. cruzi* infection was associated with the absence of IL-10-secreting B1 cells and the presence of high levels of IFN-γ [62]. In this sense, some authors suggest that B1 cells would play a pathological rather than protective role in Chagas disease. Nevertheless, the specific role of these cells in this context has not been found yet, and their association with autoimmune diseases suggests that they could be involved in self-reactive responses observed in *T. cruzi* infection.

Finally, a B cell subset with immunosuppresor function has been described, and it seems to be involved in favoring immune tolerance. In normal physiological conditions, these regulatory B cells (B_{reg}) secrete IL-10, IL-35, and TGF-β and suppress autoimmune pathologies by hampering the expansion of pathogenic T cell clones and other pro-inflammatory lymphocytes [63]. Fares et al. described that patients with chronic Chagas disease have a higher frequency of IL-10- and TGF-β-producing B cells in peripheral blood, both in basal state and upon in vitro stimulation with parasite lysate. It is yet to be determined whether, in the context of chronic Chagas disease, these cells have a benign effect by containing an overly inflammatory response or detrimental by favoring tolerance toward the parasite presence [33].

3.2 T Lymphocytes

T cell response is (generally) initiated upon signals produced by the recognition of peptide-MHC complexes on the surface of APC by the T cell receptor (TCR). The ultimate consequence of this activation is the generation of a large number of effector, pathogen-specific T cells, departing from a relatively small set of naïve T cells (T_N) with diverse specificity. The activation of a T_N cell triggers its clonal expansion, together with changes in the expression of molecules, specially membrane-anchored receptors, which enable the T cell effector capabilities [1].

In the context of an acute infection, T cell response can be dissected in three phases: (1) priming and expansion, (2) resolution and contraction, and (3) memory [64].

During the first phase, T cells divide and differentiate into effector T cells (T_E), increasing their expression of several molecules and chemokine receptors that favor activation, migration to lymphoid organs, and retention therein [1, 65]. The changes that the T cell undergoes upon activation promote rapid amplification of specific response, generation of effector and memory populations, enhancement of APC functions, and circumvention of the response within non-pathological levels.

In the case of chronic Chagas disease, it has been shown that patients have an increased frequency of circulating activated T cells and that these cells secrete pro- and anti-inflammatory cytokines [66]. T cell response is particularly important for the maintenance of the typically low parasitemia in the chronic phase of the disease [67], and if impaired (e.g., in VIH coinfection cases), it leads to rapid clinical onset [68, 69].

T_E cells carry out functions aimed at suppressing invading pathogens, but in pathological states, they may produce inflammation and tissue damage. While T_N

cells are activated mostly in lymphoid organs, T_E cells can be activated in virtually any tissue. Upon activation, they increase the expression of molecules involved in migration toward sites of infection or tissue damage, where they encounter their cognate antigen and initiate actions for the elimination of its source [1, 65]. On one hand, the different populations of effector $CD4^+$ T cells, known as helper T cells (T_H), enhance phagocyted pathogen elimination in macrophages, recruit other immune system cells, stimulate inflammation, and favor the immune functions of the mucosae and of other lymphocytes. On the other hand, $CD8^+$ T cells, also referred to as cytotoxic T lymphocytes (CTL), suppress infected or pathological cells exposing pathogen-related epitopes associated with class I MHC molecules [1]. The specific functions of each subset are discussed later.

After the elimination of the antigen source, although this is not exactly the case in chronic infections, the second phase of the T cell response, clonal contraction, takes place. The dominant process in this phase is death by apoptosis of the vast majority of the activated T_E cells. However, some of them gain a remarkable capacity to self-renew by proliferation and perpetuate in the long term, becoming memory T cells (T_M) [64]. These are present in the circulation but are specially abundant in lymphoid organs and mucosae [1]. Within this population, two major subsets can be distinguished: central memory T cells (T_{CM}) homing at the lymph nodes have a low effector capacity but high sensibility to stimulation and proliferation capability upon stimulation, especially in long-term immune protection, and effector memory T cells (T_{EM}), homing at peripheral tissue, which can rapidly produce effector profile cytokines but have a limited proliferative capacity [70, 71].

Chronic Chagas patients have been reported to have an increased frequency of circulating T_{CM} cells as compared to non-infected individuals, regardless of their clinical status [72].

In circumstances of persistent antigenic exposition, like that of tumors and chronic infections (including Chagas disease), memory T cells specific for the persistent antigen undergo loss of functionality in the T cells, in a process termed T cell exhaustion. It is characterized by the hierarchical loss of effector functions, altered expression and usage of transcription factors, metabolic disarrangements, and increased and sustained expression of inhibitory receptors, such as PD-1, CTLA-4, Tim-3, Lag-3, and TIGIT [73, 74]. Under normal physiological conditions, these molecules take part in the control of inflammation and the contraction of effector T cell populations after antigen source elimination. The conditions that lead to T cell exhaustion result in the failure of the acquisition of a homeostatic antigen-independent memory T cell response [73].

In the murine chronic *T. cruzi* infection model, lack of cytokine production was observed in T cells recovered from tissue infiltrates, suggesting some degree of dysfunctionality [75]. Despite some authors' claim that the picture observed in this model does not fully agree with the description of a T cell exhaustion process [69, 76], studies carried on chronic Chagas disease patient samples revealed the presence of exhausted T cells, exhibiting a direct relationship between their frequency and the severity of the cardiac disease [77, 78]. Furthermore, an association between therapeutic success of benznidazole (Bz) in pediatric patients and the loss of parasite-specific IFN-γ-producing T cells was observed [79], supporting the relevance of persistent antigenic stimulation during chronic *T. cruzi* infection in the development of an inefficient T cell response in the long term.

Having established this general framework of available knowledge on T cell response to infection, and in order to facilitate their comprehension, the particulars of *T. cruzi* interactions with T_H cells and with CTL are approached separately, in the next subsections.

3.2.1 Cytotoxic T Lymphocytes and Elimination of the Intracellular Forms of T. cruzi

Given that the replicative phase of *T. cruzi*'s life cycle within the human host takes place in the intracellular environment, CTL response has been, within the last few decades, the predominant subject in the study of T cell response in the context of Chagas disease. It has been demonstrated that $CD8^+$ T cells are an essential agent in the control of *T. cruzi* infection [80–82]. In the human infection, a reduction in the frequency of $CD8^+$ T cells in severe cardiac patients was observed in comparison with asymptomatic and mild cardiac patients, suggesting that CTL response might have a role in preventing the progression of the cardiac symptoms [83].

Despite the aforementioned, and given the experimental and clinical evidence, the deployment of a *T. cruzi*-specific $CD8^+$ T cell response does not imply protection against infection/reinfection. Diverse lines of research have tried to clarify the reason for which an immune memory capable of mounting an effective CTL response toward the parasite is not established. One particular feature of this system, probably common to other parasites with an intracellular replicative phase, is that expansion and contraction of $CD8^+$ T cells (experimentally assessed as cytotoxic activity and IFN-γ-producing cell frequency of T cells against a TcTS and an ASP-2 epitopes) occur with a certain delay in response to acute infection, as compared to the usual timings observed in responses toward viruses and bacteria [84]. A likely explanation for this is that the initial infection goes unnoticed by the immune system, and therefore an effective activation of the innate response is withheld until the first round of replication and reinfection, which takes place 4–5 days afterward [84, 85].

In addition to this possible difficulty to generate an effective response in the context of primary infection, the persistent antigenic stimulus takes its toll on the maintenance of a functional memory population. Of note, activated T cells may produce only one cytokine (monofunctional T cells), while others can simultaneously produce multiple cytokines (polyfunctional T cells). In general, polyfunctionality is considered a correlate of protection in the context of infection [86] and vaccination [87].

It has been observed that patients with chronic Chagas disease have an imbalanced composition of the $CD8^+$ memory T cell subset [88, 72], further supporting the hypothesis of a progressive exhaustion of the CTL response associated to cardiac function compromise. In agreement with this, an increase in the frequency of monofunctional $CD8^+$ T cells (upon in vitro stimulation with parasite lysate) was reported in patients with severe cardiomyopathy [88] and an increase in the expression of inhibitory receptor CTLA-4 on the surface of $CD8^+$ T cells in asymptomatic patients [89]. Even though CTLA-4 expression was not increased on the surface of $CD8^+$ T cells of cardiac patients, an increment in the frequency of cells positive for intracellular CTLA-4 was observed [89]. In partial discordance, an ulterior study showed a significantly increased frequency of CTLA-4-expressing $CD8^+$ T cells in Chagas disease patients, with and without cardiac symptoms [90]. Additionally, a reduction of the T_{EM} subset in favor of an augmented T_{TE} subset within circulating $CD8^+$ T cells has been demonstrated in cardiac Chagas disease patients, which suggests that CTL-mediated protection against cardiac progression depends on the existence of an elevated number of competent $CD8^+$ memory T cells, i.e., in nonterminal states of differentiation [83, 91].

Another component of this dysfunctional response panorama is the reduced secretion of IL-2 which might lead not only to a diminished effector response but also to physical deletion of specific activated T cell clones [83, 92].

Regarding the specificity of $CD8^+$ T cell response, several epitopes have been described, being those derived from proteins within the trans-sialidase proteins family the most thoroughly studied [93, 94]. Besides its participation, with the mentioned nuances, in the elimination of the parasite during infection, a remarkable amount of evidence suggests that $CD8^+$ T cells are also involved in tissue damage and inflammatory processes linked to the clinical manifestations of Chagas disease [53, 95, 96]. Inflammatory infiltrates in patients with cardiac or digestive forms of the disease are rich in activated $CD8^+$ T cells, and these express cytolytic molecules like granzymes and TIA-1 [66, 97, 98]. A predominance of pro-inflammatory cytokine-producing cells in cardiac patients was also reported [99], although cytokine profile comparative studies between asymptomatic and cardiac Chagas disease patients showed discordant results: some authors state that IFN-γ secretion has a protective effect regarding the development of the symptoms [69, 77, 100], while others propose a harmful effect for this pro-inflammatory cytokine on cardiac function [100–102]. A relationship between the T_{TE}-enriched $CD8^+$ T cell profile in cardiac patients and the cardiac damage itself has been proposed, given the enhanced cytotoxic capacity of this subset [88]. The accumulated evidence on that regards highlights the importance of the regulation of the response in order to achieve a balance between an efficient effector activity and controlling inflammatory damage.

3.2.2 Helper T Cells and Specific Immune Response Modulation

$CD4^+$ T cells are characterized by the expression of surface molecules and the secretion of cytokines which modulate the activity of other cells, mainly macrophages, DCs, and other lymphocytes [1].

It is known that in the chronic *T. cruzi* infection murine model, $CD4^+$ T cells are an important component of the cardiac lesions infiltrates, which might suggest they are of importance for the response against the parasite [103]. Nonetheless, and possibly due to the indirect nature of their physiologic role, i.e., their ultimate effect requires collaboration with other cells, little is known about their involvement in Chagas disease, especially in comparison with the extension to which $CD8^+$ T cell response has been studied. A class II MHC knockout mouse model applied to the study of the $CD4^+/CD8^+$ T cell collaboration led to the conclusion that $CD4^+$ T cells are not absolutely necessary for the establishment of a CTL response. However, the specificity profile and migratory patterns of the expanded $CD8^+$ T cells were not the same as those in wild-type control mice, and transgenic animals presented a substantially higher susceptibility to infection, perishing to it in relatively short timespans [104]. This not only limits the utility of this model for the evaluation of the role T_H cells play in the chronic *T. cruzi* infection scenario but also demonstrates that, despite not essential for the generation and expansion of parasite-specific $CD8^+$ T cells, $CD4^+$ are indeed important for the control of infection and/or inflammation.

The variety of different T_H cell profiles is tightly linked to their function and is defined according to the cytokines they produce and the expression of typical transcription factors [1, 105, 106]. Thus, T_H1 cells produce high levels of IFN-γ, while T_H2 cells produce IL-4, IL-5, IL-9, and IL-13, which activate mechanisms involved in the elimination of helminthes and other similar pluricellular parasites [106, 105]. Additionally, and in accordance with the type of pathogen they respond to, while T_H1 cells guide the production of IgM, IgA, IgG_1, IgG_2, and IgG_3, T_H2 cells favor the production of IgM, IgG_4, and IgE [105].

Results from the murine model suggest that, with regard to the level of protection that one or the other profile against *T. cruzi* infection, a coordinated response between both profiles is desirable [107, 108], with predominance, according to several authors, of T_H1 effector mechanisms for the control/elimination of the parasite [109–111]. In the human infection, Albareda et al. demonstrated an association of a lower frequency of *T. cruzi*-specific IFN-γ-producing $CD4^+$ T cells with the degree of severity in patients with Chagas chronic cardiopathy [112]. The results on this report also suggest that most parasite-specific circulating $CD4^+$ T cells in patients are newly recruited cells and they have a highly differentiated profile. Furthermore, in the most severe cases of cardiac disease, they present markers of apoptosis, while a minor fraction shows a long-term memory T cell phenotype [112]. This evidence favors the T cell response exhaustion hypothesis in the context of chronic Chagas disease. In addition, experiments in which PBMC from chronic cardiac Chagas patients were stimulated in vitro and secreted cytokines were analyzed showed that antigens from the parasite induce a response profile that does not completely fit in any of the mentioned, including T_H1 (IFN-γ, TNF-α) and T_H2 (IL-4, IL-13) cytokines, along with IL-2, IL-10 (which has an immunosuppressor function, as will be discussed later), and GM-CSF (a cytokine that favors monocytes and neutrophils production in the BM and Langerhans cells differentiation into mature DCs) [113]. GM-CSF, IL-10, and TNF-α secretion was attributed at least partially to the response against parasite ribosomal P proteins which, as mentioned before, are related to humoral self-reactivity phenomena.

There is a subset of $CD4^+$ T cells that operate tolerance mechanisms via an *in trans* action upon other T cells, thereby restricting potentially pathogenic immune responses, termed regulatory T cells (T_{reg}). Their regulatory function is mediated by IL-10 and TGF-β: IL-10 inhibits IL-12 production and class II MHC molecule expression on DCs and activated macrophages; TGF-β inhibits proliferation and effector function on T cells and macrophages, inhibits $CD4^+$ T cell differentiation into T_H1 and T_H2 cells, induces class switch to IgA on activated B cells, and promotes local tissue reparation during inflammation remission [1]. In chronic Chagas disease, T_{reg} cells are augmented in asymptomatic patients as compared to those with cardiac or digestive symptoms [114, 115]. Furthermore, the existence of IL-10-producing activated T_{reg} cells has been demonstrated in both asymptomatic and cardiac chronic Chagas disease patients [116]. In the murine model of infection, their activity and, in particular, the production of TGF-β were discarded as the cause of the lack of cytokine production observed in effector T cells isolated from inflammatory infiltrates in *T. cruzi*-infected heart tissue [117, 118].

3.3 Human Immune Response After Anti-parasitic Treatment

Treatment with nifurtimox or benznidazole (Bz) is indicated in all cases of acute phase of Chagas disease, including congenital infection and reactivation due to immunosuppression, and in children of up to 18 years of age in the chronic phase of *T. cruzi* infection [119]. In the particular case of adult patients up to 50 years old with chronic infection and without any symptoms, treatment is strongly recommended, while in those patients with cardiac alterations, the usefulness of anti-parasitic therapy is still a matter of debate [120–122]. However, and regardless of the advantages or disadvantages of treatment with these drugs in terms of clinical outcome, this section intends to review how the immune response changes after therapy in chronic Chagas disease patients.

Laucella et al. clearly demonstrated that significant changes in *T. cruzi*-specific T cell responses occur in the majority of patients in the chronic phase of infection who were treated with Bz. Thus, the specific T cell response against *T. cruzi* lysate measured as IFN-γ secretion significantly diminished over 12 months following Bz treatment, and this fall was more evident at 36 months posttreatment in three of four evaluated subjects [123]. Even more, T_H cells in the majority of treated patients showed a polyfunctional profile of responding cells against parasite antigens, by secreting not only IFN-γ but also other cytokines such as IL-2. On the other side, sera from only six out of 43 treated subjects became negative by conventional serological tests, and in five out of these six subjects, the serological negativization correlated with a decrease in IFN-γ-producing T cells to below background levels. By performing a multiplex serodiagnostic assay which included 14 recombinant *T. cruzi* proteins, the antibody titers clearly diminished in 15 out of 32 treated patients, and this correlated with T cell response decay. No difference was mentioned between the immunological status in patients without or with cardiac manifestations before and after treatment. This work was not only the first study evaluating T and B cell responses in the same subjects before and after Bz treatment but also demonstrated that changes in parasite-specific T cell responses and in antibody responses to individual parasite proteins can be useful as alternative markers of treatment efficacy [123]. Authors hypothesize that Bz treatment diminishes the parasite load and, concomitantly, the parasite antigen exposed to immune cells necessary to maintain an effective response. An extension of this study in which patients were followed for

4–12 years after treatment showed that decrease in specific antibody titers, determined by conventional serology or multiplex serological assay, strongly correlated with a drop in parasite-specific T cell response [79]. In fact, the decline in T cell activation as well as in antibody level detected by multiplex approach comes before seronegativization, as detected by conventional serological tests, demonstrating the usefulness of these assays to determine treatment success. Interestingly, some patients showed an increase in the frequency of peripheral IFN-γ-producing T cells responsive to *T. cruzi* antigens after 24–72 months posttreatment, with no changes in humoral response. Consistently with previous findings, Bz treatment also modified the functional features of T_H cells toward a profile that reflects a more effective response. In addition, treated patients showed a diminished number of single TNF-α producer T cell population. Of note, it was demonstrated that elevated concentrations of TNF are directly linked to inflammation and T cell dysfunction during chronic infection, leading to a restricted expression of helper molecules such as IL-2, IL-21, IFN-γ, and CD40L that are crucial for antimicrobial T cell immune response [124, 125]. Overall, these findings suggest that in the context of Chagas disease, Bz treatment could bias *T. cruzi*-specific T cell population toward a responder but less detrimental restricted profile.

Results from another study, carried out on a cohort of asymptomatic and cardiac adult patients, before and 1-year posttreatment with Bz, showed that treatment diminished the frequency of IL-10$^+$CD4$^+$ cells in asymptomatic patients while increasing the frequency of IL-10$^+$ monocytes in cardiac subjects. This might suggest that anti-parasitic therapy restores the balance of inflammatory and regulatory profiles of cytokine production [126].

The following studies were performed by comparing the immune response of two cohorts of subjects, treated and non-infected; however, these comparisons have not been done in the same patient before and after treatment. Sathler-Avelar et al. focused their study in the immunological response developed in six children (9–14 years old, mean 12) treated with Bz in the early onset of the asymptomatic phase of Chagas disease [127]. By insightful analysis of the results presented therein, the cytokine profile of whole blood leukocytes, isolated from infected children, before and 1 year after the end of treatment showed an increase of IFN-γ secreted by NK and CD8$^+$ T cells with a concomitant fall in the frequency of TNF-α-producing monocytes. In addition, a higher frequency of CD4$^+$ T cells secreting IL-10 were observed in the peripheral blood of treated patients. Although the authors pointed out the CD4$^+$ T and B cell subsets as the main sources of IFN-γ and IL-10 in treated patients as key element for parasite clearance in the absence of tissue damage, no differences in the secretion of these cytokines were observed in culture before and after *T. cruzi* lysate stimulation. In our opinion, this observation weakens the possibility of drawing a definitive conclusion from the exposed data. In a subsequent study performed over the same cohort of patients, Sathler-Avelar et al. observed that NK, CD4$^+$, and CD8$^+$ T cells presented an activation status after Bz treatment [128]. These authors demonstrated a positive correlation between the presence of activation markers in the surface of immune cells and the secretion of both IFN-γ and IL-10, a pro-inflammatory and an anti-inflammatory cytokine, respectively. Later, Sathler-Avelar et al. decided to extend their previous finding in adult patients with the asymptomatic form of the disease, non-treated and treated with Bz [129]. Authors observed that treatment induced low frequency of IL-10^+ neutrophils and monocytes, IFN-γ^+ NK-cells in the innate compartment while in the adaptive one, low levels of IL-12^+, TNF-α^+, IFN-γ^+ and IL-5^+ $CD4^+$ T cells, and IL-10^+ B cells and $CD8^+$ T cells were detected in treated subjects compared to non-treated group. When PBMC were incubated with epimastigote lysate, the number of IL-10^+ monocytes as well as IFN-γ^+ and TNF-α^+ $CD8^+$ T cells augmented compared with the conditions without stimulation.

Recently, a cross-sectional study [130] carried out over a cohort of Chagas disease adult patients in chronic phase before and 1 year after the end of treatment with Bz disclosed that treatment led to a clear decrease of IL-10^+ neutrophils accompanied by an increase of IL-10^+, TGF-β^+, and TLR4 expression in monocytes of cardiac patients, whereas only an impaired phagocytic capacity was observed in Bz-treated asymptomatic patients. No differences were observed in IL-12 and TNF-α secretion, NO levels, and expression of Fc receptors and TLR2 in monocytes across the groups.

In another cross-sectional study performed in a cohort of adult asymptomatic patients, Bz treatment decreased $CD4^+$ T cell activation and terminally differentiated memory T cells, while no differences were found in either T_{CM} or T_{EM} cells. Under thorough analysis, no differences were observed in the functional and phenotypic ex vivo $CD8^+$ T cells and T_{reg} lymphocytes between both groups, and only a slight increase of IFN-γ and TNF-α secretion was revealed in T cells from treated patients after stimulation with *T. cruzi* lysate [118].

Although the serological tests are widely considered as a biomarker of therapy efficacy, the dynamics of specific antibodies in the sera of patients can also be interpreted as a direct reflection of the humoral immune response in the context of anti-parasitic chemotherapy. So far, several studies demonstrate that Bz treatment induces a partial or total decrease in *T. cruzi*-specific antibodies titers determined by either conventional serological tests or multiplex assay [131, 132]. As it was previously mentioned [79, 123], the multiplex assay utilizing recombinant proteins from *T. cruzi* showed that the antibody levels declined as early as 2–24 months posttreatment, whereas with conventional serological tests using whole *T. cruzi* lysate, the decay is visible at 24–48 months posttreatment. Similar results were observed when antibody levels against *T. cruzi* B13, IF8, and JL7 proteins were compared with *T. cruzi* lysate all assessed by ELISA and Chagas test, a commercial recombinant protein-based ELISA [133]. Other examples of antibody response during treatment come from data obtained with other recombinant proteins or parasite fractions (see references in [134] and [135]). As a whole, these results clearly show that antibodies with certain specificities diminish with Bz treatment, while others remain stable in time. In this regard, it has been demonstrated that antibody-secreting plasma cells become long-lived in survival niches, like the bone marrow, spleen, lymph nodes, or mucosal-associated lymphoid tissues of the gut, producing molecules critical for their survival, even after pathogen clearance or in the absence of memory B cells [136–138]. So far, we can speculate that *T. cruzi*-specific long-lived plasma cells residing in survival niches could contribute to the persistence of humoral immunity observed in treated patients even in the absence of repeated *T. cruzi* antigen exposure. To our knowledge, no investigation in this field has been carried out in the context of Chagas disease.

References

1. Abbas AK, Lichtman AH, Pillai S. Cellular and molecular immunology. 7th ed. Amsterdam: Elsevier; 2012.
2. Ferreira V, Valck C, Sánchez G, Gingras A, Tzima S, Molina MC, Sim R, Schwaeble W, Ferreira A. The classical activation pathway of the human complement system is specifically inhibited by calreticulin from Trypanosoma cruzi. J Immunol. 2004;172(5):3042–50.
3. Sosoniuk E, Vallejos G, Kenawy H, Gaboriaud C, Thielens N, Fujita T, Schwaeble W, Ferreira A, Valck C. Trypanosoma cruzi calreticulin inhibits the complement lectin pathway activation by direct interaction with L-Ficolin. Mol Immunol. 2014;60(1):80–5.
4. Ribeiro CH, Lynch NJ, Stover CM, Ali YM, Valck C, Noya-Leal F, Schwaeble WJ, Ferreira A. Deficiency in mannose-binding lectin-associated serine protease-2 does not increase susceptibility to Trypanosoma cruzi infection. Am J Trop Med Hyg. 2015;92(2):320–4.
5. Cardoso MS, Reis-Cunha JL, Bartholomeu DC. Evasion of the immune response by trypanosoma cruzi during acute infection. Front Immunol. 2016;6:1–15.
6. Geiger A, Bossard G, Sereno D, Pissarra J, Lemesre JL, Vincendeau P, Holzmuller P. Escaping deleterious immune response in their hosts: lessons from trypanosomatids. Front Immunol. 2016;7:1–21.
7. Lidani KCF, Bavia L, Ambrosio AR, de Messias-Reason IJ. The complement system: a prey of Trypanosoma cruzi. Front Microbiol. 2017;8:1–14.
8. Norris KA, Bradt B, Cooper NR, So M. Characterization of a Trypanosoma cruzi C3 binding protein with functional and genetic similarities to the human complement regulatory protein, decay-accelerating factor. J Immunol. 1991;147(7):2240–7.
9. a Norris K, Schrimpf JE, Szabo MJ. Identification of the gene family encoding the 160-kilodalton Trypanosoma cruzi complement regulatory protein. Infect Immun. 1997;65(2):349–57.
10. Henrique PM, Marques T, da Silva MV, Nascentes GAN, de Oliveira CF, Rodrigues V, Gómez-Hernández C, Norris KA, Ramirez LE, Meira WSF. Correlation between the virulence of T. cruzi strains, complement regulatory protein expression levels, and the ability to elicit lytic antibody production. Exp Parasitol. 2016;170:66–72.
11. Ramírez-Toloza G, Ferreira A. Trypanosoma cruzi evades the complement system as an efficient strategy to survive in the mammalian host: the specific roles of host/parasite molecules and Trypanosoma cruzi calreticulin. Front Microbiol. 2017;8:1–13.
12. Romano PS, Cueto JA, Casassa AF, Vanrell MC, Gottlieb RA, Colombo MI. Molecular and cellular mechanisms involved in the Trypanosoma cruzi/host cell interplay. IUBMB Life. 2012;64(5):387–96.
13. Stempin C, Giordanengo L, Gea S, Cerbán F. Alternative activation and increase of Trypanosoma cruzi survival in murine macrophages stimulated by cruzipain, a parasite antigen. J Leukoc Biol. 2002;72(4):727–34.
14. Vitelli-Avelar DM, Sathler-Avelar R, Massara RL, Borges JD, Lage PS, Lana M, Teixeira-Carvalho A, Dias JCP, Elói-Santos SM, Martins-Filho OA. Are increased frequency of macrophage-like and natural killer (NK) cells, together with high levels of NKT and CD4+CD25high T cells balancing activated CD8+ T cells, the key to control Chagas' disease morbidity? Clin Exp Immunol. 2006;145(1):81–92.
15. Chen L, Watanabe T, Watanabe H, Sendo F. Neutrophil depletion exacerbates experimental Chagas' disease in BALB/c, but protects C57BL/6 mice through modulating the Th1/Th2 dichotomy in different directions. Eur J Immunol. 2001;31(1):265–75.

16. Luna-Gomes T, Filardy AA, Rocha JDB, Decote-Ricardo D, LaRocque-de-Freitas IF, Morrot A, Bozza PT, Castro-Faria-Neto HC, DosReis GA, Nunes MP, Freire-de-Lima CG. Neutrophils increase or reduce parasite burden in trypanosoma cruzi-infected macrophages, depending on host strain: role of neutrophil elastase. PLoS One. 2014;9(3):3–10.
17. Medeiros NI, Fares RC, Franco EP, Sousa GR, Mattos RT, Chaves AT, Nunes MD, Dutra WO, Correa-Oliveira R, Rocha MO, Gomes JA. Differential expression of matrix metalloproteinases 2, 9 and cytokines by neutrophils and monocytes in the clinical forms of Chagas Disease. PLoS Negl Trop Dis. 2017;11(1):e0005284.
18. Sousa-Rocha D, Thomaz-Tobias M, Diniz LFA, Souza PSS, Pinge-Filho P, Toledo KA. Trypanosoma cruzi and its soluble antigens induce NET release by stimulating toll-like receptors. PLoS One. 2015;10(10):1–16.
19. Gil-Jaramillo N, Motta FN, Favali CBF, Bastos IMD, Santana JM. Dendritic cells: a double-edged sword in immune responses during Chagas disease. Front Microbiol. 2016;7:1–12.
20. Da Costa TA, Silva MV, Mendes MT, Carvalho-Costa TM, Batista LR, Lages-Silva E, Rodrigues V, Oliveira CJ, Ramirez LE. Immunomodulation by Trypanosoma cruzi: toward understanding the association of dendritic cells with infecting TcI and TcII populations. J Immunol Res. 2014;2014:962047.
21. Van Overtvelt L, Vanderheyde N, Verhasselt V, Ismaili J, De Vos L, Goldman M, Willems F, Vray B, Van Overtvelt L, Vanderheyde N, Verhasselt V, Ismaili J, De Vos L, Goldman M, Willems F, Vray B. Trypanosoma cruzi infects human dendritic cells and prevents their maturation: inhibition of cytokines, HLA-DR, and costimulatory molecules. Infect Immun. 1999;67(8):4033–40.
22. Van Overtvelt L, Andrieu M, Verhasselt V, Connan F, Choppin J, Vercruysse V, Goldman M, Hosmalin A, Vray B. Trypanosoma cruzi down-regulates lipopolysaccharide-induced MHC class I on human dendritic cells and impairs antigen presentation to specific CD8+ T lymphocytes. Int Immunol. 2002;14(10):1135–44.
23. Brodskyn C, Patricio J, Oliveira R, Lobo L, Arnholdt A, Mendonça-previato L, Barral A, Barral-netto M. Glycoinositolphospholipids from Trypanosoma cruzi Interfere with Macrophages and Dendritic Cell Responses. Society. 2002;70(7):3736–43.
24. Erdmann H, Steeg C, Koch-Nolte F, Fleischer B, Jacobs T. Sialylated ligands on pathogenic Trypanosoma cruzi interact with Siglec-E (sialic acid-binding Ig-like lectin-E). Cell Microbiol. 2009;11(11):1600–11.
25. Cardillo F, Voltarelli JC, Reed SG, Silva JS. Regulation of Trypanosoma cruzi infection in mice by gamma interferon and interleukin 10: role of NK cells. Infect Immun. 1996;64(1):128–34.
26. Lieke T, Graefe SEB, Klauenberg U, Fleischer B, Jacobs T. NK cells contribute to the control of Trypanosoma cruzi infection by killing free parasites by perforin-independent mechanisms NK cells contribute to the control of Trypanosoma cruzi infection by killing free parasites by perforin-independent mechanisms. Infect Immun. 2004;72(12):6817–25.
27. Batalla EI, Pino Martínez AM, Poncini CV, Duffy T, Schijman AG, González Cappa SM, Soto CDA. Impairment in natural killer cells editing of immature dendritic cells by infection with a virulent trypanosoma cruzi population. J Innate Immun. 2013;5(5):494–504.
28. Lieke T, Steeg C, Graefe SEB, Fleischer B, Jacobs T. Interaction of natural killer cells with Trypanosoma cruzi-infected fibroblasts. Clin Exp Immunol. 2006;145(2):357–64.
29. Ferreira LRP, Ferreira FM, Nakaya HI, Deng X, da D, Cândido S, Campos de Oliveira L, Billaud J-N, Lanteri MC, Oliveira-Carvalho R, Seielstad M, Kalil J, Fernandes F, Pinho Ribeiro AL, Sabino EC, Cunha-Neto E. Blood gene signatures of Chagas disease cardiomyopathy with or without ventricular dysfunction. J Infect Dis. 2015;3(215):387–95.
30. Guilmot A, Carlier Y, Truyens C. Differential IFN-γ production by adult and neonatal blood CD56+ natural killer (NK) and NK-like-T cells in response to *Trypanosoma cruzi* and IL-15. Parasite Immunol. 2014;36(1):43–52.
31. Hoffman W, Lakkis FG, Chalasani G. B cells, antibodies, and more. Clin J Am Soc Nephrol. 2016;11(1):137–54.
32. Acosta Rodriguez EV, Zuniga EI, Montes CL, Merino MC, Bermejo DA, Amezcua Vesely MC, Motran CC, Gruppi A. Trypanosoma cruzi infection beats the B-cell compartment favouring parasite establishment: can we strike first? Scand J Immunol. 2007;66(2–3):137–42.

33. Fares RCG, Correa-Oliveira R, de Araújo FF, Keesen TSL, Chaves AT, Fiuza JA, Ferreira KS, Rocha MOC, Gomes JAS. Identification of phenotypic markers of B cells from patients with Chagas disease. Parasite Immunol. 2013;35(7–8):214–23.
34. Sathler-Avelar R, Lemos EM, Reis DD, Medrano-Mercado N, Araujo-Jorge TC, Antas PRZ, Correa-Oliveira R, Teixeira-Carvalho A, Eloi-Santos SM, Favato D, Martins-Filho OA, Corrêa-Oliveira R, Teixeira-Carvalho A, Elói-Santos SM, Favato D, Martins-Filho OA, Correa-Oliveira R, Teixeira-Carvalho A, Eloi-Santos SM, Favato D, Martins-Filho OA. Phenotypic Features of Peripheral Blood Leucocytes during Early Stages of Human Infection with Trypanosoma cruzi. Scand J Immunol. 2003;58(6):655–63.
35. Fernández ER, Olivera GC, Quebrada Palacio LP, González MN, Hernandez-Vasquez Y, Sirena NM, Morán ML, Ledesma Patiño OS, Postan M. Altered distribution of peripheral blood memory B cells in humans chronically infected with Trypanosoma cruzi. PLoS One. 2014;9(8):e104951.
36. Cherukuri A, Cheng PC, Pierce SK. The role of the CD19/CD21 complex in B cell processing and presentation of complement-tagged antigens. J Immunol. 2001;167(1):163–72.
37. Krautz GM, Kissinger JC, Krettli AU. The targets of the lytic antibody response against Trypanosoma cruzi. Parasitol Today. 2000;16(1):31–4.
38. Almeida IC, Milani SR, Gorin PA, Travassos LR. Complement-mediated lysis of Trypanosoma cruzi trypomastigotes by human anti-alpha-galactosyl antibodies. J Immunol. 1991;146(7):2394–400.
39. de Oliveira Mendes TA, Reis Cunha JL, de Almeida Lourdes R, Rodrigues Luiz GF, Lemos LD, dos Santos ARR, da Câmara ACJ, da Cunha Galvao LM, Bern C, Gilman RH, Fujiwara RT, Gazzinelli RT, Bartholomeu DC. Identification of strain-specific B-cell epitopes in Trypanosoma cruzi using genome-scale epitope prediction and high-throughput immunoscreening with peptide arrays. PLoS Negl Trop Dis. 2013;7(10):e2524.
40. Carmona SJ, Nielsen M, Schafer-Nielsen C, Mucci J, Altcheh J, Balouz V, Tekiel V, Frasch AC, Campetella O, Buscaglia CA, Agüero F. Towards high-throughput immunomics for infectious diseases: use of next-generation peptide microarrays for rapid discovery and mapping of antigenic determinants. Mol Cell Proteomics. 2015;14(7):1871–84.
41. Kumar S, Tarleton RL. The relative contribution of antibody production and CD8+ T cell function to immune control of *Trypanosoma cruzi*. Parasite Immunol. 1998;20(5):207–16.
42. Pitcovsky TA, Buscaglia CA, Mucci J, Campetella O. A functional network of intramolecular cross-reacting epitopes delays the elicitation of neutralizing antibodies to Trypanosoma cruzi trans-sialidase. J Infect Dis. 2002;186(3):397–404.
43. Buscaglia C a, Campo V a, Frasch ACC, Di Noia JM. Trypanosoma cruzi surface mucins: host-dependent coat diversity. Nat Rev Microbiol. 2006;4(3):229–36.
44. Zuniga E, Acosta-Rodriguez E, Merino MC, Montes C, Gruppi A. Depletion of immature B cells during Trypanosoma cruzi infection: involvement of myeloid cells and the cyclooxygenase pathway. Eur J Immunol. 2005;35(6):1849–58.
45. Ortiz-ortiz L, Parks DE, Rodriguez M, Weigle W. Polyclonal B lymphocyte activation during Trypanosoma cruzi infection. J Immunol. 1980;124(1):121–6.
46. Minoprio P, Burlen O, Pereira P, Guilbert B, Andrade L, Hontebeyrie-Joskowciz M, Coutinho A. Most B cells in acute Trypanosoma cruzi infection lack parasite specificity. Scand J Immunol. 1988;28(5):553–61.
47. Bermejo DA, Amezcua Vesely MC, Khan M, Acosta Rodriguez EV, Montes CL, Amezcua Vesely MC, Toellner KM, Mohr E, Taylor D, Cunningham AF, Gruppi A. Trypanosoma cruzi infection induces a massive extrafollicular and follicular splenic B-cell response which is a high source of non-parasite-specific antibodies. Immunology. 2011;132(1):123–33.
48. Bryan MA, Guyach SE, Norris KA. Specific humoral immunity versus polyclonal B Cell activation in trypanosoma cruzi infection of susceptible and resistant mice. PLoS Negl Trop Dis. 2010;4(7):e733.
49. Grauert MR, Houdayer M, Hontebeyrie-Joskowciz M. Trypanosoma cruzi infection enhances polyreactive antibody response in an acute case of human Chagas' disease. Clin Exp Immunol. 1993;93(1):85–92.

50. Montes CL, Acosta-Rodríguez EV, Mucci J, Zuniga EI, Campetella O, Gruppi A. A Trypanosoma cruzi antigen signals CD11b+ cells to secrete cytokines that promote polyclonal B cell proliferation and differentiation into antibody-secreting cells. Eur J Immunol. 2006;36(6):1474–85.
51. Reina-San-Martín B, Degrave W, Rougeot C, Cosson a, Chamond N, Cordeiro-Da-Silva a, Arala-Chaves M, Coutinho a, Minoprio P. A B-cell mitogen from a pathogenic trypanosome is a eukaryotic proline racemase. Nat Med. 2000;6(8):890–7.
52. Gao W, Wortis HH, Pereira M a. The Trypanosoma cruzi trans-sialidase is a T cell-independent B cell mitogen and an inducer of non-specific Ig secretion. Int Immunol. 2002;14(3):299–308.
53. Bonney KM, Engman DM. Autoimmune pathogenesis of Chagas heart disease: looking back, looking ahead. Am J Pathol. 2015;185(6):1537–47.
54. Gironès N, Cuervo H, Fresno M. *Trypanosoma cruzi*-induced molecular mimicry and Chagas' disease. In: Molecular mimicry: infection-inducing autoimmune disease. New York, NY: Springer; 2005. p. 89–123.
55. Gómez K, Longhi S, Levin M. The genesis of anti-cardiac G protein coupled receptor antibodies in Chagas disease. In: Peter J-C, editor. Immunology of G-protein coupled receptors. Ketala: Dipak Haldar, S.G.; 2006. p. 45–63.
56. Labovsky V, Smulski C, Gómez K, Levy G, Levin M. Anti-β-adrenergic receptor autoantibodies in patients with chronic Chagas heart disease. Clin Exp Immunol. 2007;140:440–9.
57. Cunha-Neto E, Teixeira PC, Fonseca SG, Bilate AM, Kalil J. Myocardial gene and protein expression profiles after autoimmune injury in Chagas' disease cardiomyopathy. Autoimmun Rev. 2011;10(3):163–5.
58. Leon JS, Engman DM. Autoimmunity in Chagas heart disease. Int J Parasitol. 2001;31(5–6):555–61.
59. Ternynck T, Bleux C, Gregoire J, Avrameas S, Kanellopoulos LC. Comparison between autoantibodies arising during Trypanosoma cruzi infection in mice and natural autoantibodies. J Immunol. 1990;144(4):1504–11.
60. Cardillo F, Postol E, Nihei J, Aroeira LS, Nomizo A, Mengel J. B cells modulate T cells so as to favour T helper type 1 and CD8+ T-cell responses in the acute phase of Trypanosoma cruzi infection. Immunology. 2007;122(4):584–95.
61. Bermejo D a, Jackson SW, Gorosito-Serran M, Acosta-Rodriguez EV, Amezcua-Vesely MC, Sather BD, Singh AK, Khim S, Mucci J, Liggitt D, Campetella O, Oukka M, Gruppi A, Rawlings DJ. Trypanosoma cruzi trans-sialidase initiates a program independent of the transcription factors RORγt and Ahr that leads to IL-17 production by activated B cells. Nat Immunol. 2013;14(5):514–22.
62. Minoprio P, el Cheikh MC, Murphy E, Hontebeyrie-Joskowicz M, Coffman R, Coutinho A, O'Garra A. Xid-associated resistance to experimental Chagas' disease is IFN-gamma dependent. J Immunol. 1993;151(8):4200–8.
63. Rosser EC, Mauri C. Regulatory B cells: origin, phenotype, and function. Immunity. 2015;42(4):607–12.
64. Laidlaw BJ, Craft JE, Kaech SM. The multifaceted role of CD4+ T cells in CD8+ T cell memory. Nat Rev Immunol. 2016;16(2):102–11.
65. Appay V, Van Lier RAW, Sallusto F, Roederer M. Phenotype and function of human T lymphocyte subsets: consensus and issues. Cytom Part A. 2008;73(11):975–83.
66. Dutra WO, Gollob KJ. Current concepts in immunoregulation and pathology of human Chagas disease. Curr Opin Infect Dis. 2008;21(3):287–92.
67. Sartori AMC, Neto JE, Nunes EV, Braz LMA, Caiaffa HH, Oliveira OD, Neto VA, Shikanai-Yasuda MA. Trypanosoma cruzi parasitemia in chronic Chagas disease: comparison between human immunodeficiency virus (HIV)-positive and HIV-negative patients. J Infect Dis. 2002;186:872–5.
68. Vaidian AK, Weiss LM, Tanowitz HB. Chagas' disease and AIDS. Kinetoplastid Biol Dis. 2004;3(1):2.
69. Virgilio S, Pontes C, Dominguez MR, Ersching J, Rodrigues MM, Vasconcelos JR, dos Santos Virgilio F, Pontes C, Dominguez MR, Ersching J, Rodrigues MM, Vasconcelos JR. CD8+ T

Cell-mediated immunity during Trypanosoma cruzi infection: a path for vaccine development? Mediators Inflamm. 2014;2014:243786.
70. Sallusto F, Geginat J, Lanzavecchia A. Central memory and effector memory T cell subsets: function, generation, and maintenance. Annu Rev Immunol. 2004;22:745–63.
71. Lanzavecchia A, Sallusto F. Understanding the generation and function of memory T cell subsets. Curr Opin Immunol. 2005;17(3):326–32.
72. Fiuza JA, Fujiwara RT, Gomes JAS, Rocha MODC, Chaves AT, De Araújo FF, Fares RCG, Teixeira-Carvalho A, Martins-Filho ODA, Cançado GGL, Correa-Oliveira R. Profile of central and effector memory T cells in the progression of chronic human Chagas disease. PLoS Negl Trop Dis. 2009;3(9):e512.
73. Wherry EJ, Kurachi M. Molecular and cellular insights into T cell exhaustion. Nat Publ Gr. 2015;15(8):486–99.
74. Anderson AC, Joller N, Kuchroo VK. Lag-3, Tim-3, and TIGIT: co-inhibitory receptors with specialized functions in immune regulation. Immunity. 2016;44(5):989–1004.
75. Leavey JK, Tarleton RL. Cutting edge: dysfunctional CD8+ T cells reside in nonlymphoid tissues during chronic Trypanosoma cruzi infection. J Immunol. 2003;170(5):2264–8.
76. Tarleton RL. CD8+ T cells in Trypanosoma cruzi infection. Semin Immunopathol. 2015;37(3):233–8.
77. Laucella SA, Postan M, Martin D, Hubby Fralish B, Albareda MC, Alvarez MG, Lococo B, Barbieri G, Viotti RJ, Tarleton RL. Frequency of interferon- gamma -producing T cells specific for Trypanosoma cruzi inversely correlates with disease severity in chronic human Chagas disease. J Infect Dis. 2004;189(5):909–18.
78. Albareda MC, De Rissio AM, Tomas G, Serjan A, Alvarez MG, Viotti R, Fichera LE, Esteva MI, Potente D, Armenti A, Tarleton RL, Laucella SA. Polyfunctional T cell responses in children in early stages of chronic trypanosoma cruzi infection contrast with monofunctional responses of long-term infected adults. PLoS Negl Trop Dis. 2013;7(12):e2575.
79. Alvarez MG, Bertocchi GL, Cooley G, Albareda MC, Viotti R, Perez-Mazliah DE, Lococo B, Castro Eiro M, Laucella SA, Tarleton RL. Treatment success in Trypanosoma cruzi infection is predicted by early changes in serially monitored parasite-specific T and B cell responses. PLoS Negl Trop Dis. 2016;10(4):e0004657.
80. Junqueira C, Caetano B, Bartholomeu DC, Melo MB, Ropert C, Rodrigues MM, Gazzinelli RT. The endless race between Trypanosoma cruzi and host immunity: lessons for and beyond Chagas disease. Expert Rev Mol Med. 2010;12:e29.
81. Tarleton RL. Depletion of CD8+ T cells increases susceptibility and reverses vaccine-induced immunity in mice infected with Trypanosoma cruzi. J Immunol. 1990;144:717–24.
82. Tarleton RL, Koller BH, Latour A, Postan M. Susceptibility of β2-microglobulin-deficient mice to Trypanosoma cruzi infection. Nature. 1992;356:338–40.
83. Albareda MC, Laucella SA, Alvarez MG, Armenti AH, Bertochi G, Tarleton RL, Postan M. Trypanosoma cruzi modulates the profile of memory CD8+ T cells in chronic Chagas' disease patients. Int Immunol. 2006;18(3):465–71.
84. Tzelepis F, de Alencar B, Penido M, Gazzinelli R, Persechini P, Rodrigues MM. Distinct kinetics of effector CD8+ cytotoxic T cells after infection with Trypanosoma cruzi in Naıve or vaccinated mice. Infect Immun. 2006;74(4):2477–81.
85. Padilla AM, Bustamante JM, Tarleton RL. CD8+ T cells in Trypanosoma cruzi infection. Curr Opin Immunol. 2009;21(4):385–90.
86. Lewinsohn DA, Lewinsohn DM, Scriba TJ. Polyfunctional CD4+T cells as targets for tuberculosis vaccination. Front Immunol. 2017;8:1262.
87. Thakur A, Pedersen LE, Jungersen G. Immune markers and correlates of protection for vaccine induced immune responses. Vaccine. 2012;30(33):4907–20.
88. Mateus J, Lasso P, Pavia P, Rosas F, Roa N, Valencia-Hernández CA, González JM, Puerta CJ, Cuéllar A. Low frequency of circulating CD8+ T stem cell memory cells in chronic chagasic patients with severe forms of the disease. PLoS Negl Trop Dis. 2015;9(1):e3432.
89. Souza PEA, Rocha MOC, Menezes CAS, Coelho JS, Chaves ACL, Gollob KJ, Dutra WO. Trypanosoma cruzi infection induces differential modulation of costimulatory mole-

cules and cytokines by monocytes and T cells from patients with indeterminate and cardiac Chagas' disease. Infect Immun. 2007;75(4):1886–94.
90. Argüello RJ, Albareda MC, Alvarez MG, Bertocchi G, Armenti AH, Vigliano C, Meckert PC, Tarleton RL, Laucella SA. Inhibitory receptors are expressed by Trypanosoma cruzi-specific effector T cells and in hearts of subjects with chronic Chagas disease. PLoS One. 2012;7(5):e35966.
91. Costa RP, Gollob KJ, Fonseca LL, Rocha MOC, Chaves ACL, Medrano-mercado N, Arau Â Jo-jorge TC, Antas PRZ, Colley DG, Correa-oliveira R, Gazzinelli G, Carvalho-parra J, Dutra WO. T-cell repertoire analysis in acute and chronic human Chagas' disease: differentail frequencies of Vb5 expressing T cells. Scand J Immunol. 2000;51:511–9.
92. Alvarez MG, Postan M, Weatherly DB, Albareda MC, Sidney J, Sette A, Olivera C, Armenti AH, Tarletona RL, Laucella SA, Tarleton RL, Laucella SA, Tarletona RL, Laucella SA, Tarleton RL, Laucella SA. HLA class I-T cell epitopes from trans-sialidase proteins reveal functionally distinct subsets of CD8+ T cells in chronic Chagas disease. PLoS Negl Trop Dis. 2008;2(9):e288.
93. Martin DL, Weatherly DB, Laucella SA, Cabinian M a, Crim MT, Sullivan S, Heiges M, Craven SH, Rosenberg CS, Collins MH, Sette A, Postan M, Tarleton RL. CD8+ T-cell responses to Trypanosoma cruzi are highly focused on strain-variant trans-sialidase epitopes. PLoS Pathog. 2006;2(8):0731–40.
94. Tzelepis F, de Alencar BCG, Penido MLO, Claser C, Machado AV, Bruna-Romero O, Gazzinelli RT, Rodrigues MM. Infection with Trypanosoma cruzi restricts the repertoire of parasite-specific CD8+ T cells leading to immunodominance. J Immunol. 2008;180(3):1737–48.
95. Engman DM, Leon JS. Pathogenesis of Chagas heart disease: role of autoimmunity. Acta Trop. 2002;81(2):123–32.
96. Scharfstein J, Gomes J d AS, Correa-Oliveira R. Back to the future in Chagas disease: from animal models to patient cohort studies, progress in immunopathogenesis research. Mem Inst Oswaldo Cruz. 2009;104(Suppl 1):187–98.
97. Lannes-Vieira J. Trypanosoma cruzi-elicited CD8+ T cell-mediated myocarditis: chemokine receptors and adhesion molecules as potential therapeutic targets to control chronic inflammation? Mem Inst Oswaldo Cruz. 2003;98(3):299–304.
98. da Silveira ABM, Lemos EM, Adad SJ, Correa-Oliveira R, Furness JB, D'Avila Reis D. Megacolon in Chagas disease: a study of inflammatory cells, enteric nerves, and glial cells. Hum Pathol. 2007;38(8):1256–64.
99. Fonseca SG, Reis MM, Coelho V, Nogueira LG, Monteiro SM, Mairena EC, Bacal F, Bocchi E, Guilherme L, Zheng XX, Liew FY, Higuchi ML, Kalil J, Cunha-Neto E. Locally produced survival cytokines IL-15 and IL-7 may be associated to the predominance of CD8+ T cells at heart lesions of human chronic Chagas disease cardiomyopathy. Scand J Immunol. 2007;66(2–3):362–71.
100. Dutra WO, Menezes C a S, Magalhães LMD, Gollob KJ. Immunoregulatory networks in human Chagas disease. Parasite Immunol. 2014;36(8):377–87.
101. Gomes JAS, Rocha MOC, Gazzinelli G. Evidence that development of severe cardiomyopathy in human Chagas' disease is due to a Th1-specific immune response. Infect Immun. 2003;71(3):1185–93.
102. Gomes JAS, Bahia-oliveira LMG, Rocha OC, Busek SCU, Teixeira MM, Silva JS, Correa-oliveira R, Ota M, Rocha C. Type 1 chemokine receptor expression in Chagas' disease correlates with morbidity in cardiac patients. Infect Immun. 2005;73(12):7960–6.
103. Ben Younes-Chennoufi A, Said G, Eisen H, Durand A, Hontebeyrie-Joskowicz M. Cellular immunity to Trypanosoma cruzi is mediated by helper T cells (CD4+). Trans R Soc Trop Med Hyg. 1988;82(1):84–9.
104. Padilla A, Xu D, Martin D, Tarleton R. Limited role for CD4+ T-cell help in the initial priming of Trypanosoma cruzi-specific CD8+ T cells. Infect Immun. 2007;75(1):231–5.
105. Annunziato F, Romagnani S. Heterogeneity of human effector CD4+ T cells. Arthritis Res Ther. 2009;11(6):257.

106. Sallusto F. Heterogeneity of human CD4+ T cells against microbes. Annu Rev Immunol. 2016;34(1):317–34.
107. Silva J, Morrisey P, Grabstein K, Mohler K, Anderson D, Reed S. Interleukin 10 and Interferon γ regulation of experimental Trypanosoma cruzi infection. J Exp Med. 1992;175:169–74.
108. Petray PB, Rottenberg ME, Bertot G, Corral RS, Diaz A, Örn A, Grinstein S. Effect of anti-γ-interferon and anti-interleukin-4 administration on the resistance of mice against infection with reticulotropic and myotropic strains of Trypanosoma cruzi. Immunol Lett. 1993;35(1):77–80.
109. Rodrigues MM, Ribeirão M, Pereira-Chioccola V, Renia L, Costa F. Predominance of CD4 Th1 and CD8 Tc1 cells revealed by characterization of the cellular immune response generated by immunization with a DNA vaccine containing a Trypanosoma cruzi gene. Infect Immun. 1999;67(8):3855–63.
110. Kumar S, Tarleton RL. Antigen-specific Th1 but not Th2 cells provide protection from lethal Trypanosoma cruzi infection in mice. J Immunol. 2001;166(7):4596–603.
111. Hoft DF, Eickhoff CS. Type 1 immunity provides both optimal mucosal and systemic protection against a mucosally invasive, intracellular pathogen. Infect Immun. 2005;73(8):4934–40.
112. Albareda MC, Olivera GC, a Laucella S, Alvarez MG, Fernandez ER, Lococo B, Viotti R, Tarleton RL, Postan M. Chronic human infection with Trypanosoma cruzi drives CD4+ T cells to immune senescence. J Immunol. 2009;183(6):4103–8.
113. Longhi SA, Atienza A, Perez Prados G, Buying A, Balouz V, Buscaglia C a, Santos R, Tasso LM, Bonato R, Chiale P, Pinilla C, Judkowski V a, Gómez KA, Prados GP, Buying A, Balouz V, Buscaglia C a, Santos R, Tasso LM, Bonato R, Chiale P, Perez Prados G, Buying A, Balouz V, Buscaglia C a, Santos R, Tasso LM, Bonato R, Chiale P, Pinilla C, Judkowski V a, Gómez K a, Prados GP, Buying A, Balouz V, Buscaglia C a, Santos R, Tasso LM, Bonato R, Chiale P. Cytokine production but lack of proliferation in peripheral blood mononuclear cells from chronic Chagas' Disease cardiomyopathy patients in response to T. cruzi ribosomal P proteins. PLoS Negl Trop Dis. 2014;8(6):e2906.
114. Vitelli-Avelar DM, Sathler-Avelar R, Dias JCP, Pascoal VPM, Teixeira-Carvalho A, Lage PS, Elói-Santos SM, Corrêa-Oliveira R, Martins-Filho OA. Chagasic patients with indeterminate clinical form of the disease have high frequencies of circulating CD3 +CD16 -CD56 + natural killer T cells and CD4 +CD25 High regulatory T lymphocytes. Scand J Immunol. 2005;62(3):297–308.
115. da Silveira ABM, Fortes de Araújo F, Freitas MAR, Gomes JAS, Chaves AT, de Oliveira EC, Neto SG, Luquetti AO, da Cunha Souza G, Bernardino Júnior R, Fujiwara R, d'Ávila Reis D, Correa-Oliveira R. Characterization of the presence and distribution of Foxp3+ cells in chagasic patients with and without megacolon. Hum Immunol. 2009;70(1):65–7.
116. de Araújo FF, Vitelli-Avelar DM, Teixeira-Carvalho A, Antas PRZ, Gomes JAS, Sathler-Avelar R, Rocha MOC, Elói-Santos SM, Pinho RT, Correa-Oliveira R, Martins-Filho OA. Regulatory T cells phenotype in different clinical forms of Chagas' disease. PLoS Negl Trop Dis. 2011;5(5):1–8.
117. Kotner J, Tarleton R. Endogenous CD4+ CD25+ regulatory T cells have a limited role in the control of Trypanosoma cruzi infection in mice. Infect Immun. 2007;75(2):861–9.
118. Martin DL, Postan M, Lucas P, Gress R, Tarleton RL. TGF-β regulates pathology but not tissue CD8+ T cell dysfunction during experimental Trypanosoma cruzi infection. Eur J Immunol. 2007;37(10):2764–71.
119. Guhl F, Lazdins-Helds J. Reporte sobre la enfermedad de Chagas. Geneva: Grupo de Trabajo Científico, WHO; 2007. p. 104.
120. Morillo CA, Marin-Neto JA, Avezum A, Sosa-Estani S, Rassi A, Rosas F, Villena E, Quiroz R, Bonilla R, Britto C, Guhl F, Velazquez E, Bonilla L, Meeks B, Rao-Melacini P, Pogue J, Mattos A, Lazdins J, Rassi A, Connolly SJ, Yusuf S. Randomized trial of benznidazole for chronic Chagas' cardiomyopathy. N Engl J Med. 2015;373(14):1295–306.
121. Viotti R, Vigliano C, Armenti H, Segura E. Treatment of chronic Chagas' disease with benznidazole: clinical and serologic evolution of patients with long-term follow-up. Am Heart J. 1994;127(1):151–62.

122. Fabbro DL, Streiger ML, Arias ED, Bizai ML, Del Barco M, Amicone NA. Trypanocide treatment among adults with chronic Chagas disease living in Santa Fe City (Argentina), over a mean follow-up of 21 years: parasitological, serological and clinical evolution. Rev Soc Bras Med Trop. 2007;40(1):1–10.
123. Laucella S a, Mazliah DP, Bertocchi G, Alvarez MG, Cooley G, Viotti R, Albareda MC, Lococo B, Postan M, Armenti A, Tarleton RL. Changes in Trypanosoma cruzi-specific immune responses after treatment: surrogate markers of treatment efficacy. Clin Infect Dis. 2009;49(11):1675–84.
124. Day CL, Abrahams DA, Lerumo L, Janse van Rensburg E, Stone L, O'rie T, Pienaar B, de Kock M, Kaplan G, Mahomed H, Dheda K, Hanekom WA. Functional capacity of mycobacterium tuberculosis-specific T cell responses in humans is associated with mycobacterial load. J Immunol. 2011;187(5):2222–32.
125. Beyer M, Abdullah Z, Chemnitz JM, Maisel D, Sander J, Lehmann C, Thabet Y, Shinde PV, Schmidleithner L, Köhne M, Trebicka J, Schierwagen R, Hofmann A, Popov A, Lang KS, Oxenius A, Buch T, Kurts C, Heikenwalder M, Fätkenheuer G, Lang PA, Hartmann P, Knolle PA, Schultze JL. Tumor-necrosis factor impairs CD4+ T cell–mediated immunological control in chronic viral infection. Nat Immunol. 2016;17(5):593–603.
126. Vitelli-Avelar DM, Sathler-Avelar R, Teixeira-Carvalho A, Pinto Dias JC, Gontijo ED, Faria AM, Elói-Santos SM, Martins-Filho OA. Strategy to assess the overall cytokine profile of circulating leukocytes and its association with distinct clinical forms of human Chagas disease. Scand J Immunol. 2008;68(5):516–25.
127. Sathler-Avelar R, Vitelli-Avelar DM, Massara RL, Borges JD, Lana M, Teixeira-Carvalho A, Dias JCP, Elói-Santos SM, Martins-Filho OA. Benznidazole treatment during early-indeterminate Chagas' disease shifted the cytokine expression by innate and adaptive immunity cells toward a type 1-modulated immune profile. Scand J Immunol. 2006;64(5): 554–63.
128. Sathler-Avelar R, Vitelli-Avelar DM, Massara RL, de Lana M, Pinto Dias JC, Teixeira-Carvalho A, Elói-Santos SM, Martins-Filho OA. Etiological treatment during early chronic indeterminate Chagas disease incites an activated status on innate and adaptive immunity associated with a type 1-modulated cytokine pattern. Microbes Infect. 2008;10(2): 103–13.
129. Sathler-Avelar R, Vitelli-Avelar DM, Elói-Santos SM, Gontijo ED, Teixeira-Carvalho A, Martins-Filho OA. Blood leukocytes from benznidazole-treated indeterminate Chagas disease patients display an overall type-1-modulated cytokine profile upon short-term in vitro stimulation with trypanosoma cruzi antigens. BMC Infect Dis. 2012;12(1):123.
130. Campi-Azevedo AC, Gomes JAS, Teixeira-Carvalho A, Silveira-Lemos D, Vitelli-Avelar DM, Sathler-Avelar R, Peruhype-Magalhães V, Béla SR, Silvestre KF, Batista MA, Schachnik NCC, Correa-Oliveira R, Eloi-Santos SM, Martins-Filho OA. Etiological treatment of Chagas disease patients with benznidazole lead to a sustained pro-inflammatory profile counterbalanced by modulatory events. Immunobiology. 2015;220(5):564–74.
131. Rassi A, Marin-Neto JA, Rassi A. Chronic Chagas cardiomyopathy: a review of the main pathogenic mechanisms and the efficacy of aetiological treatment following the BENznidazole evaluation for interrupting trypanosomiasis (BENEFIT) trial. Mem Inst Oswaldo Cruz. 2017;112(3):224–35.
132. Viotti R, Vigliano C, Álvarez MG, Lococo B, Petti M, Bertocchi G, Armenti A, de Rissio AM, Cooley G, Tarleton R, Laucella S. Impact of aetiological treatment on conventional and multiplex serology in chronic Chagas disease. PLoS Negl Trop Dis. 2011;5(9):e1314.
133. Niborski LL, Grippo V, Lafón SO, Levitus G, García-Bournissen F, Ramirez JC, Burgos JM, Bisio M, Juiz NA, Ayala V, Coppede M, Herrera V, López C, Contreras A, Gómez KA, Elean JC, Mujica HD, Schijman AG, Levin MJ, Longhi SA. Serological based monitoring of a cohort of patients with chronic Chagas disease treated with benznidazole in a highly endemic area of northern Argentina. Mem Inst Oswaldo Cruz. 2016;111(6):365–71.
134. Balouz V, Buscaglia CA, Aires B. Chagas disease diagnostic applications: present knowledge and future steps. In: Advances in parasitology. London: Academic Press; 2017. p. 1–45.

135. Schijman A, Burgos J, Marcet P. Molecular tools and strategies for diagnosis of Chagas Disease and leishmaniasis. In: Santos da Silva M, Cano MIN, editors. Frontiers in parasitology. Sharjah: Bentham Books; 2017. p. 394–453.
136. Fairfax KA, Kallies A, Nutt SL, Tarlinton DM. Plasma cell development: from B-cell subsets to long-term survival niches. Semin Immunol. 2008;20(1):49–58.
137. Tangye SG. Staying alive: regulation of plasma cell survival. Trends Immunol. 2011;32(12):595–602.
138. Hammarlund E, Thomas A, Amanna IJ, Holden LA, Slayden OD, Park B, Gao L, Slifka MK. Plasma cell survival in the absence of B cell memory. Nat Commun. 2017;8(1):1781.

Part III
Epidemiology

Epidemiology of Chagas Disease

Roberto Chuit, Roberto Meiss, and Roberto Salvatella

Abstract Chagas disease has been described more than 100 years ago and has existed or coexisted with man for millennia.

In the last 30 years, control programs have had a significant impact on the primary transmission routes, achieving notable reductions in *T. cruzi* infection prevalence in vast regions of the continent and even ended the vector transmission by *Triatoma infestans* in Uruguay, Chile, and Brazil households.

Current transmission of *T. cruzi* and the perspectives toward the future are analyzed, as well as the economic impact of the interventions to have a better understanding of the burden of the disease.

We believe that the transmission of *T. cruzi* was initiated at the beginning by digestive contagion, and, after the application of all vector control measures, mother to child, transfusion, and transplant are the essential routes remaining.

This new transmission scenario will force health structures to prepare themselves to face new challenges. In this way, they will be able to keep new infections, controlled through adequate surveillance systems to block and prevent household transmission.

R. Chuit (✉) · R. Meiss
Institute of Epidemiological Research, National Academy of Medicine (Bs. As.), Buenos Aires, Argentina
e-mail: chuit@aya.yale.edu

R. Salvatella
Pan American Health Organization/World Health Organization, Washington, DC, USA

J. M. Altcheh, H. Freilij (eds.), *Chagas Disease*, Birkhäuser Advances in Infectious Diseases, https://doi.org/10.1007/978-3-030-00054-7_4

1 Introduction

Chagas disease or American trypanosomiasis is a parasitic disease caused by the protozoan *Trypanosoma cruzi* (*T. cruzi*). The parasite lives in the blood and tissues of different mammals, man, and intestines of blood-sucking bugs of the family Reduviidae, subfamily Triatominae, known in Argentina, Chile, Uruguay, and Bolivia as "vinchucas" (Quechua voice: wikchukuy means "throw") [1].

In 1909 [2] and subsequent years, Carlos Chagas characterized the disease in humans, as well as causing the parasite and the transmitting vector as a unit of transmission in Argentina [3], Venezuela [4], and other American countries. Coexistence among the different actors of the infection process in the American continent has occurred since much earlier, and in the beginning, this was an enzootic infection, and the parasite was restricted to wild animals. It became a zoonosis when contact between humans and domestic and synanthropic animals started human transmission in dwellings (domestic cycle), with a high number of vector colonies and elevated parasite infection [5, 6].

The following transmission routes of *T. cruzi* have been reported:

(a) Vectorial (when the vector insects feed and deposit their feces simultaneously)
(b) Transplacental or congenital (when the parasite crosses the placenta of the seropositive mother and infects the child during pregnancy or childbirth)
(c) Oral route of trypomastigotes (by parasite-contaminated food)
(d) Blood transfusion through *T. cruzi*-contaminated blood
(e) Organ transplants

The acute stage, connected with the first infection and particularly in children, is often unapparent and not opportunely diagnosed. If there are symptoms, they are generally unspecific, characterized by fever, localized or generalized edema, localized and/or generalized lymphatic adenopathy, myocarditis, and encephalitis, and the diagnosis is made by parasite evidence. The chronic phase may induce nerve diseases with different forms of manifestation, neurological disorders, or massive organ dilatation (megacolon and megaesophagus).

2 The Vectors

Most of the 110 species of triatomine have strictly wild habits, and they live in association with birds, edentates, lizards, and mammals. As wild enzootic, it extends from 42° N (North Carolina and Maryland, USA) to 49° S (south of Argentina and Chile) including the Caribbean islands [7]. The wild reservoirs reported infected by the *T. cruzi* parasite are armadillos (Dasypus), opossum (*Didelphis* sp.), bats, raccoons, squirrels, edentates, and primates. These wild foci do not include humans. It has been reported that Andean cultures came into contact with guinea pig species approximately 7000–8000 years ago in the regions of Peru and Bolivia, and between 5000 and 3500 years ago, they began to keep them in the houses and to use them as a food source. The precondition of the use of wild animals in the ancients' diet suggests that the ample contact between *T. cruzi* and humans gave rise to the oral contagion of the disease, showed by:

(a) Parasite findings in some mummies in the Rio Grande Valley (Rio Bravo) (1150 BP), to the north of the states of Chihuahua and Coahuila (Mexico) and south of the Rio Grande (Texas, USA)
(b) The presence of hairs and bones of wild rodents without cooking in human coprolites associated with megacolon and very large fecal pellets that filled the pelvic cavity [8–11]

This practice was also described for the last Incas and is common in the current populations of Latin America [9, 12].

It has been proposed that the guinea pig, with its wild habits, is still a reservoir of *T. cruzi*. It once took part in the parasitic cycle, since, at the time of its domestication, it was able to attract wild triatomines, among other insects, thus originating the domestic circle. This presumption is based on the finding of mummies with megacolon in rooms of Chiribaya, dating from 900 AD to 1350 AD. It is possible to establish this relationship today, thanks to the finding of mummies with megacolon [13]. An analogous situation can be described for the Chinchorro culture (6000–2000 BC) that inhabited the valleys of Azapa, Camarones, and Lluta (Chile) [14]. There, populations were exposed to vectors when they spent the night in the mountain slopes, in their small huts made of sticks, skins, and vegetable mats (Fig. 1).

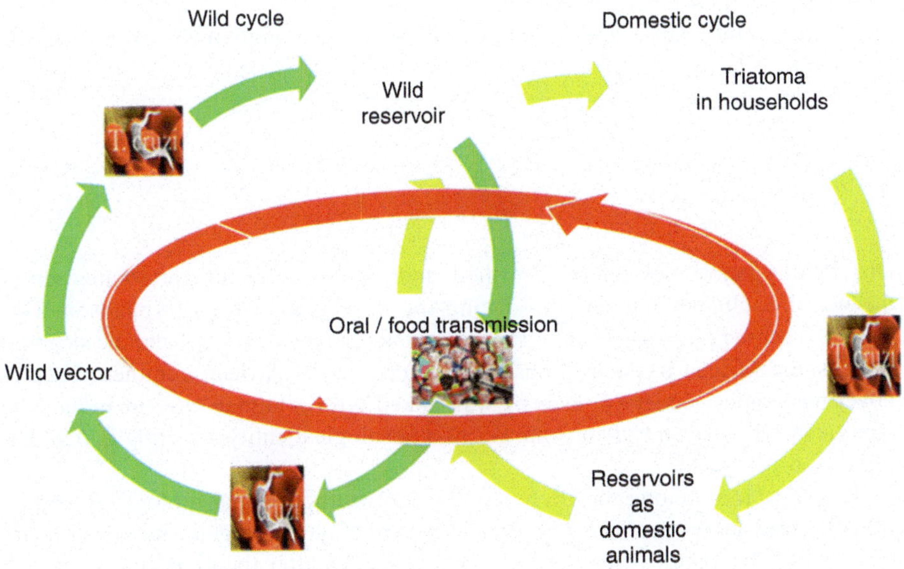

Fig. 1 Representation of vector-associated transmission. *Source* Elaborated by the author

Different investigations have shown that the *T. cruzi* complete cycle can develop in marsupials, and infectious forms were found in their anal glands as well as in their peripheral blood [15]. Due to the presence of the complete cycle in these reservoirs, it can be assumed that they have been significantly relevant in parasite contribution regarding the installation of an intradomiciliary cycle [16–18].

Today, it is common to use construction materials such as mud, straw, and palm leaves that offer a physical medium equivalent to the nests and dens of wild animals, where the vector can have access to a blood meal. *T. infestans* (main vector domiciled in southern Peru, Bolivia, Chile, Argentina, Paraguay, Brazil, and Uruguay) and *Panstrongylus megistus* (main vector domiciled in large areas of Brazil) live in wall cracks near the roofs [19]. *R. prolixus*, on the other hand, prefers the characteristic palm walls and ceilings of Venezuela, Colombia, and Central America. *Triatoma dimidiata*, the main vector in Ecuador and the Pacific coast of Colombia and Central America to Mexico, prefers cracks in mud walls, but at a low level from the floor or on the ground floor, hiding by means of a phenomenon known as "camouflage." It is also possible that while performing their subsistence activities, individuals were exposed to triatomines while shucking, manufacturing hooks, and preparing "totora" and jonquil to make mats [20]. In Chile, *Mepraia spinolai* is distributed in the regions of Arica and Parinacota, Tarapacá, Antofagasta, Atacama, Coquimbo, and Valparaiso [21].

Each vector species listed has specific conditions and/or attributes that make it biologically or behaviorally different, but as domiciliary species, we can prove that they all have the following characteristics:

- Lack of mobility and little ability for active distribution, that is, a high degree of stability of domiciled populations.
- Population replacement is slow, since, on average, the new generations develop in a year approximately.
- All the evolutionary stages of the vector are present simultaneously in the same ecotope.

3 The Parasite

Different trypanosomes can be identified, and they are infecting vertebrates across the world, including humans, producing the disease known as trypanosomiasis. Chagas disease (*T. cruzi*) is the most representative in the Americas and the sleeping sickness or African trypanosomiasis (*T. brucei*) in the African continent [22]. A novel nosological entity (*T. evansi*) was linked with a human case in India. The American and African trypanosomiasis as a nosological entity affecting human has more than 100 years.

Trypanosomes belong to different subgenera with particular biological aspects; the American are intracellular parasites of the vertebrate host, deposited in situ with the stool of the vector (triatomine), which defecates after the blood inlet, and the African live and replicate in the salivary glands and are inoculated with the slobber through the vector bite (tsetse fly) [23].

This parasitic differentiation could have occurred 475 million years ago, when the breakup of the supercontinent Pangea separated the family Trypanosomatidae, and the ancestor of *T. cruzi* diverged from the ancestor of the salivary parasites (*T. gambiense, T. rangeli, T. rhodesiense, T brucei*) to become dregs. *T. cruzi* would have risen 280–150 million years ago in America, and, between 88 and 37 million years ago, the subpopulations of *T. cruzi* I and *T. cruzi* II would have been separated as genotypes [24]. The *T. cruzi* I would be autochthonous of South America and would have coevolved with primates and rodents. *T. cruzi* II would have entered from North America 5 million years ago with the great exchange of mammals in the small islands that gave origin to the Isthmus of Panama [25, 26]. At present, both groups would be circulating in different environments [27, 28].

4 Possible Routes of Dissemination of Domestic Vector Species

Paleoparasitological studies found *T. cruzi* DNA in Chinchorro mummies (+9000 years old), found in the Chilean coastal desert, that is, before the domestication of guinea pigs (8000 years ago [29]). Therefore, Chagas disease is present since ancient times in the Americas, and it is probably older than animal domestication and human presence in any part of America where wild vectors and reservoirs are present [30]. The dispersal of host mammals in South America and, perhaps, that of *Triatoma infestans* could have occurred in times prior to the uplift of the Andes Mountains [31]. In this way, the native species of the western slope of the Andes (degu or mouse of the pircas—*Octodon degus* and *Triatoma spinolai*) remained genetically isolated, living together inside the caves and multiple galleries excavated by the degu [32].

The reciprocal relationship would have favored the wild cycle of *T. cruzi* between vectors and reservoirs, long before the appearance of man. Men would have been incorporated into that cycle when they displaced mammals from their shelters or when they domesticated some of them. This scenario would have favored the introduction of infected vectors in the peridomiciliary areas and later in the dwellings. Evidences of this contact are found in mummified remains in different parts of the continent (Chihuahua desert in Mexico, Inca area of Peru, Minas Gerais in Brazil, and Atacama Desert in Chile), with an age of 4000–9000 years. Other mummies found in the Tarapacá gorge (Chile) at 1500 m high, corresponding to indigenous Wankarani, with an antiquity extending back to 1500 BC, who emigrated from Bolivia 3500 years ago, have shown a sign compatible with megacolon and cardiomegaly, characteristic conditions of this disease. As primitive populations went from nomad (hunter-gatherers) to sedentary (farmers) and, therefore, brought wild species (vectors and animals) into the interior of dwellings, they gave rise to the domestic cycle of transmission. With the complete cycle installed in the housing – parasite, vector, and human reservoir in the same reduced unit – human relationships and their migratory movements were enough for this biotic unit to install in other regions, far away from those considered initially.

We can conclude that the geographical distribution of the most important vectors of the disease in the Americas before the implementation of vector control programs was as follows [33]:

4.1 *Triatoma infestans*

- Argentina (except the provinces of Chubut, Santa Cruz, and Tierra del Fuego)
- Bolivia (Beni, Chuquisaca, Cochabamba, La Paz, Potosí, Santa Cruz, Tarija)
- Brazil (Alagoas, Bahia, Goiás, Mato Grosso, Mato Grosso do Sul, Minas Gerais, Paraíba, Paraná, Pernambuco, Piauí, Rio de Janeiro, Rio Grande do Sul, São Paulo, Sergipe, Tocantins)
- Chile (Regions I–VI and the Metropolitan Santiago area)
- Paraguay (Alto Paraguay, Boquerón, Caaguazú, Caazapá, Central, Chaco, Concepción, Cordillera, Guairá, Misiones, Nueva Asunción, Paraguarí, Presidente Hayes, San Pedro)
- Peru (Arequipa, Ica, Moquegua, Tacna)
- Uruguay

4.2 *Panstrongylus megistus*

- Argentina (Corrientes, Jujuy, Misiones, Salta)
- Brazil (Alagoas, Bahia, Ceará, Espirito Santo, Goiás, Maranhão, Mato Grosso, Mato Grosso do Sul, Minas Gerais, Pará, Paraíba, Paraná, Pernambuco, Piauí, Rio de Janeiro, Rio Grande del Norte, Rio Grande do Sul, Santa Catarina, São Paulo, Sergipe)
- Paraguay (Amambay, Cordillera)
- Uruguay

4.3 *Rhodnius prolixus*

- Colombia (Antioquia, Arauca, Boyacá, Caquetá, Casanare, César, Cundinamarca, Guajira, Huila, Magdalena, Meta, Norte de Santander, Putumayo, Santander, Tolima, Vichada)
- El Salvador
- Guatemala (in five of the 22 departments)
- Honduras (in 11 of the 18 departments)
- México (Chiapas, Oaxaca)
- Nicaragua
- Venezuela (Aragua, Carabobo, Cojedes, Miranda, Portuguesa, Yaracuy)

4.4 *Triatoma brasiliensis*

- Brazil (widespread in all the semiarid northeast of the country Alagoas, Bahia, Ceará, Maranhão, Paraíba, Piauí, Rio Grande del Norte, Sergipe, Tocantins—and the north of Minas Gerais)

4.5 *Triatoma dimidiata*

- Belize
- Colombia
- Costa Rica
- Ecuador
- El Salvador
- Guatemala
- Honduras (in 16 of the 18 departments)
- Mexico (Campeche, Chiapas, Guerrero, Jalisco, Nayarit, Oaxaca, Puebla, Quintana Róo, San Luis Potosi, Tabasco, Veracruz, Yucatan)
- Nicaragua
- Panama
- Peru (Tumbes)
- Venezuela

At present, the geographical distribution of these species as transmission risk has been modified as an effect of the control actions developed by the countries.

5 Transmission and Infection

Chagas disease is not homogeneous in the Americas, and it has a varying epidemiology associated with distribution, transmission mechanisms, clinical manifestations, and predominant pathologies. There are regional forms in which acute cases can be clinically lethal, frequent and florid, or asymptomatic and benign. The chronic forms show cardiomyopathy, megaesophagus, megacolon, or cardiomegaly according to geographic areas; mother-to-child transmission varies, and it is frequent in some regions and extremely low in others.

If untreated, *T. cruzi* infection is lifelong; 90% of new infections occur before 10 years of age [34]. During the disease evolution, 25% of infected cases would develop some alteration, where 18% would develop cardiomyopathy without heart failure, 4% cardiomyopathy with heart failure, and 3% megavisceras [35–40]. High mortality appears in infected individuals aged between 20 and 59, increasing among those with electrocardiographic alterations [41].

The transmission ways showed in Fig. 2 can be defined differently with the various weights already stated.

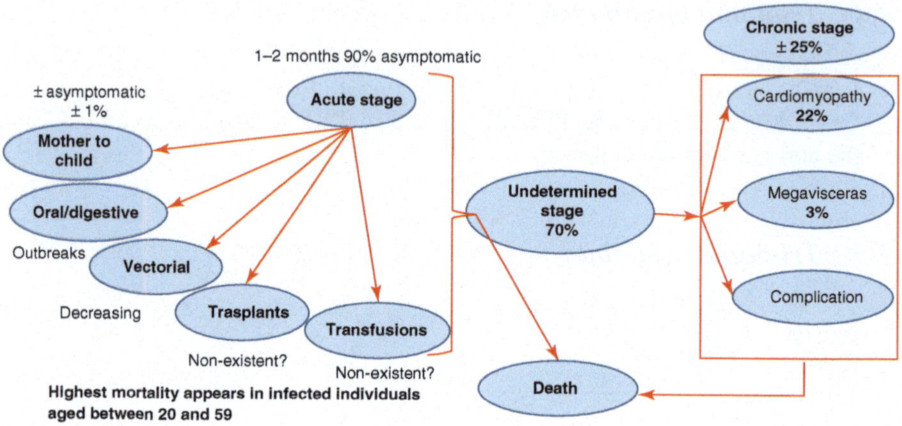

Fig. 2 Transmission of *T. cruzi* infection. *Source* Elaborated by the author

Historically, the classic model of transmission can be defined as that associated with the presence of the vector in the domiciliary units in an environment conducive to its development. This environment, associated with wild or domestic reservoirs (cats and/or dogs) [42] infected by *T. cruzi*, established the condition of intradomiciliary or domestic transmission. This cycle was for years responsible for the occurrence of new infections in the Americas. Also, it was considered the classic model of vector transmission, with abundant rural population above urban, precarious housing and coexistence of different transmission routes, associated with bioclimatic factors of the regions of poor areas characterized by subsistence agriculture. This model no longer exists; it has changed from rural areas to marginal urban areas, where the domestic cycle is maintained and the peridomiciliary influence is low or inexistent for transmission maintenance. Population is no longer dispersed, but it is grouped in such a way that it gives rise to a new Chagas transmission model (Fig. 3).

This model changes depending on vector control interventions developed by the countries and shows an impact on the prevalence of *T. cruzi* in the populations of the Americas. These figures can be explained by the reduced number of cases that started at more than 20,000,000 positive cases in the middle of the twentieth century (result of estimations made according to the estimated value of exposed population and possibly infected people that arises from mid-century publications of the different countries) [43, 44] and decreased to less than 7,000,000 in 2015 (Fig. 4).

The reduction of new infections associated with vectorial transmission led to a decrease in prevalence in all the countries of the region, as a direct result of the regional initiatives to "Eliminate Intradomiciliary Vector Transmission and Control of Blood Banks" launched in 1991 under the executive secretary of the PAHO/

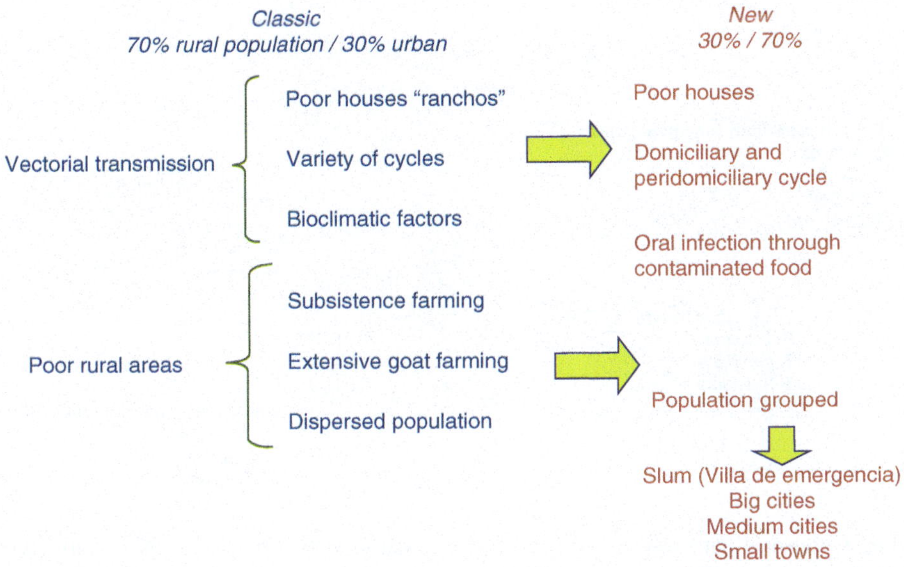

Fig. 3 Change of the vectorial transmission model. *Source* Elaborated by the author

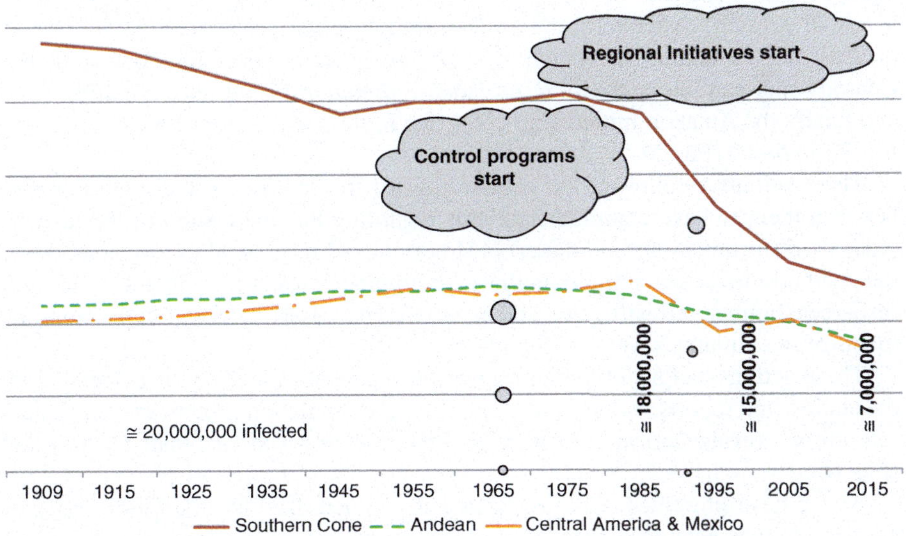

Fig. 4 Estimated prevalences for *T. cruzi* infection. Region of the Americas. 1909–2016. *Source* Elaborated by the author

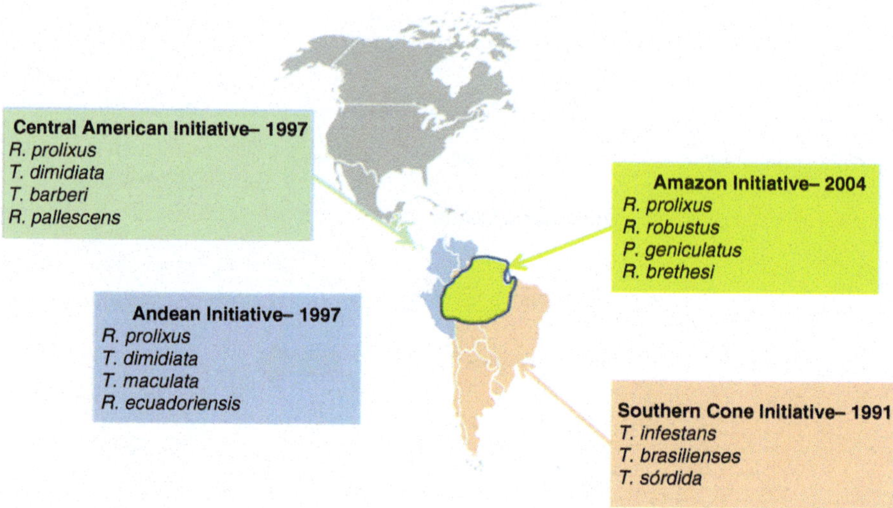

Fig. 5 Regional initiatives for the control of vectorial transmission. *Source* Elaborated by the author based on countries' report to OPS/OMS

WHO. The first one was the Southern Cone Initiative in 1991 (*T. infestans*, *T. brasiliensis*, *T. sordida*). Afterward, the Andean Initiative was organized in 1997 (for *R. prolixus*, *T. dimidiata*, *T. maculata*, *R. ecuadoriensis*), in conjunction with the Central American Initiative (for *R. prolixus*, *T. dimidiata*, *T. barberi*, *R. pallescens*), and finally the Amazon Initiative in 2004 (for *R. prolixus*, *R. robustus*, *P. geniculatus*, *R. brethesi*) (Fig. 5).

These initiatives allowed the actions toward the interior of the countries and between them to have common standards regarding the application of the insecticides used (formulations and doses) [45] normalized from local experiences. Also, the methods to detect intradomiciliary triatomine infestation and the minimum resources (human, infrastructure, and economic) necessary to carry out the commitments of the initiatives were agreed upon.

These actions had a substantial impact on vector transmission, as evidenced by the studies and reports, where studies in different population groups and ages show that initial prevalence averages were around 10% down to less than 1% [46–49] (Fig. 6).

As we have presented, Chagas disease is characterized by a large diversity of epidemiological data, resulting from the variety of vectors and reservoirs that serve as sources of infection. In this epidemiological scenario, it is important to distinguish national and/or local conditions.

For the control of the domiciled vector, it is necessary to know the species present or the species that intervene in the transmission of the infection in the home

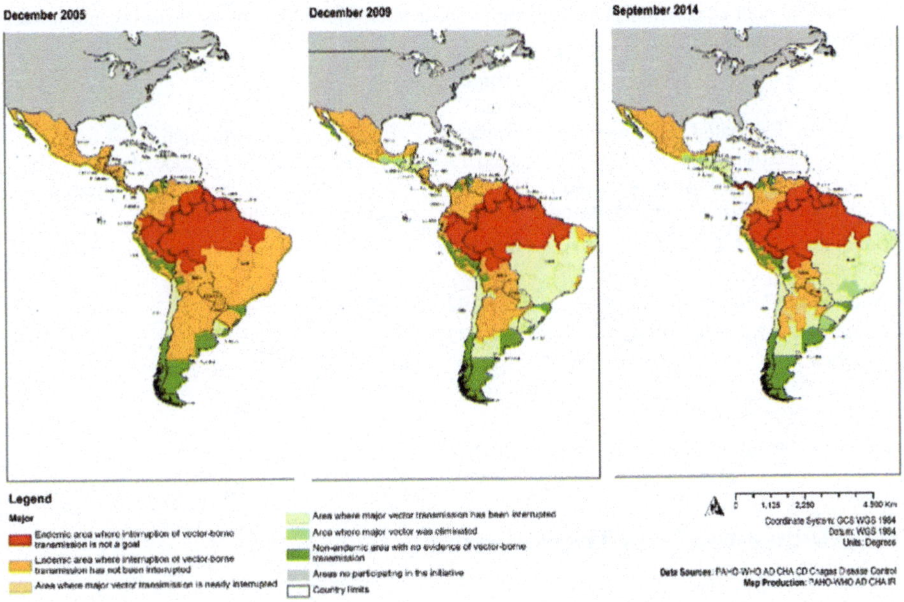

Fig. 6 Chagas disease, major vector transmission 2005–2014. *Source* Elaborated by the author of PAHO information

environment, the degree of vulnerability to the control measures that depend on vulnerability (with greater adaptation to human housing) and human behaviors. The removal of a vector from a determined area depends primarily on whether the species is invasive or introduced and, as such, strictly domiciliary.

The maximum degree of vector control achieved in the case of autochthonous species is the extinction of intradomiciliary colonies through chemical treatment with insecticides. Also, the absence of vector colonies in the interior of the house could represent the interruption of the transmission or its transformation into a fortuitous or accidental event.

The WHO Report [50] updated the epidemiological information on Chagas disease in Latin American countries, based on the available 2010 demographic and epidemiologic information, and it is possible to calculate the population infected, new cases due to vector transmission, number of women infected, and positive children due to mother-to-child transmission. For countries included in the Southern Cone Initiative (Argentina, Bolivia, Brazil, Chile, Paraguay, and Uruguay), the average prevalence is 2.17% (0.03–6.14%); in the Andean Initiative (Colombia, Ecuador, Peru, and Venezuela), it is 1.08% (0.043–1.38%); and in the Central American Initiative with Mexico (Belize, Costa Rica, El Salvador, Guatemala, Honduras, Mexico, Nicaragua, and Panama), it is 0.72% (0.16–1.30%) (Table 1).

Table 1 Number of cases of positive population, women, and cases due to vectorial and mother-to-child transmission

Initiative	Estimated # of people infected by *T. cruzi*	Estimated # of cases due to vectorial transmission	Estimated # of women aged 15–44 infected by *T. cruzi*	Estimated # of cases from congenital transmission	Estimated prevalence of *T. cruzi* infection—%	Max./min. prevalence (%)
Southern Cone	3,581,423	8430	606,765	3304	2.17	0.03–6.14
Andean	971,053	10,524	251,292	2657	1.08	0.43–1.38
Central America and Mexico	1,263,024	9893	266,873	2707	0.72	0.16–1.30

Source by the author of the WHO Report 2015

6 Mother-to-Child Transmission

Due to population movements (migration) or due to old infections transmitted by mothers, by the vector, or by transfusions, the distribution of congenital Chagas disease is not restricted to rural areas only. Therefore, congenital transmission goes beyond traditional transmission areas and occurs in remote areas where the vector does not exist. In this way, it becomes a world disease where the transmission factor is the person (mother to child or transfusions), in such remote places as Europe or Japan [51].

The most significant number and case studies of congenital Chagas disease have been reported from Argentina, Bolivia, Brazil, Chile, Colombia, Guatemala, Honduras, Paraguay, Uruguay, and Venezuela. The risk of congenital transmission varies according to the confluence of epidemiological factors, and perhaps here the strain of the parasite, the level of parasitemia of the mother, and the existence of placental lesions are critical. It is estimated that, under these conditions, the risk would vary between 1% and 7%, or more, in some regions of Argentina, Bolivia, Chile, and Paraguay. The number of cases to be detected caused by congenital transmission depends on the prevalence of infection in fertile women who become pregnant, so to achieve control and cutting maternal-fetal transmission, it is necessary to develop actions that allow women to reach their delivery with a proper diagnosis to follow and treat newborns properly.

With vectorial transmission controlled, it can be said that congenital transmission is becoming the primary way of occurrence of new cases in many countries. The absence of sensitive and adequate systems for the detection of infection in newborns leads to the assumption that these will be the new chronic cases in the future.

We can make an estimate, through a theoretical and modeling exercise, of the situation in Argentina from official data [52]. Based on these official data, in

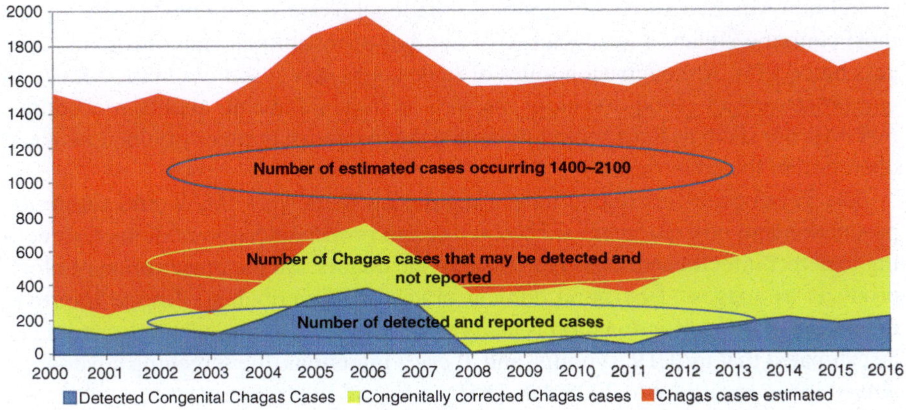

Fig. 7 Notified cases of congenital Chagas disease, correction of really detected non-notified and estimated that should be detected. Argentina 2017. *Source* Elaborated by the author

Argentina, there is an average of 700,000 deliveries per year, and the prevalence of infection by *T. cruzi* in pregnant women is 2.8% (12–0.5%). According to this information, between 1400 and 2100 cases of *T. cruzi*-positive newborns would occur in the country. If this estimated value is compared with detected values, and corrections are made for those possibly treated (detected and not notified), a final result is obtained showing that in Argentina, only for this cause, more than 1000 new chronic cases per year would be produced. These numbers, compared with those reported by vector transmission, are widely superior (Fig. 7).

7 Transfusions and Blood Banks

As we described, in the last century, there were important population movements from rural areas to urban areas in the region of the Americas. This motivated a change not only in the pattern of vector transmission, since in areas that were not defined as risky, infected persons were blood suppliers generating the risk of transmission by blood transfusions. Although the risk is present in all regions, the prevalence of blood givers is not the same in all countries.

Since 2005, Spain has a clear regulation regarding the disease of Chagas, which indicates that people with positive *T. cruzi* serology should not donate blood and requires serological testing of all donors who have lived in countries where Chagas disease is endemic or of children from a mother who has lived in those countries [53].

In Latin America, legislation and regulations or regulations related to blood transfusion began to appear between 1939 and the 1950s–1960s in some countries like Argentina, Brazil, and Chile. In others, they started appearing in the 1970s (Bolivia, Colombia, Costa Rica, Ecuador, Paraguay, Venezuela) and 1980s (Honduras, Mexico, Nicaragua, Uruguay) and in the 1990s, in Guatemala, Panama, and Peru.

In Latin America, there are varieties of national blood system modalities, but the most common situation is an excessive number of institutions that obtain and process blood [54]. Although economies of scale could save resources in the collection and processing of blood donations, as well as in guaranteeing quality procedures, this does not seem to have served to reduce the excessive number of blood banks in operation in the countries.

The 37 states of this region have a regulatory framework that would assure the production and utilization of safe blood through proper selection of donors, screening to detect infectious diseases of 100% of the donors and the prescription of blood products as suggested by good clinical practices. Nevertheless, despite improvements, not all blood for donation is analyzed as required by the standards, and it can be estimated that for HIV, syphilis, and *T. cruzi* infections, the screening coverage is almost 99%, according to 2005 data [55].

Since regional initiatives, the danger of being infected after a blood transfusion was modified. Most of the states of the area have proven schemes and mechanisms to study all blood to be transfused, achieving around 100% of control of the stock to be transfused in blood banks.

Thus, the prevalence of infection by *T. cruzi* in blood banks cannot be used as an indicator of prevalence in the population.

8 Oral Transmission

As we have stated, perhaps in its origins, oral transmission had a greater importance than the other routes. As progress is being made in the control of vectorial, transfusional, and mother-to-child transmission, oral transmission becomes increasingly important, being currently responsible for localized foci of risk and fundamentally associated with food [56, 57].

In the Amazon, there have been frequent and epidemiologically significant outbreaks between 1968 and 2005. The Evandro Chagas Institute [58] reported 442 cases in 62 different outbreaks associated mainly with the consumption of açai juice. As a hypothesis, it was suggested that the fruit would have been transported with triatomine and squeezed together to prepare the juice and thus contaminated the drinking.

Diaz presents a summary of the contamination routes for the occurrence of oral transmission. It includes the contamination of food by the reservoirs either directly by the vector in the preparation or directly by the consumption of infective animals [59]. Recently, Rueda Karina [60] published, in a review of oral transmission mechanisms and records, outbreaks, case number, and sources of infection showing that by 2013 Argentina, Bolivia, Brazil, Colombia, French Guiana, Ecuador, and Venezuela reported outbreaks with varying numbers of affected people between 2 and 217 cases.

Of all the routes of transmission, the oral route is the most difficult to control because it is associated with the habits and customs of the population, infected vectors of wild origin, and infected reservoirs present in the diet of many populations of the Americas.

9 Economic Impact

The control program is both costly and cost-effective as control intervention. This is shown in the Southern Cone Initiative that has spent more than USD 345 million from its national budget between 1991 and 2000 to finance vector control and blood banks (transmission through blood transfusion) activities in its territories since the launch of the initiative [44].

Research to measure the impact of interventions by different researchers in the region's countries [61] showed that the actions developed were effective to achieve an interruption of vectorial transmission, shown by the reduction of *T. cruzi* prevalence in humans.

Effectiveness was defined using various parameters, with the main one being the measurement of the burden of disease prevented, in DALYs (disability-adjusted life years), potential disease transmission, the overall burden of Chagas disease, etc.

With data from different studies of *T. cruzi* infection (prevalence), mathematical models were developed trying to determine the trend of the disease in the region countries, giving an economic value to each of the stages of infection, disease, or death [62, 63].

If we carried out an exercise using different serological data of the region and summarized it in the present chapter, associated with average economic values emerging from the various models, we could construct a theoretical economic impact scenario of money savings produced by the countries' interventions. In 1909, it was estimated that the total expenditure on Chagas disease was almost USD 900 million, reduced by half to the present day with a downward trend for the future (Fig. 8).

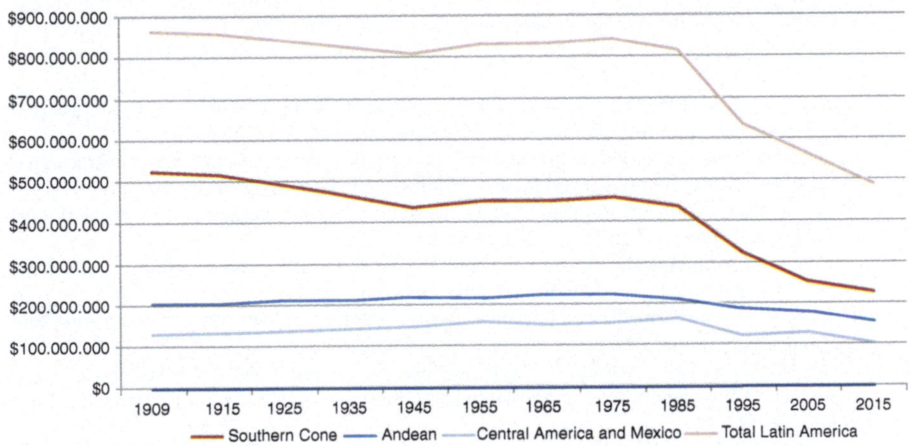

Fig. 8 Total expenditure in USD dollars by American region and year. 1909–2015. *Source* Elaborated by the author

In current times, it is necessary to ensure sustainability of the control program in an unsteady epidemiological context with low *T. cruzi* infection rates and a political-institutional context of health sector reforms, in which the decentralization of operations may result in the risk of control activities losing priority.

The new scenario requires Chagas disease control activities to be integrated into other programs as EMTCT Plus (PAHO) [64] and become part of a broader scheme for meeting the health needs of the population.

We can say that in terms of transmission, after millennial of evolution, we are where we started.

References

1. Lenz R. Diccionario etimológico de las voces chilenas derivadas de lenguas indígenas Americanas. Santiago de Chile: Cervantes; 1910.
2. Chagas C. Nova tripanozomiaze humana: estudos sobre a morfología e o ciclo evolutivo do Schizotrypanum cruzi n. gen., n. sp., agente etiológico de nova entidade mórbida do homem. Mem Inst Oswaldo Cruz. 1909;1(2):159–218. ISSN 0074-0276.
3. Mazza S. Casos agudos benignos de enfermedad de Chagas comprobados en la Provincia de Jujuy. MEPRA. 1934;17:3–11.
4. La TE. Trypanosomose americaine ou maladie de Chagas au Venezuela. Bull Soc Pathol Exot. 1919;12:509–13.
5. Walsh J, Molinex D, Birley M. Deforestation: effects on vector-borne disease. Parasitology. 1993;106:55–75.
6. Walter A. Human activities and American trypanosomiasis. Review of the literature. Parasite. 2003;10:191–204.
7. Dias JC. Epidemiologia. In: Brener Z, Andrade Z, Barral-Netto M, Diotaiut L, Pereira A, Loiola C, editors. Trypanosoma cruzi e doença de Chagas. Rio de Janeiro: Guanabara-Koogan; 2000. p. 48–74.
8. Dittmar K, Jansen AM, Araújo A, Reinhard K. Molecular diagnosis of prehistoric Trypanosoma cruzi in the Texas-Coahuila border region. Paleopathology Newsletter 4. Thirteenth Annual Meeting of the Paleopathology Association, Tempe, AZ, USA. 2003.
9. Rodríguez-Morales A. Chagas disease: an emerging food-borne entity? J Infect Dev Ctries. 2008;2:149–50.
10. Reinhard KJ, Fink TM, Skiles J. Case of megacolon in Rio Grande Valley as a possible case of Chagas Disease. Mem Inst Oswaldo Cruz. 2003;98(Suppl 1):165–72.
11. Aufderheide A, Salo W, Madden M, Streitz J, Dittmar K. Aspects of ingestion transmission of Chagas identified in mummies and their coprolites. Chungara. 2005;37:85–90.
12. Pinto J. Notes on Trypanosoma cruzi and its bio-ecological characteristics, as an agent of food-borne diseases. RevSoc Bras Trop Med. 2006;39:370–5.
13. Martinson E, Reinhard KJ, Buikstra JE, Dittmar de la Cruz K. Pathoecology of Chiribaya parasitism. Mem Inst Oswaldo Cruz. 2003;98:195–205.
14. Arriaza B. Cultura Chinchorro: las momias más antiguas del mundo. Santiago de Chile: Editorial Universitaria; 2003.
15. Deane MP, Lenzi HL, Jansen A. Trypanosoma cruzi: vertebrate and invertebrate cycles in the same mammal host, the opossum Didelphis marsupial. Mem Inst Oswaldo Cruz. 1984;79(4):513–5.
16. McKeever S, Gorman GW, Norman L. Occurrence of a Trypanosoma cruzi- like organism in some mammals from southwestern Georgia and northwestern Florida. J Parasitol. 1958;44:583–7.

17. Jansen AM, Deane MP. Trypanosoma cruzi infection of mice by ingestion of food contaminated with material from the anal glands of the opossum Didelphis marsupialis. In: XII Annual Meeting on Basic Research in Chagas Disease BI-09, Caxambu, MG, Brazil. 1985.
18. Schofield CJ. Trypanosoma cruzi. The vector parasite paradox. Mem Inst Oswaldo Cruz. 2000;95:535–44.
19. Carcavallo RU, Martínez A. Biología, ecología y distribución geográfica de los triatominos Americanos. In: Carcavallo RU, Rabinovich JE, Tonn RJ, editors. Factores biológicos y ecológicos en la enfermedad de Chagas. Buenos Aires: OPS/ECO, MSAS, SNCH; 1985. p. 149–208.
20. Orellana-Halkyer N, Arriaza-Torres B. Enfermedad de Chagas en poblaciones prehistóricas del norte de Chile. Rev Chil Hist Nat. 2010;83(4):531–41.
21. Cattan PE, Pinochet A, Botto-Mahan C, Acuña MI, Canals M. Abundance of Mepraia spinolai in a periurban zone of Chile. Mem Inst Oswaldo Cruz. 2002;97(3):285–7.
22. Sleeping sickness. http://www.who.int/mediacentre/factsheets/fs259/en/. Accessed Oct 2017.
23. Trypanosomiasis, human African (sleeping sickness). 2017. http://www.who.int/mediacentre/factsheets/fs259/en/.
24. Miles MA, Feliciangeli MD, Arias AR. American trypanosomiasis (Chagas disease) and the role of molecular epidemiology in guiding control strategies. BMJ. 2003;326:1444–8.
25. Fernández O, Souto RP, Castro JA, Pereira JB, Fernandes NC, Junqueira AC. Brazilian isolates of Trypanosoma cruzi from humans and triatomines classified into two lineages mini-exon andribosomal RNA sequences. Am J Trop Med Hyg. 1998;58:807–11.
26. Anez N, Crisante G, da Silva FM, Rojas A, Carrasco H, Umezawa E. The predominance of lineage I among Trypanosoma cruzi isolates from Venezuelan patients with different clinical profiles of acute Chagas disease. Trop Med Int Health. 2004;9:1319–26.
27. Briones M, Souto R, Stolf B, Zingales B. The evolution of Two Trypanosoma cruzi subgroups inferred from rRNA genes can be correlated with the interchange of American mammalian faunas in the Cenozoic and has implications for pathogenicity and host specificity. Mol Biochem Parasitol. 1999;104:219–32.
28. Stevens J, Gibson W. The molecular evolution of trypanosomes. Parasitol Today. 1999;15:432–7.
29. Aufderheide A, Salo W, Madden M, Streitz J, Buikstra J, Guhl F, Arriaza B, Renier C, Wittmers L Jr, Fornaciari G, Allison M. A 9,000-year record of Chagas' disease. Proc Natl Acad Sci. 2004;101:2034–9.
30. Araújo A, Jansen AM, Reinhard K, Ferreira LF. Paleoparasitology of Chagas disease: a review. Mem Inst Oswaldo Cruz. 2009;104(Suppl):S9–S16.
31. Difusión de la Enfermedad de Chagas en América del Sur. Curto, Susana Isabel; Ling, Claudia Marcela; Chuit Roberto. Centro de Investigaciones Precolombinas. Anti. Latinoamérica: una mirada desde el presente hacia el pasado/María Teresita de Haro... [et al.]; compilado por María Teresita de Haro... [et al.]. - 1a ed. - Ciudad Autónoma de Buenos Aires: Aspha, Pag: 285 - 306. 2017
32. Mann Fischer G. Los pequeños mamíferos de Chile. Guyana Zoología. 1978;40:1.
33. WHO - World Health Organization. Report of the Expert Committee on the control of Chagas disease. Technical report series 905. Geneva: WHO; 2002. p. 42.
34. Chuit R, Subias E, Perez AC, Paulone I, Wisnivesky-Colli C, Segura EL. Usefulness of serology for the evaluation of Trypanosome cruzi transmission in endemic areas of Chagas' Disease. Rev Soc Brazil Trop Med. 1989;22(3):119–29. ISSN: 0037-8682-1989.
35. Laranja FS, Dias E, Nobrega G, Miranda A. Chagas Disease, A clinical, epidemiologic and pathologic study. Circulation. 1956;14:1035–60.
36. Mazza S. La enfermedad de Chagas en la República Argentina. Mem Inst Oswaldo Cruz. 1949;47:273.
37. Rosenbaum MB. Chagasic myocardiopathy. Prog Cardiovasc Dis. 1964;7:199–225.
38. Pinto Dias JC, Kloetzel K. The prognostic value of the electrocardiographic features of chronic Chagas' disease. Rev Inst Med Trop São Paulo. 1968;10(3):158–62.

39. Maguire JH, Hoff R, Sherlock I, Guimaraes AC, Sleigh A, Ramos NB, Mott KE, Weller T. Cardiac morbidity and mortality due to Chagas' disease: prospective electrocardiographic study of a Brazilian community. Circulation. 1987;75:1140.
40. Manzullo E. Epidemiology of Chagas disease in Argentina. Rev Fed Arg Cardiol. 1988;17:141.
41. Manzullo EC, Chuit R. Risk of death due to chronic chagasic cardiopathy. Mem Inst Oswaldo Cruz. 1999;94(1):317–20. ISSN 0074¬0276-1999.
42. Chuit R, Gürtler RE, Mac Dougall L, Segura EL, Singer B. Chagas disease: risk assessment by an environmental approach in northern Argentina. Rev Patol Trop (Goias). 2001;30(2):193–207.
43. Estimación cuantitativa de la enfermedad de Chagas en las Américas. OPS/HDM/CD/425-06. OPS/OMS.
44. Moncayo Á, Silveira AC. Current epidemiological trends for Chagas disease in Latin America and future challenges in epidemiology, surveillance and health policy. Mem Inst Oswaldo Cruz. 2009;104(Suppl I):17–30.
45. WHO. OMS protocolo de evaluación de efecto insecticida sobre tríatominos [WHO protocol for the evaluation of insecticidal effect on triatomines]. Acta Toxicol Argent, 1994;2: 29–32.
46. Reports of the Intergovernmental Commission of the Southern Cone Initiative. Washington, DC: Pan American Health Organization. 1998–1999.
47. WHO. Chagas disease: interruption of transmission in Brazil. Wkly Epidemiol Rec. 2000;75:153–5.
48. WHO. Chagas disease: interruption of transmission in Chile. Wkly Epidemiol Rec. 2000;75:10–2.
49. WHO. Chagas disease: interruption of transmission in Uruguay. Wkly Epidemiol Rec. 1998;73:1–4.
50. WHO. Chagas disease in Latin America: an epidemiological update based on 2010 estimates. Wkly Epidemiol Rec. 2015;90(6):33–44.
51. Schmunis GA, Yadon ZE. Chagas disease: a Latin American health problem becoming a world health problem. Acta Trop. 2010;115:14–21.
52. Boletines Integrados Epidemiológicos de Argentina. http://www.msal.gob.ar/index.php/home/boletin-integrado-de-vigilancia.
53. Real Decreto España 1088/2005. Boletín Oficial del Estado del 20 de Septiembre. 225:31288–304. 2005.
54. Organización Panamericana de la Salud. Taller sobre control de calidad de sangre en serología de bancos de sangre. Documento OPS/HPC/HCT/96/79. Washington, DC: Organización Panamericana de la Salud; 1996.
55. Organización Panamericana de la Salud. Elegibilidad para la donación de sangre. Washington, DC: Organización Panamericana de la Salud; 2009.
56. Cabello C, Cabello F. Zoonosis con reservorios silvestres: Amenazas a la salud pública y a la economía. Rev Med Chil. 2008;136:385–93.
57. Silveira A. Factores de riesgo implica-dos en la transmisión oral de la Enfermedad de Chagas. In: Informe Final Consulta Técnica e Epidemiología, Prevención y Manejo de la Transmisión de la Enfermedad de Chagas como Enfermedad Transmitida por Alimentos (ETA), Rio de Janeiro, 4–5 Mayo 2006, pp. 16–9.
58. Valente SA, Valente VC, Pinto A. Epidemiologia e transmissão oral da doença de Chagas na Amazonia Brasileira. Instituto Evandro Chagas. Rodovia. In: Informe Final Consulta Técnica y Epidemiología, Prevención y Manejo de la Transmisión de la Enfermedad de Chagas como Enfermedad Transmitida por Alimentos (ETA), Rio de Janeiro, 4–5 Mayo 2006, pp. 21–6.
59. Díaz ML, González CI. Enfermedad de Chagas agudo: transmisión oral de Trypanosoma cruzi como una vía de transmisión re-emergente. Rev Univ Ind Santander Salud. 2014;46(2):177–88.
60. Rueda K, Trujillo JE, Carranza JC, Vallejo GA. Transmisión oral de Trypanosoma cruzi: una nueva situación epidemiológica de la enfermedad de Chagas en Colombia y otros países suramericanos. Biomedica. 2014;34:631–41. https://doi.org/10.7705/biomedica.v34i4.2204.

61. Basombrío A, Schofield CJ, Rojas CL, del Rey EC. A cost-benefit analysis of Chagas disease control in northwestern Argentina. Trans R Soc Trop Med Hyg. 1998;92(2):137–43. https://doi.org/10.1016/S0035-9203(98)90720-9.
62. Wilson LS, Strosberg AM, Barrio K. Cost-effectiveness of chagas disease interventions in Latin America and the Caribbean: Markov models. Am J Trop Med Hyg. 2005;73(5):901–10.
63. Lee BY, Bacon K-t M, Bottazzi ME, Hotez PJ. Global economic burden of Chagas disease: a computational simulation model. Lancet Infect Dis. 2013;13:342–8.
64. Organización Panamericana de la Salud. ETMI Plus. Marco para la eliminación de la transmisión maternoinfantil del VIH, la sífilis, la hepatitis y la enfermedad de Chagas. Washington, DC: OPS; 2017.

Chagas Disease in Europe

Julio Alonso-Padilla, María Jesús Pinazo, and Joaquim Gascón

Abstract Chagas disease is an infectious disease caused by the parasite *Trypanosoma cruzi*. It affects approximately seven million people worldwide, most of them in Latin America, where insect vectors that transmit the infection are endemic. Besides, *T. cruzi* can also be transmitted through blood transfusion, organ transplant, and from mother to child. The infection is chronic in a majority of cases and remains asymptomatic for years. It is estimated that ~30% of those chronically infected will end up developing the life-threatening symptoms characteristic of the disease: heart and/or gastrointestinal tract tissue disruptions. In the last decades, large migratory flows between Latin American countries and non-endemic regions like Europe have spread Chagas disease impact. Its silent clinical progression and vector-independent transmission routes entail a health challenge in non-endemic countries too. In this chapter we present the epidemiological status of Chagas disease in Europe as well as the measures being taken to downsize its public health risk and to control the disease.

J. Alonso-Padilla · M. J. Pinazo · J. Gascón (✉)
Centre for Research in International Health (CRESIB), Barcelona Institute for Global Health (ISGlobal), Hospital Clinic-University of Barcelona, Barcelona, Spain
e-mail: julio.a.padilla@isglobal.org; mariajesus.pinazo@isglobal.org; jgascon@clinic.cat

J. M. Altcheh, H. Freilij (eds.), *Chagas Disease*, Birkhäuser Advances in Infectious Diseases, https://doi.org/10.1007/978-3-030-00054-7_5

1 Introduction

Chagas disease is a parasitic infection caused by the protozoan parasite *Trypanosoma cruzi* (*T. cruzi*). Although originally circumscribed to the Americas, where the vectors that generally transmit the infection are endemic, migratory flows in recent decades have spread the disease to non-endemic regions like Europe.

It is estimated that three million people arrived into Europe originating from Latin America (LA) [1]. The distribution of Latin American migrants among European countries has not been homogeneous. In addition to economic factors (chances of finding a job), political factors (ease of entry to countries, old colonial relations, current relationships between origin and reception states), and cultural features (shared language and/or customs) have been very important for migrant distribution [2]. Possibly that is why Spain and to a lesser extent Italy are the countries that have received a greater flow of people from LA (Fig. 1).

Prevalence of Chagas disease in endemic countries is not homogeneous. This has certainly contributed to shape important differences in the prevalence of Chagas disease in European receptor countries accordingly to the origin of migrants. Furthermore, the typology of migratory flows has also varied over time. Most recent migratory flows from LA are basically economic and come from rural areas that are highly endemic for Chagas disease [3].

Emergence of Chagas disease in Europe is manifest from the beginning of this century, as it has been evidenced by several studies [4–6]. Unlike other tropical diseases such as malaria or schistosomiasis, known through previous migratory flows originating in other latitudes and also through traveler's medicine, Chagas disease was unknown to European health professionals. The clinical characteristics of this disease and its variety of forms of transmission have involved new challenges that, especially in those countries that have received a lower flow of people from LA, are still not completely solved. One of the characteristics of this migration is the tendency to feminization, which is relevant in the context of Chagas disease due to the possibility of congenital transmission.

The onset of the economic crisis in Europe in 2008 and the economy improvements seen in some Latin American countries have led to the return of a percentage of this immigration to their countries of origin. Nonetheless, part of this population still remains in Europe, and a percentage of it continued its journey within the European Union (EU), basically from Spain to richer northern countries, less affected by the economic crisis [7]. In any case, this phenomenon has not substantially changed the Chagas disease problem in Europe. It has rather made it more complex, as the preparedness of health systems and the knowledge to manage the disease are not equally set in all European countries. Thereof the importance of generalizing already acquired knowledge to reaching a consensus position for the management and control of Chagas disease in the continent.

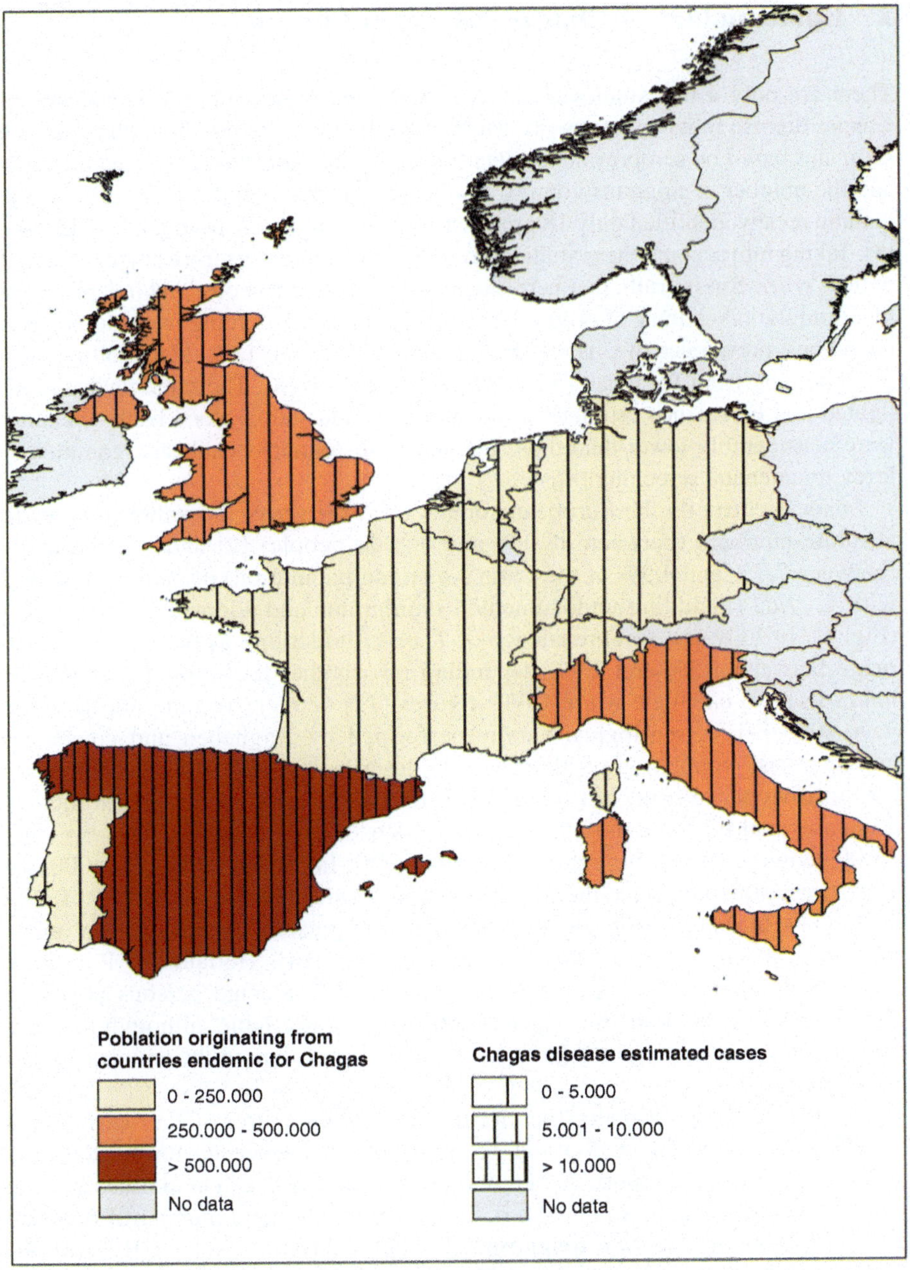

Fig. 1 Map of Europe: countries that have received migrant population originating from Chagas disease endemic countries shaded according to the legend details; stripes pattern within each country limits indicates the number of estimated cases of Chagas disease per country. Data were extracted from reference [11] to plot the figure. [Photo Credit: Carme Subirà]

2 Epidemiology of Chagas Disease in Europe

There are only a few studies conducted in Europe to measure the prevalence of Chagas disease in its countries [8]. Most of the figures currently being handled are estimates based on seroprevalence data from the countries of origin of the migrants and the number of migrants coming from each endemic country [2, 9, 10]. A systematic review identified only 18 prevalence studies as having been made in Europe [8]. Taking into account these studies, around 4.2% of migrants from LA are infected with *T. cruzi*. But in truth, that percentage is very heterogeneous, and it depends on the immigrants' country of origin. For instance, migrants coming from Bolivia had the highest prevalence of Chagas disease (18.1%, 95% CI: 13.9–22.7), followed by those coming from Paraguay (5.5%, 95% CI: 3.5–7.9) [8]. The same review highlighted that prevalence estimates from studies conducted in blood bank screening were considerably lower than those derived from primary healthcare, community level, or antenatal screening [8].

Spain is currently the European country with the greatest number of cases in absolute numbers (between 48,000 and 86,000 people) [2] and in percentage (between 2.7% and 4.9% of the Latin American population) of patients infected with *T. cruzi* (including undocumented immigrants and adopted children) [11] (Fig. 1). In Italy, the seroprevalence of *T. cruzi* infection has been estimated to range between 1.5% and 2.9% depending on whether the seroprevalence estimates used to calculate it are 1990s figures [12] or more recent data from the year 2005 [9]. A serological survey performed by Angheben and coworkers among at-risk population residing in Italy described a 4.3% seroprevalence rate (36 positive participants out of 867) [6]. In Switzerland, up to 2009, a total of 258 cases had been diagnosed, although it is estimated that there may be some 3,000 people infected throughout the country [7]. In the UK, between 6,000 and 12,000 people could have the disease, which would mean a prevalence of 1.3–2.4% [11]. In other European countries that also present Latin American immigration to a lesser extent (Belgium, France, Germany, Holland, or Portugal), absolute numbers are estimated to be below 3,000 infected persons [11] (see Fig. 1). Data from other European countries is not available, although the estimated number of immigrants from LA is much lower than in the countries mentioned above.

In summary, it is estimated that in Europe absolute figures of *T. cruzi*-infected people range between 68,000 and 123,000 [11]. However, up until 2009 only 4,290 cases had been reported [11]. A study carried out in England illustrates the degree of infra-diagnosis that occurs. In this work, the total number of reported cases of *T. cruzi* infection diagnosed in London from 2001 to 2014 was 41, which yielded a prevalence of 0.043% among the Latin American migrants in the city. However, the ratio between the observed and the expected prevalence of *T. cruzi* infection was 3.34%, resulting in an index of underdiagnosis of 96.6% [13].

3 Routes of Transmission of *T. cruzi*

The triatomine vectors (order Hemiptera; family Reduviidae) that generally transmit the disease in America are not present in Europe [14]. Vector-independent transmission routes, like organ transplant, blood transfusion, and from mother to child, are of relevance in endemic and non-endemic regions, such as Europe [15].

3.1 *Blood Banks and Transplants Recipients*

In Europe there have been a few cases of Chagas disease acquired through blood transfusion [16–18]. Although disease acquisition through organ transplant has also been reported [19], no prevalence studies have been published in organ donors.

Regarding blood bank surveillance, a study performed in Spain reported that 0.62% (11/1,777) of blood donors from LA were seropositive to *T. cruzi* antigens [20]. The highest rate (10.2%) was observed in Bolivian people. Other studies from France and Italy showed figures of 0.3% (3/972) and 1.0% (1/102) positive donors, respectively [21, 22]. In contrast, a work performed in the Netherlands showed 0.0% seropositive samples out of 1,333 at-risk donors tested, which mostly were from Suriname and Brazil [23]. Results from these studies come to illustrate the heterogeneous parasite prevalence rates found between different European countries in relation to the immigrants' countries of origin.

3.2 *Congenital Transmission*

Several studies in pregnant women of Latin American origin have shown that prevalence rates of *T. cruzi* infection range between 1.5% and 4.7% of women [24–29]. In a study performed between 2005 and 2007 at two maternity hospitals in Barcelona (Spain), 3.4% of the LA women were positive for Chagas disease (46 out of 1,350 tested) [27]. Furthermore, a 7.5% rate of *T. cruzi* congenital transmission was found [27]. The incidence of Chagas disease clinical cases due to vertical transmission have been published in several European countries [28–32].

4 Chagas Control in Europe and Current Challenges

Chagas disease has a number of connotations that go beyond a simple parasitic infection. In many areas of LA, it is stigmatizing to endure Chagas disease, which makes of it a forgotten disease. The late onset of symptoms, linked to the fact that they are not pathognomonic of infection and are confused with cardiac or gastrointestinal symptoms of other etiologies, has historically led to a great deal of ignorance. When symptoms do exist, patients' quality of life is impaired. Besides, *T. cruzi* infection does sometimes co-occur with other morbidities and affects other pathological processes. However, despite the high number of people that has arrived from endemic countries, studies on the health status of LA migrants are scarce [33].

In Europe, a major challenge posed by Chagas disease to public health systems and healthcare professionals is the generalized lack of knowledge of the disease, which may preclude an appropriate clinical management of patients. Another big issue is that *T. cruzi* infection is underdiagnosed [11, 13]. Poor access to diagnosis is an acknowledged massive hurdle toward disease control in endemic regions, which is most frequently observed in rural areas that are distantly located from microbiological reference laboratories [34]. Motivated by other features perhaps, but it is a phenomenon that also occurs in Europe despite the availability of wealthier healthcare systems.

4.1 T. cruzi *Infection Diagnosis*

Similarly to what is made in endemic countries, the diagnostic algorithms applied in Europe differ depending on whether congenital (acute) or chronic infection is to be diagnosed. In the former, due to potential false-positive confounders from parasite-specific mother-derived immunoglobulins, diagnosis in Europe is largely performed by molecular methods like that described by Piron et al. [35]. Commercial polymerase chain reaction methodologies are also available [29, 36] although at high prices. Since the sensitivity of molecular methods is not perfect, newborns to seropositive mothers (and their kin) must be serologically assayed when maternally derived antibody levels decline. In this regards, an algorithm to reduce the number of tests and restrict serological testing to months 9 and 12 of age of the child has been proposed in order to save costs [37].

At the chronic stage diagnosis is made serologically. At this stage parasitemia is low, and sensitivity of molecular detection is much poorer than indirect detection of anti-*T. cruzi* immunoglobulins in sera. Serological diagnosis involves two assays based on different antigenic sets due to the parasite high antigenic variability. If discordant results are obtained, then a third assay must be performed for tipping the scales. A recent work has questioned this procedure as it reported that a single highly specific and sensitive chemiluminescent assay (Chagas Architect, Abbott) would suffice to discard negative cases and only doubtful positive results ("gray zone") should need to be confirmed by another serological test [38].

In general, the inconveniences faced to get access to Chagas disease diagnosis in Europe are not as cumbersome as those encountered in many areas of endemic regions. However, unawareness of the disease and its characteristic silent clinical progression involves that a large percentage of patients are not timely diagnosed. Thus, specific programs have been set in place to directly bring information and promote disease screening to target populations like immigrants coming from Chagas disease endemic countries [39, 40].

On the other hand, a feature observed upon talking to experts from several European countries was the high level of heterogeneity among diagnostic algorithms used in each place. Certainly, arrival to a consensus could be of great help to standardize the diagnosis and ulterior access to treatment of patients, but also, very importantly, to save costs in the process.

4.2 Treatment and Management of Patients

The two anti-parasitic drugs used to date (benznidazole and nifurtimox) to treat *T. cruzi* infection are available in Europe. However, the routes of acquisition of these drugs may vary from country to country depending on whether benznidazole or nifurtimox is prescribed. Mirroring what occurs in endemic areas, there is also an open debate in Europe about whether all patients infected with *T. cruzi* should or should not be treated. In general, international consensus is followed, which means that anti-parasitic drug treatment is recommended for patients in the acute stage, for those at chronic stage with infection reactivation, and for chronic patients under 50 years of age without clinical symptoms or mild cardiologic compromise (Kushnir level I) [41]. It is especially relevant to treat women at child-bearing age as it has been shown that benznidazole treatment of women before pregnancy significantly reduces the risk of transmission of the infection to their newborns [42, 43]. Whether older patients may receive treatment or not depends on each clinician's judgment. The lack of biomarkers of therapeutic efficacy is certainly a handicap when it comes to establishing more solid consensuses [44].

Benznidazole is the most widespread drug due to its availability. The regime indicated for adults involves a 5 mg/kg daily dose (up to a maximum of 400 mg per day) administered in two doses for 60 days. In children, benznidazole should be indicated with an 8–10 mg/kg daily administered in two or three doses for 60 days as well. A pediatric formulation of benznidazole has been successfully assessed in a clinical trial and will be produced soon in Argentina [45, 46]. Nifurtimox should be prescribed at a 15 mg/kg daily dose for children and 8–10 mg/kg for adults in three doses for 60 days [15]. Nifurtimox daily accumulated dose should not surpass 600 mg. Both drugs are well tolerated by children, and even a more specific age-related dosing has been proposed [47]. Once treatment is initiated, patients are regularly observed for the onset of adverse drug reactions (ADRs) which are mostly skin-related manifestations, digestive disorders, and general ADRs like headache, asthenia, and fever [48]. ADRs such as muscular-articular and neurological complains are less common [48, 49]. Nonetheless, in a very low percentage of cases, hospitalization is required, and updated clinical guidelines are of major importance to closely monitor these events [48, 49].

Patients' access to diagnosis and treatment within Europe differs accordingly to the health systems of each country and the personal status of immigrants (legal entitlements). For instance, in Spain universal access to the healthcare system facilitates the entry of patients into the system. Despite this, there are other barriers (work schedules, permits, language, unfamiliarity with rights, entitlements, and the overall health system gaps in health literacy, social exclusion, and direct and indirect discrimination) that hinder their care [50]. On the other hand, health systems of recipient countries should ensure that health professionals are aware of the existence of Chagas disease and have adequate clinical guidelines. In Spain, the most affected European country, a series of clinical guidelines and consensus documents have been produced and published in national and international journals with the aim to help health professionals to know about Chagas disease and to provide protocols for chronic Chagas cardiologic and digestive disease [51–53]. The management of Chagas disease has been as well documented in the context of primary healthcare [54] and under immunosuppression conditions like in patients with HIV/AIDS [55] or organ and tissue transplants recipients [56].

Patient management after access to diagnosis and treatment is not easy. In one study focusing on process of care for Chagas disease in Italy, less than 30% of patients completed treatment with dropouts along the cascade of care. The authors concluded that there is an urgent need to involve affected communities and local regional health authorities to take part in the model of care, adapting it to the local needs [57]. Probably similar facts occur in other European countries. In complex cases with advanced disruption of heart and/or gastrointestinal tract tissues, the referral to specialists in cardiology or gastroenterology should follow the usual circuits of the different health systems.

5 Efforts to Control Transmission

5.1 Blood Banks

Most European countries follow the EU Directive 2004/23/EC on safety and quality of blood. In this document, an antecedent of Chagas disease is specified as a permanent exclusion criterion for homologous donors. But there are many patients at risk of *T. cruzi* infection who have never had a screening test and therefore do not know whether or not they carry and may transmit the parasite. Only France, Spain, and the UK currently have a legal regulation that makes explicit the screening of *T. cruzi* prior to donation; this includes not only migrants from endemic areas but also children born to mothers of endemic areas and persons who have received transfusions in endemic countries [58–60].

Italy is in the process of approving a new law in the parliament that allows systematic screening in patients at risk of infection [61]. The legislation in Sweden directly excludes people who have lived more than 5 years in countries endemic to the disease, although they do not refer to children of mothers born in endemic areas [62]. As a rule, donation is excluded in Switzerland in case of diagnosis of Chagas disease, but some cantons such as Geneva and Vaud have now implemented unofficial screening measures at the hospital level. There is no data from other European countries, although the Latin American presence in these countries is practically nonexistent.

5.2 Transplants

The use of donor organs with acute infection is contraindicated, and the use of a donor heart with chronic infection is also contraindicated. However, the use of other organs from donors with chronic infection has a relative contraindication. If transplantation is decided, periodic monitoring of the recipient should be recommended using parasitological and serological methods [56].

There are few European countries with a current legislation that considers transplants and Chagas disease. But in the EU directives on organ transplantation, there is no mention of Chagas disease [63]. It only points out that it is necessary to investigate certain epidemiological situations that may affect the suitability of the transplant and that may imply a risk in the transmission of some disease. In Italy, since 2012, a legal regulation has been approved obliging the screening of *T. cruzi* in donors at risk [64]. In Spain, although the legislation concerning this issue is vague [65], the National Transplant Organization (ONT) has made some official recommendations [66].

5.3 Congenital Transmission

It is of special interest the management of *T. cruzi* infection in pregnancy, during which, although it is of vital importance to carry out the diagnosis, it is not possible to administer treatment to the pregnant woman. Treatment in newborns is highly effective, and the early treatment during the first months of life will prevent future complications of the disease, thereby the great relevance of adequately diagnosing mothers before delivery. Diagnosis of *T. cruzi* infection during pregnancy will allow careful monitoring of the affected women and early control of the newborns, which should be immediately treated in case the parasite is transmitted. Several studies have shown that congenital transmission control programs are cost-effective in endemic countries [67]. In European countries, where health systems are widely established and health economics less stringent, timely screening of pregnant women suspected of at risk of infection should be mandatory. Furthermore, preventive widespread diagnosis and treatment of *T. cruzi*-infected women in child-bearing age has been shown to be beneficial to control transmission of the infection during pregnancy [42, 43]. For this particular group of patients, it would then be very advisable to implement diagnostic algorithms to limit the transmission and save newborns from receiving treatment.

In some areas of Spain, specifically in Catalonia and Valencia, and in Tuscany in Italy, control measures for *T. cruzi* infection in pregnant women at risk of infection and control programs of newborns have already been approved by regional governments [68–70]. In other regions of several European countries (at least four in Spain, three in Italy, one in Germany, two in Switzerland, and two in Portugal, and there might be more the authors do not currently know about), there are local initiatives, generally promoted by hospitals or research centers, implemented for the early detection of *T. cruzi* infection in pregnant women and the screening of newborns born to positive mothers. However, up to now there is yet no official recommendation or guide at national or EU levels.

6 Conclusions

1. Population movements during the last decades between Chagas disease endemic countries in Latin America and Europe have contributed to extend the impact of the disease, which should now be considered an emerging infectious disease due to the number of cases registered and its relevance as public health threat.
2. There are between 68,000 and 123,000 people infected with *T. cruzi* in Europe, a majority of them residing in Spain, Italy, the UK, and France.
3. The distribution and epidemiology of the infection in Europe is very heterogeneous and depends on the origin of the immigrants received by each country.
4. There is a lack of knowledge of the disease and how to manage it clinically, which entails a public health risk in countries where it is a new challenge.
5. Access to diagnosis is still shaded by the stigma that accompanies this disease, which along with miscommunication and unawareness complicate widespread testing of at-risk populations.
6. Diagnostic algorithms are diverse and may lead to delays in treatment administration to congenital cases as well as to excessive costs due to cost-ineffectiveness.
7. Although treatment with benznidazole and nifurtimox is generally widely available, there are still issues that preclude access to it, most importantly the huge level of underdiagnosed cases.
8. Treatment is highly effective and well tolerated by children, and it should therefore be administered to them as soon as a positive diagnosis is known.
9. Transmission routes in non-endemic regions are vector-independent (blood transfusion, organ transplant, and from mother to child), and control measures must be put on place for each of them correspondingly.
10. Blood bank screening in European countries most affected by Chagas disease is well established. Serological testing of at-risk organ donors is not that obvious.
11. Control of congenital transmission should be particularly enforced due to the great benefits it provides. Both by early identifying potentially infected newborns and immediately treating them, as well as preventively treating women at child-bearing age to reduce chances of vertical transmission of the parasite.

References

1. Yépez del Castillo I. Las migraciones entre América Latina y Europa: una dimensión de las relaciones entre estas dos regiones. In: Yépez del Castillo I, Herrera G, editors. Nuevas migraciones latinoamericanas a Europa. Balances y desafíos. Quito: Biblioteca FLACSO; 2007. p. 19–30.
2. Gascon J, Bern C, Pinazo MJ. Chagas disease in Spain, the United States and other non-endemic countries. Acta Trop. 2010;115:22–7.
3. López de Lera D, Oso Casas L. La inmigración latinoamericana en España. Tendencias y estado de la cuestión. In: Yépez del Castillo I, Herrera G, editors. Nuevas migraciones latinoamericanas a Europa. Balances y desafíos. Quito: Biblioteca FLACSO; 2007. p. 31–68.

4. Roca C, Pinazo MJ, López-Chejade P, Bayó J, Posada E, López-Solana J, Gállego M, Portús M, Gascón J. Chagas disease among the Latin American adult population attending in a primary care center in Barcelona, Spain. PLoS Negl Trop Dis. 2011;5:e1135.
5. Jackson Y, Chappuis F. Chagas disease in Switzerland: history and challenges. Euro Surveill. 2011;16:pii 19963.
6. Angheben A, Anselmi M, Gobbi F, et al. Chagas disease in Italy: breaking an epidemiological silence. Euro Surveill. 2011;16:pii 19969.
7. Jackson Y, Herrera MV, Gascon J. Economic crisis and increased immigrant mobility: new challenges in managing Chagas disease in Europe. Bull World Health Organ. 2014;92:771–2.
8. Requena-Méndez A, Aldasoro E, de Lazzari E, Sicuri E, Brown M, Moore DAJ, Gascon J, Muñoz J. Prevalence of Chagas disease in Latin-American migrants living in Europe: a systematic review and meta-analysis. PLoS Negl Trop Dis. 2015;9:e0003540.
9. Jannin J, Salvatella R. Estimación cuantitativa de la enfermedad de Chagas en las Américas. Organización Panamericana de la Salud; OPS/HDM/CD/425-06. Geneva: Department of Control of Neglecled Tropical Diseases (NTD). Innovative and Intensified Disease Management (IDM). WHO/NTD/IDM; 2006.
10. WHO. Control and prevention of Chagas disease in Europe. Issue December 2009; 69 p.
11. Basile L, Jansa JM, Carlier Y, et al. Chagas disease in European countries: the challenge of a surveillance system. Euro Surveill. 2011;16:pii 19968.
12. Schmunis GA, Yadon ZE. Chagas disease: a Latin American health problem becoming a world health problem. Acta Trop. 2010;115:14–21.
13. Requena-Méndez A, Moore DAJ, Subirà C, Muñoz J. Addressing the neglect: Chagas disease in London, UK. Lancet Glob Heal. 2016;4:e231–3.
14. Bern C, Kjos S, Yabsley MJ, Montgomery SP. *Trypanosoma cruzi* and Chagas' disease in the United States. Clin Microbiol Rev. 2011;24:655–81.
15. Gascon J, Pinazo MJ. Chagas disease: from Latin America to the world. Re Parasitol. 2015;2015:7–14.
16. Villalba R, Fornes G, Alvarez MA, Roman J, Rubio V, Fernandez M, Garcia JM, Vinals M, Torres A. Acute Chagas' disease in a recipient of a bone marrow transplant in Spain: case report. Clin Infect Dis. 1992;14:594–5.
17. Forés R, Sanjuán I, Portero F, et al. Chagas disease in a recipient of cord blood transplantation. Bone Marrow Transplant. 2007;39:127–8.
18. Valls FV, Puy RV, Bassany EE, Cano ML, Sánchez JL, Lozoya R, Azem J, Naval X, Armengol M. Caso clínico de enfermedad de Chagas transfusional. Emf Emerg. 2008;10(Suppl 1):14–8.
19. Rodriguez-Guardado A, González ML, Rodriguez M, Flores-Chavez M, Boga JA, Gascon J. *Trypanosoma cruzi* infection in a Spanish liver transplant recipient. Clin Microbiol Infect. 2015;21:687.
20. Piron M, Vergés M, Muñoz J, et al. Seroprevalence of *Trypanosoma cruzi* infection in at-risk blood donors in Catalonia (Spain). Transfusion. 2008;48:1862–8.
21. El Ghouzzi MH, Boiret E, Wind F, Brochard C, Fittere S, Paris L, Mazier D, Sansonetti N, Bierling P. Testing blood donors for Chagas disease in the Paris area, France: first results after 18 months of screening. Transfusion. 2010;50:575–83.
22. Gabrielli S, Girelli G, Vaia F, Santonicola M, Fakeri A, Cancrini G. Surveillance of Chagas disease among at-risk blood donors in Italy: preliminary results from Umberto I Polyclinic in Rome. Blood Transfus. 2013;11:558–62.
23. Slot E, Hogema BM, Molier M, Bart A, Zaaijer HL. Risk factors and screening for *Trypanosoma cruzi* infection of Dutch blood donors. PLoS One. 2016;11:e0151038.
24. Martinez de Tejada B, Jackson Y, Paccolat C, Irion O. Congenital Chagas disease in Geneva: diagnostic and clinical aspects. Rev Méd Suisse. 2009;5:2091–2–4–6.
25. Muñoz-Vilches MJ, Salas J, Cabezas T, Metz D, Vázquez J, Soriano MJ. Chagas screening in pregnant Latin-American women. Experience in Poniente Almeriense (Almeria, Spain). Enferm Infecc Microbiol Clin. 2012;30:380–2.
26. Paricio-Talayero JM, Benlloch-Muncharaz MJ, Collar-del-Castillo JI, et al. Epidemiological surveillance of vertically-transmitted Chagas disease at three maternity hospitals in the Valencian Community. Enferm Infecc Microbiol Clin. 2008;26:609–13.

27. Muñoz J, Coll O, Juncosa T, et al. Prevalence and vertical transmission of *Trypanosoma cruzi* infection among pregnant Latin American women attending 2 maternity clinics in Barcelona, Spain. Clin Infect Dis. 2009;48:1736–40.
28. Francisco-González L, Gastañaga-Holguera T, Jiménez Montero B, Daoud Pérez Z, Illán Ramos M, Merino Amador P, Herráiz Martínez MÁ, Ramos Amador JT. Seroprevalencia y transmisión vertical de enfermedad de Chagas en una cohorte de gestantes latinoamericanas en un hospital terciario de Madrid. An Pediatr (Barc). 2018;88:122. pii: S1695-4033(17)30148-0.
29. Barona-Vilar C, Gimenez-Marti MJ, Fraile T, et al. Prevalence of *Trypanosoma cruzi* infection in pregnant Latin American women and congenital transmission rate in a non-endemic area: the experience of the Valencian Health Programme (Spain). Epidemiol Infect. 2012;140:1896–903.
30. Jackson Y, Myers C, Diana A, Marti HP, Wolff H, Chappuis F, Loutan L, Gervaix A. Congenital transmission of Chagas disease in Latin American immigrants in Switzerland. Emerg Infect Dis. 2009;15:601–3.
31. Muñoz J, Portús M, Corachan M, Fumadó V, Gascon J. Congenital *Trypanosoma cruzi* infection in a non-endemic area. Trans R Soc Trop Med Hyg. 2007;101:1161–2.
32. Riera C, Guarro A, El Kassab H, et al. Congenital transmission of *Trypanosoma cruzi* in Europe (Spain): a case report. Am J Trop Med Hyg. 2006;75:1078–81.
33. Roura M, Domingo A, Leyva-Moral JM, Pool R. Hispano-Americans in Europe: what do we know about their health status and determinants? A scoping review. BMC Public Health. 2015;15:472.
34. Egüez KE, Alonso-Padilla J, Terán C, Chipana Z, García W, Torrico F, Gascon J, Lozano-Beltran D-F, Pinazo M-J. Rapid diagnostic tests duo as alternative to conventional serological assays for conclusive Chagas disease diagnosis. PLoS Negl Trop Dis. 2017;11:e0005501.
35. Piron M, Fisa R, Casamitjana N, López-Chejade P, Puig L, Vergés M, Gascón J, Gómez i Prat J, Portús M, Sauleda S. Development of a real-time PCR assay for *Trypanosoma cruzi* detection in blood samples. Acta Trop. 2007;103:195–00.
36. Seiringer P, Pritsch M, Flores-Chavez M, et al. Comparison of four PCR methods for efficient detection of *Trypanosoma cruzi* in routine diagnostics. Diagn Microbiol Infect Dis. 2017;88:225–32.
37. Abras A, Muñoz C, Ballart C, et al. Towards a new strategy for diagnosis of congenital *Trypanosoma cruzi* infection. J Clin Microbiol. 2017;55:1396–07.
38. Abras A, Llovet T, Tebar S, Herrero M, Berenguer P, Ballart C. Serological diagnosis of chronic Chagas disease: is it time for a change? J Clin Microbiol. 2016;54:1566–72.
39. ISGlobal site. Barcelona's Bolivian community spreads the word about Chagas disease. 2016. http://www.isglobal.org/en/-/la-comunidad-boliviana-en-barcelona-pasa-la-voz-contra-el-chagas. Accessed 10 Nov 2017.
40. Navarro M, de los Santos JJ. Access to Chagas disease treatment in non-endemic countries: the case of Spain. Lancet Glob Health. 2017;5:e577.
41. Bern C, Montgomery SP, Herwaldt BL, et al. Evaluation and treatment of Chagas disease in the United States: a systematic review. JAMA. 2007;298:2171–81.
42. Alvarez MG, Vigliano C, Lococo B, Bertocchi G, Viotti R. Prevention of congenital Chagas disease by benznidazole treatment in reproductive-age women. An observational study. Acta Trop. 2017;174:149–52.
43. Fabbro DL, Danesi E, Olivera V, Codebó MO, Denner S, Heredia C, Streiger M, Sosa-Estani S. Trypanocide treatment of women infected with *Trypanosoma cruzi* and its effect on preventing congenital Chagas. PLoS Negl Trop Dis. 2014;8:e3312.
44. Viotti R, Alarcón De Noya B, Araujo-Jorge T, et al. Towards a paradigm shift in the treatment of chronic Chagas disease. Antimicrob Agents Chemother. 2014;58:635–9.
45. DNDi site. Paediatric dosage form of benznidazole (Chagas). https://www.dndi.org/diseases-projects/portfolio/paediatric-benznidazole/. Accessed 15 Nov 2017.
46. MundoSano site. An Argentine laboratory will develop the pediatric formula for Chagas disease treatment. https://www.mundosano.org/en/an-argentine-laboratory-will-develop-the-pediatric-formula-for-chagas-disease-treatment/. Accessed 15 Nov 2017.
47. Wagner N, Jackson Y, Chappuis F, Posfay-Barbe KM. Screening and management of children at risk for Chagas disease in nonendemic areas. Pediatr Infect Dis J. 2016;35:335–7.

48. Pinazo MJ, Muñoz J, Posada E, López-Chejade P, Gállego M, Ayala E, del Cacho E, Soy D, Gascon J. Tolerance of benznidazole in treatment of Chagas' disease in adults. Antimicrob Agents Chemother. 2010;54:4896–9.
49. Sperandio da Silva GM, Felix Mediano MF, Hasslocher-Moreno AM, et al. Benznidazole treatment safety: the Médecins Sans Frontières experience in a large cohort of Bolivian patients with Chagas' disease. J Antimicrob Chemother. 2017;72:2596–601.
50. Rechel B, Mladovsky P, Ingleby D, Mackenbach JP, McKee M. Migration and health in an increasingly diverse Europe. Lancet. 2013;381:1235–45.
51. Pinazo MJ, Cañas E, Elizalde JI, et al. Diagnosis, management and treatment of chronic Chagas' gastrointestinal disease in areas where *Trypanosoma cruzi* infection is not endemic. Gastroenterol Hepatol. 2010;33:191–00.
52. Gascon J, Albajar P, Canas E, et al. Diagnosis, management and treatment of chronic Chagas' heart disease in areas where *Trypanosoma cruzi* infection is not endemic. Enferm Infecc Microbiol Clin. 2008;26:99–106.
53. González-Tomé MI, Rivera M, Camaño I, et al. Recommendations for the diagnosis, treatment and follow-up of the pregnant woman and child with Chagas disease. Enferm Infecc Microbiol Clin. 2013;31:535–42.
54. Roca Saumell C, Soriano-Arandes A, Solsona Díaz L, Gascón Brustenga J. Consensus document for the detection and management of Chagas disease in primary health care in a non-endemic areas. Aten Primaria. 2015;47:308–17.
55. Pérez-Molina J, Rodríguez-Guardado A, Soriano A, et al. Guidelines on the treatment of chronic coinfection by *Trypanosoma cruzi* and HIV outside endemic areas. HIV Clin Trials. 2011;12:287–98.
56. Pinazo MJ, Miranda B, Rodríguez-Villar C, et al. Recommendations for management of Chagas disease in organ and hematopoietic tissue transplantation programs in nonendemic areas. Transplant Rev. 2011;25:91–101.
57. Repetto EC, Zachariah R, Kumar A, et al. Neglect of a neglected disease in Italy: the challenge of access-to-care for Chagas disease in Bergamo area. PLoS Negl Trop Dis. 2015; 9:e0004103.
58. Ministerio de Sanidad y Consumo. Real Decreto 1088/2005. Boletín Oficial del Estado. 2005. pp. 31288–304.
59. National Blood Service. Guidelines for the blood services in the United Kingdom. 2005.
60. Ministère de la Santé et des Sports. Arrêté du 12 janvier 2009 fixant les critères de sélection des donneurs de sang. NOR: SJSP0901086A.
61. Ministerio della Salute. Protocolli per l'accertamento della idoneita del donatore di sangue e di emocomponenti. GU n.85 del 13-04-2005. 2005. https://infoleges.it. Accessed 5 Nov 2017.
62. Källstrand Nord E. Socialstyrelsens författningssamling. SOSFS 2009:28 (M).
63. Directiva 2010/45/UE. Diario Oficial de la Unión Europea. 2010; 1–8.
64. Criteri generali per la valutazione di idoneita del donatore. Allegato E. Centro Nazionale Trapanti. 9 Agosto 2012.
65. Ministerio de Sanidad, Servicios Sociales e Igualdad. Real Decreto 1732/2012. Boletín Oficial del Estado. pp. 89315–48.
66. Criterios de Selección de donantes de órganos respecto a la transmisión de infecciones. 2004. Organización Nacional de Trasplantes. 2ª Ed. pp. 1–37.
67. Sicuri E, Muñoz J, Pinazo MJ, Posada E, Sanchez J, Alonso PL, Gascon J. Economic evaluation of Chagas disease screening of pregnant Latin American women and of their infants in a non endemic area. Acta Trop. 2011;118:110–7.
68. Protocol de cribatge i diagnòstic de malaltia de Chagas en dones embarassades llatinoamericanes i en els seus nadons. Planificació i avaluació Salut Pública - Protocol Chagas. Departamente de Salut. Generalitat de Catalunya. 2010.
69. Regulación del control de las infecciones congénitas y perinatales en la Comunidad Valenciana. Circ. 3/2007/8/1. Conselleria de Sanitat de la Comunitat Valenciana. 2007.
70. Programma regionale per la prevenzione e il controllo della malattia di Chagas congenita: indicazione per l'assistenza in gravidanza. Delibera Regione Toscana n. 489 del 04/0672012.

Chagas Disease in the United States (USA)

Melissa S. Nolan, Kyndall Dye-Braumuller, and Eva Clark

Abstract In recent years, Chagas disease has become an emerging public health interest, and new evidence suggests that a significant disease burden exists in the United States. Implementation of national blood donor screening and regionalized community screening projects have provided novel insight into the at-risk populations residing in the country. Despite the presence of triatomines in the United States being known to the scientific community for over a century, very little is known about the distribution of vectors and their influence on autochthonous human cases. This chapter reviews the 11 triatomine species naturally found in the United States and provides a summary of the clinical and epidemiologic characteristics of human disease.

1 Triatomine Vector Biology and Ecology

Kissing bugs belong to the family Reduviidae, subfamily Triatominae. There are 138 described species, and all triatomine bugs have the potential to transmit the pathogen responsible for Chagas disease, *Trypanosoma cruzi* [1, 2]. The relationship between triatomine bugs, their hosts, and the *Trypanosoma cruzi* parasite is ancient, thought to have evolved over millions of years ago in the New World,

M. S. Nolan (✉)
Section of Tropical Medicine, Department of Pediatrics, Baylor College of Medicine, Houston, TX, USA

Arnold School of Public Health, University of South Carolina, Columbia, SC, USA
e-mail: msnolan@mailbox.sc.edu

K. Dye-Braumuller
Mosquito and Vector Control Division, Harris County Public Health, Houston, TX, USA

E. Clark
Section of Infectious Disease, Department of Medicine, Baylor College of Medicine, Houston, TX, USA

J. M. Altcheh, H. Freilij (eds.), *Chagas Disease*, Birkhäuser Advances in Infectious Diseases, https://doi.org/10.1007/978-3-030-00054-7_6

millennia before humans arrived [3, 4]. In fact, morphometric evidence suggests that triatomine bugs originated in the New World, after which they spread throughout other regions of the globe [3, 5]. Triatomine bugs are thought to have been described first by Fray Reginaldo de Lizarraga around 1590 in Peru or Chile, and the first valid species described in the United States was *Conorhinus sanguisugus* by John LeConte in Georgia in 1855, which is known as *Triatoma sanguisuga* today [6, 7]. Since 1855, an additional ten species of kissing bugs have been described and documented in the United States: *Triatoma gerstaeckeri, Paratriatoma hirsuta, T. incrassata, T. indictiva, T. lecticularia, T. neotomae, T. protracta, T. recurva, T. rubida,* and *T. rubrofasciata.* These bugs are found throughout the central to southern states, roughly south of the Great Lakes.

Relative to the species found in Latin, Central, and South America, there are fewer studies describing the biology, behavior, and ecology of the kissing bugs in the United States. Thus, knowledge is somewhat lacking regarding the trypanosome disease ecology, sylvatic cycles, and human risk in the United States. A recent uptick in research has brought more attention to this neglected tropical disease and triatomines in the United States in the past decade, especially around the United States-Mexico border. The US kissing bug species' biology and history are described below according to the most recent research on these 11 species.

1.1 **T. gerstaeckeri** ***(Ståhl)***

As the most frequently studied and collected kissing bug in the United States, *T. gerstaeckeri* has contributed to much of the current knowledge of triatomines in the southwest and central United States, and it has only been recorded in New Mexico and Texas [8]. Historically, this species was recorded as a pest of livestock and humans—even invading rural homes in Texas [7, 9–11]. In more recent decades, *T. gerstaeckeri* is found in various peridomestic and sylvatic habitats including bird nests, dog kennels, livestock pens, chicken coops, woodrat nests, lights on human structures at night, and caves [7, 8, 12–14]. Hosts for this species can be wide-ranging as well from dogs, chickens, amphibians, livestock, and humans; however woodrats are considered their primary host [7, 13–15]. *T. gerstaeckeri* has been found to be naturally infected with *T. cruzi* anywhere from 26% to 64% of specimens tested [7, 8, 12, 14, 15]. Some work has been conducted on understanding the feeding to defecation interval of this species in order to estimate risk of *T. cruzi* parasite transmission. Martinez-Ibarra et al. recorded an average defecation time of 11.5 min ($N = 733$) for this species [16].

1.2 **T. incrassata** ***(Usinger)***

This species of triatomine is largely understudied compared to *T. gerstaeckeri. T. incrassata* has only been collected on lights at night in two southern Arizona counties: Santa Cruz and Pima [7, 8, 17]. Even though little is known about this species, Klotz et al. inferred that the favored habitat of this species is woodrat nests based on similar species and their primary hosts would be woodrats and squirrels [13] *T. incrassata* has not been found naturally infected with *T. cruzi*, so its infection status is unknown in the United States [7, 8].

1.3 T. indictiva *(Neiva)*

Little is known about this US species of kissing bug as well; it has been found only in Arizona, New Mexico, and Texas in relatively small numbers [7, 8]. *T. indictiva* has been collected from woodrat nests and from lights at night [11, 18], and woodrats are known to be their primary host [13, 15]. However, Wozniak et al. collected one *T. indictiva* specimen inside of a house [14]. Of the relatively small number of bugs collected, between 0% and 50% test positive for *T. cruzi* parasites [14, 15, 18].

1.4 T. lecticularia *(Ståhl)*

T. lecticularia has been collected in multiple states including New Mexico, Texas, Kansas, Missouri, Tennessee, Georgia, South Carolina, and Florida; its distribution most likely includes the adjacent states of Oklahoma, Arkansas, Louisiana, Mississippi, and Alabama as well [8]. This species has been collected from human dwellings as early as 1940 to now, dog kennels, hollowed logs, lights at night, rock squirrel burrows, and woodrat nests [7, 8, 14, 15, 19–23]. Woodrats and squirrels are thought to be the primary hosts for *T. lecticularia* [13, 24], but like its sister species, these are generalist ectoparasites; a study in 2017 investigating triatominae blood meals found multiple *T. lecticularia* specimens had fed on turkey vultures (Georgieva et al. [30]). Studies investigating the *T. cruzi* infection prevalence have shown that this kissing bug is commonly infected with the parasite, with results ranging anywhere from 38% to 83% [11, 14, 15, 25]. Martinez-Ibarra et al. recorded an average defecation time of 8.3 min ($N = 368$) for this species [16].

1.5 T. protracta *(Uhler)*

This triatomid was first reported in the United States in 1860 as a pest of humans in central California [26] and has since been recorded in Nevada, Utah, Arizona, Colorado, New Mexico, and Texas [8, 27–29]. Because this species is attracted to lights at night, it has been commonly found in and around human dwellings, and it has also been collected from woodrat nests [7, 8, 14, 29] and potentially dog kennels [30]. The woodrat and closely related rodents are believed to be *T. protracta*'s primary hosts [13]. The infection prevalence for this species varies widely through the literature from 18% to 100% of bugs tested [8, 14, 15]. Results of defecation time experiments have varied: Wood reported a defecation time of 30.6 min ($N = 10$) [31], while Martinez-Ibarra et al. reported 6.7 min ($N = 475$) [16]. Klotz et al.

recorded that *T. protracta* defecate less often and typically further away from their host compared to their Latin American cousins [32].

1.6 T. recurva *(Ståhl)*

T. recurva, the largest kissing bug in the United States, has been recorded only in Arizona and Texas; however it has not been found in Texas since the single report in 1984 [8, 33, 34]. In 1940s Arizona, it was described as a frequent pest invader of human dwellings and miner tents [11, 35] and continues to invade homes as recently as this decade [13, 36, 37]. This species can be found in rodent nests; however the primary host is still unknown although *T. recurva* can be found associated with rock squirrels and readily feeds on various reptiles and guinea pigs in laboratory settings [7, 8, 13, 17, 35, 38–40]. This species has been found naturally infected with *T. cruzi* [7, 30]. Wood recorded an average defecation time of 75.7 min following a blood meal from this species ($N = 3$) [31].

1.7 T. rubida *(Uhler)*

In the United States, this species has been recorded in Texas, New Mexico, Arizona, and California [7, 8]. Commonly collected in woodrat nests throughout this range [7, 10, 11, 13, 20, 35], it has also been known to infest or be attracted to human houses and other peridomestic structures where it has bitten people [36, 37, 41]. In addition, *T. rubida* has been sometimes collected from fish-eating bat refuges [7, 13]. The infection prevalence of this triatomine ranges from 0% to 41% in the published literature [8, 15, 30, 37]. Of the few studies on *T. rubida*'s defecation time, most have documented relatively fast defecation times relative to feeding. Wood recorded an average defecation time of 1.6 min ($N = 5$) [31], and Reisenman et al. demonstrated that 93% of adult females ($N = 15$) defecated while feeding and 62% of immature stages ($N = 75$) defecated within 10 min of feeding less than 3 cm from the feeding site [42]. Klotz et al. reported that *T. rubida* defecated less often compared to Latin American species [32].

1.8 Paratriatoma hirsuta *(Barber)*

The documented range in the United States of this species covers Nevada, California, and Arizona where it has been collected from human dwellings and lights at night; however the natural setting most often associated with *P. hirsuta* is woodrat nests [8, 35, 43, 44]. In fact, no other known hosts are associated with this kissing bug in the United States [7, 8]. To date, no naturally *T. cruzi*-infected specimen has been collected in the United States [7, 8, 30, 35]. Wood recorded an average defecation time of 35 min ($N = 2$) following a blood meal for this species [31].

1.9 T. neotomae *(Neiva)*

Named for its primary host, *Neotoma* spp. woodrats, *T. neotomae* has been found almost exclusively in woodrat nests, with one unique report from a dog kennel [7, 8, 13, 15]. This species is believed to only inhabit Texas, and previous reports of this kissing bug found in other states are thought to be in error [8]. The literature is lacking in the reported infection prevalence of this species, with only two reports: 0% and 76% of specimens testing positive for *T. cruzi* [12, 15].

1.10 T. sanguisuga *(LeConte)*

Triatoma sanguisuga has one of the widest ranges of kissing bugs in the United States; this species has been found in Texas, Oklahoma, Kansas, Louisiana, Arkansas, Missouri, Mississippi, Tennessee, Kentucky, Illinois, Indiana, Alabama, Florida, Georgia, South Carolina, North Carolina, Virginia, Ohio, Pennsylvania, Maryland, and New Jersey [7, 8]. It is most likely also found in West Virginia, but no published records of this exist. In parallel with its wide range, *T. sanguisuga* can be found in a diverse amount of habitats ranging from sylvatic to domestic: human dwellings, lights at night, armadillo burrows, chicken coops, raccoon and opossum nests, dog kennels, horse stables, cotton rat nests, etc. [7, 8, 10, 11, 14, 15, 45–48]. Klotz et al. identified their primary hosts as raccoons, armadillos, opossums, frogs, woodrats, dogs, squirrels, and humans [13]. Historically, this species has been associated with human biting activity since the 1800s [6, 14, 15, 36, 49]. Infection prevalence for this species can vary from 17% to 70%; however in every state where *T. sanguisuga* has been tested, at least one specimen is positive for *T. cruzi* [8, 12, 14, 15, 50].

1.11 T. rubrofasciata *(De Geer)*

This is the only kissing bug species found in both the Eastern and Western Hemispheres; it is largely associated with human developments and was most likely distributed around the world through the shipping industry [8, 13]. In the United States, it has been found in Florida and Hawaii, commonly collected in chicken and pigeon coops [7, 8, 43, 51–54]. This species has been found naturally infected with *T. cruzi* parasites; however infection prevalence data is not published in the United States [7].

2 Clinical Aspects of Chagas Disease in the United States

In recent years, there has been growing awareness of the significant burden of Chagas disease in the United States. The National Notifiable Diseases Surveillance System (NNDSS) does not require the mandatory notification of Chagas disease to public health authorities, and only a few states, including Arizona, Arkansas, Louisiana, Mississippi, Utah, Texas, and Tennessee, consider Chagas disease a reportable diagnosis, making it difficult to accurately evaluate the prevalence and incidence of infection.

Historically, *Trypanosoma cruzi* transmission was concentrated in rural Latin America. However, in recent decades migration has brought infected individuals to cities in Latin America, as well as to the United States and other countries [55]. In contrast to countries outside of North America, Chagas disease has been able to take root in the United States because the disease vector is already established. The southern states provide habitats to several triatomine species and animal reservoirs such as woodrats, raccoons, opossums, and dogs that influence transmission to human hosts living in these areas. In fact, human disease has likely been occurring in the United States for more than a century [56]. The first official reports of human autochthonous *T. cruzi* infection occurred in 1955, when two Texas infants were diagnosed [57, 58]. The next report was from California in 1982 of an acute infection in an adult woman. Over the following 34 years, another four cases were reported. In five of the total seven cases, triatomines were found in or near the patient's dwelling.

US blood banks began testing for Chagas disease in 2007. Subsequently, two studies investigated *T. cruzi*-positive blood donors [59, 60]. These studies were designed to exclude positive blood donors whose suspected infection had likely been acquired outside the United States, and they identified another 21 cases of likely autochthonous *T. cruzi* infection, bringing the total number of documented infections acquired in the United States to 28 during 1955–2015. Nevertheless, most individuals infected with *T. cruzi* are immigrants from endemic countries in Latin America.

The number of Chagas disease cases varies from state to state and reflects the distribution of Latin American immigrant populations. Manne-Goehler et al. provided an updated national estimate of Chagas disease prevalence in their 2016 study, which showed the first state-level estimates of cases of *Trypanosoma cruzi* infection in the United States using data from the American Community Survey and from the American Association of Blood Banks (AABB) [61]. Their study estimated 238,091 cases of *T. cruzi* infection in the United States as of 2012, a number which excludes undocumented immigrants who may account for as many as 109,000 additional cases. The state-level results show that four states (California, Texas, Florida, and New York) had more than 10,000 cases, and an additional seven states had over 5000 cases. In addition, the AABB has reported 1908 cases of *T. cruzi* infection between 2007 and 2016, identified through screening of blood donations [61].

Screening of blood donors for *T. cruzi* infection has led to increased awareness of Chagas disease because of the identification of chronic infections among asymptomatic individuals [11]. In addition, it has improved understanding of the geographic distribution of chronically infected people in the United States and may help to direct future public health efforts to improve diagnosis and management for those at risk for the manifestations of chronic Chagas disease. Currently, the Ortho *T. cruzi* enzyme-linked immunosorbent assay (ELISA) test system is the initial screening assay, and it has been approved by the US Food and Drug Administration. The ELISA result is subsequently confirmed using a radioimmune precipitation assay [62]. Between 2007 and 2015, 1908 confirmed seropositive donations were detected, with the largest numbers found in California, Florida, and Texas [61]. At this time individuals who test positive for *T. cruzi* in the United States are not permitted to donate blood in their lifetime, regardless of treatment status.

2.1 Congenital Chagas Disease in the United States

In the United States, there is great concern for potential congenital transmission of *T. cruzi* from infected mothers to infants [63]. The first documented case of congenital transmission in the United States occurred in 2012 [64]. The patient's mother was a Bolivian woman who had been diagnosed with Chagas disease in Bolivia but never treated. As in this case, most individuals born in the United States with congenital Chagas disease are the children of foreign-born parents [65]; thus the prevalence of congenital Chagas disease varies. Two important congenital risk studies were performed in Texas, which is one of the states with the highest prevalence [66, 67]. The first, conducted from 1993 to 1996, found that 0.3% (11 out of 3765) of predominantly Hispanic pregnant women in Southern Texas were *T. cruzi*-positive [66]. Nine of these 11 women were of Hispanic origin. A similar study was repeated between 2011 and 2012 [66] and found a consistent perinatal infection rate of 0.25% (ten out of 4000). These studies indicate that Chagas disease occurs with sufficient frequency (0.25%) in the United States and warrants a need for perinatal screening to identify infected mothers and infants at risk for congenital infection.

2.2 Identifying and Treating Chagas Disease in the United States

Although Chagas disease is one of the five neglected parasitic infections in the United States that have been targeted for public health action, most healthcare providers are not familiar enough with Chagas disease to routinely include it among their differential diagnoses for cardiac and intestinal disease, even with epidemiologically at-risk patients [17, 18]. Improved provider education on diagnosis and treatment of Chagas disease is imperative; one study showed that only 11% of *T. cruzi*-positive blood donors in the United States sought or were offered treatment [68]. This finding was supported by another study that found that only 25% of positive blood donors received additional follow-up care for their disease [60]. Increasing provider awareness will lead to better diagnosis and management for all patients, regardless of where the infection was acquired.

Approximately 30,000–45,000 persons in the United States are estimated to have Chagas cardiomyopathy [69, 70]. A New York study of Latin American immigrants diagnosed with dilated cardiomyopathy found that 13% of these patients had Chagas disease [71]. A larger California study followed 135 Latin American immigrants with advanced non-ischemic cardiomyopathy [72]. Chagas disease was diagnosed in 19% of those patients.

Identifying Chagas disease in patients with cardiomyopathy not only helps with prognostication but also aids in clinical treatment decisions. Many treatment options are available for patients with *T. cruzi* infection that develop heart disease in the United States that are not available in most other endemic countries, ranging from anti-arrhythmic medication to advanced cardiac devices to heart transplantation. Several studies have been done in the United States evaluating various interventions in Chagas disease patients. Many investigators have noted that these patients have a higher burden of malignant ventricular arrhythmias, which is a crucial factor when considering implantation of cardioverter-defibrillator devices or radiofrequency ablation [73]. One study demonstrated that the anti-arrhythmic amiodarone has direct anti-*T. cruzi* effects [74]. Chagas cardiomyopathy patients in the United States seem to be more likely to be offered ionotropes as well as electrical (ICDs and cardiac resynchronization therapy) and ventricular support devices, especially while awaiting heart transplant, than in other endemic countries, including countries with significant medical resources such as Brazil [75].

Heart transplantation for patients with Chagas cardiomyopathy has been done since the 1990s, even though early on, given the potential for reactivation of *T. cruzi* with immunosuppression, Chagas cardiomyopathy was initially considered to be a relative contraindication to heart transplantation. Subsequently, studies have shown that the outcome after heart transplantation for Chagas cardiomyopathy is acceptable [76], and it is now recognized that survival in patients with Chagas cardiomyopathy after heart transplant may be better than those patients with other forms of non-ischemic cardiomyopathy. *T. cruzi* post-transplantation reactivation rates are thought to be as high as 26.5–42.9% [72]; thus, screening for *T. cruzi* infection in all patients born in a Chagas disease endemic country and undergoing transplant evaluation for dilated cardiomyopathy is critical.

Reactivation of *T. cruzi* infection is a major concern because of the risk of allograft dysfunction [77]. One study of Chagas cardiomyopathy patients who underwent heart transplantation in the United States identified reactivation in five patients (45%), detected by clinical signs of reactivation with accompanying allograft dysfunction by echocardiography in two cases and whole blood PCR testing in three patients [75]. The *T. cruzi* reactivation rate in the Kransdorf study is higher than the rates of 21–39% that were reported by transplant centers in Brazil [76–78]. This difference is likely due to the use of the more potent immunosuppressive agents tacrolimus and mycophenolate mofetil in the United States, as compared to predominant use of cyclosporine and azathioprine in Brazil. Mycophenolate mofetil in particular has been associated with a higher rate of *T. cruzi* reactivation [78, 79].

Conversely, testing of potential organ donors for chronic *T. cruzi* infection is quite important due to the risk of donor-transmitted infection. Given the prevalence of *T. cruzi* infection in blood donors discussed previously, universal testing is necessary to prevent donor-transmitted *T. cruzi* infections, which led to the death of two heart transplant recipients in 2006 [80]. Kransdorf et al. found that only four of 11 organ donors underwent testing for *T. cruzi* infection [75]. The most common reason for omission of *T. cruzi* testing in their cohort was that the Organ Procurement Organization (OPO) did not routinely perform *T. cruzi* testing, even on at-risk donors. Recent data indicate that only 19% of OPOs in the United States performs testing for *T. cruzi* [79].

Another barrier to improving control of Chagas disease in the United States includes diagnostic testing. Better diagnostic tests are needed so that effective screening can be performed outside of formal laboratory settings. Currently, when a healthcare provider suspects Chagas disease, most health departments in endemic states recommend that specimens should first be screened at a commercial laboratory. Several types of serologic tests are used among commercial laboratories in the United States, including indirect hemagglutination (IHA) tests, indirect immunofluorescence (IIF) tests, and ELISAs. Most of these tests use a complex mixture of parasite antigens (IHA and ELISA) or the whole-parasite lysate (IIF). This increases the likelihood that the infection will be diagnosed, even when the antibody level is low; however, false-positive results can occur with *Leishmania* species or *Trypanosoma rangeli*. Samples that test positive should then be forwarded to the Centers for Disease Control and Prevention (CDC) for confirmatory testing. At the CDC, two serologic tests are performed to confirm the diagnosis, the Wiener recombinant antigen Chagatest ELISA and the TESA (trypomastigote excreted-secreted antigen) immunoblot [81, 82]. Although it is commonly accepted that polymerase chain reaction (PCR)-based tests for Chagas disease are more sensitive in acute Chagas infection, their results are highly variable in chronic infection. Thus, at this time PCR-based assays are only used as a clinical tool when transmission via blood transfusion, organ transplant, or congenital or laboratory exposure is suspected [83, 84].

Treatment is now recommended for asymptomatic patients of all ages who are diagnosed with Chagas disease [81]. Ideally, treatment should be started before the patient begins to develop complications of the disease. Two medications currently exist for the treatment of Chagas disease, benznidazole and nifurtimox. Benznidazole has been approved by the US Food and Drug Administration (FDA) for pediatric applications (aged 2–12 years); however, nifurtimox (all ages) and benznidazole for adult use are only available through investigational protocols from the CDC. Administrative requirements for participation under the protocols often are a barrier for the typical busy healthcare provider; thus FDA approval of these drugs would allow clinicians to access these drugs more easily for their patients.

Although benznidazole is the preferred treatment due to its lower incidence of adverse events and shorter treatment duration [81], due to supply chain issues, the most often prescribed treatment for Chagas disease in the United States is nifurtimox [85–87]. Only one study has been done to date to assess the safety of nifurtimox in patients with Chagas disease in the United States [88]. Forsyth et al. evaluated 53 adult Latin American immigrants with Chagas disease who underwent treatment with nifurtimox (8–10 mg/kg in three daily doses for 12 weeks) between 2008 and 2012. They recorded 435 adverse events among these 53 individuals, most (93.8%) of which were mild. The most common adverse events were gastrointestinal symptoms and weight loss, which has been shown in prior studies performed outside of the United States. Thus, while nifurtimox frequently causes side effects, the majority is mild and can be managed with dose reduction or temporary suspension of medication. Importantly, patients undergoing treatment should be closely monitored during the first few weeks of treatment due to the increased frequency of potentially severe adverse effects, including depression, rash, and anxiety.

3 Concluding Remarks

As an estimated 99% of patients in the United States diagnosed with Chagas disease remain untreated, there is a significant need for increased national attention to strategies to increase treatment availability for this disease. Investigation and production of new drugs targeting *Trypanosoma cruzi*, as well as increasing the supply of the existing effective medications, are essential to expand treatment of Chagas disease in the United States. With increased treatment of infected, asymptomatic individuals, many lives will be saved due to prevention of deaths and morbidity from the sequelae of Chagas disease, such as heart failure and arrhythmias.

References

1. Schofield CJ, Dolling WR. Bedbugs and kissing-bugs (bloodsucking Hemiptera). In: Lane RP, Crosskey RW, editors. Medical Insects and Arachnids. Dordrecht: Springer; 1993.
2. Krinsky WL. True Bugs (Hemiptera). In: Mullen GR, Durden LA, editors. Medical and veterinary entomology. Burlington, MA: Elsevier; 2009.
3. Schofield CJ. The biosystematics of Triatominae. In: Service MW, editor. Biosystematics of haematophagous insects. Systematics Association. Oxford: Clarendon Press; 1988. p. 284–312.
4. Briones MR, Souto RP, Stolf BS, Zingales B. The evolution of two Trypanosoma cruzi subgroups inferred from rRNA genes can be correlated with the interchange of American mammalian faunas in the Cenozoic and has implications to pathogenicity and host specificity. Mol Biochem Parasitol. 1999;104(2):219–32.
5. Gorla DE, Dujardin JP, Schofield CJ. Biosystematics of old world triatominae. Acta Trop. 1997;63(2-3):127–40.
6. LeConte J. Remarks on two species of American Cimex. Proc Acad Nat Sci. 1855;7:404.
7. Lent H, Wygodzinsky P. Revision of the Triatominae (Hemiptera, Reduviidae) and their significance as vectors of Chagas' Disease. Bull Am Museum Nat Hist. 1979;163(3):123–520.
8. Bern C, Kjos S, Yabsley MJ, Montgomery SP. Trypanosoma cruzi and Chagas' Disease in the United States. Clin Microbiol Rev. 2011;24(4):655–81.
9. Packchanian A. Natural infection of *Triatoma gerstaeckeri* with *Trypanosoma cruzi* in Texas. Public Health Rep. 1939;54:1547–54.
10. Wood SF. New localities for Trypanosoma cruzi Chagas in southwestern United States. Am J Trop Med Hyg. 1941;17:85–94.
11. Wood SF. Notes on the distribution and habits of reduviid vectors of Chagas' disease in the southwestern United States, part I. Pan-Pacific Entomol. 1941;17:85–94.
12. Burkholder JE, Allison TC, Kelly VP. Trypanosoma cruzi (Chagas) (Protozoa: Kinetoplastida) in invertebrate, reservoir, and human hosts of the lower Rio Grande valley of Texas. J Parasitol. 1980;66(2):305–11.

13. Klotz SA, Dorn PL, Mosbacher M, Schmidt JO. Kissing bugs in the United States: risk for vector-borne disease in humans. Environ Health Insights. 2014;8(Suppl 2):49–59.
14. Wozniak EJ, Lawrence G, Gorchakov R, Alamgir H, Dotson E, Sissel B, et al. The biology of the triatomine bugs native to South Central Texas and assessment of the risk they pose for autochthonous Chagas Disease exposure. J Parasitol. 2015;101(5):520–8.
15. Kjos SA, Snowden KF, Olson JK. Biogeography and Trypanosoma cruzi infection prevalence of Chagas disease vectors in Texas, USA. Vector Borne Zoonotic Dis. 2009;9(1):41–50.
16. Martinez-Ibarra JA, Alejandre-Aguilar R, Paredes-Gonzalez E, Martinez-Silva MA, Solorio-Cibrian M, Nogueda-Torres B, et al. Biology of three species of North American Triatominae (Hemiptera: Reduviidae: Triatominae) fed on rabbits. Mem Inst Oswaldo Cruz. 2007;102(8):925–30.
17. Ryckman RE. The vertebrate hosts of the Triatomine of North and Central America and the West Indies (Hemiptera: Reduviidae: Triatominae). Bull Soc Vector Ecol. 1986;11:221–41.
18. Pippin WF, Law PF, Gaylor MJ. Triatoma sanguisuga Texana Usinger and Triatoma sanguisuga indictive Neiva naturally infected with Trypanosoma cruzi Chagas in Texas (Hemiptera: Triatominae) (Kinetoplastida: Trypanosomidae). J Med Entomol. 1968;5(1):134.
19. Packchanian A. Natural infection of *Triatoma heidemanni* with *Trypanosoma cruzi* in Texas. Public Health Rep. 1940;55:1300–6.
20. Ryckman RE, Ryckman JV. Epizootiology of Trypanosoma cruzi in southwestern North America. XII. Does Gause's rule apply to the ectoparasitic Triatominae? (Hemiptera: Reduviidae) (Kinetoplastidae: Trypanosomidae) (Rodentia: Cricetidae). J Med Entomol. 1967;4(3):379–86.
21. Williams GD, Adams LG, Yaeger RG, McGrath RK, Read WK, Bilderback WR. Naturally occurring trypanosomiasis (Chagas' disease) in dogs. J Am Vet Med Assoc. 1977;171(2):171–7.
22. Yabsley MJ, Noblet GP. Seroprevalence of Trypanosoma cruzi in raccoons from South Carolina and Georgia. J Wildl Dis. 2002;38(1):75–83.
23. Moffett A, Strutz S, Guda N, Gonzalez C, Ferro MC, Sanchez-Cordero V, et al. A global public database of disease vector and reservoir distributions. PLoS Negl Trop Dis. 2009;3(3):e378.
24. Kjos SA, Gillespie JJ, Olson JK, Snowden KF. Detection of Blastocrithidia spp. (Kinetoplastida: Trypanosomatidae) in Chagas disease vectors from Texas, USA. Vector Borne Zoonotic Dis. 2009;9(2):213–6.
25. Sullivan TD, Mc GT, et al. Incidence of Trypanosoma cruzi, Chagas, in Triatoma (Hemiptera, Reduviidae) in Texas. Am J Trop Med Hyg. 1949;29(4):453–8.
26. Mortenson EW, Walsh JD, editors. Review of the *Triatoma protracta* problem in the Sierra Nevada foothills of California. 31st Annual Conference of the California Mosquito Control Association; 1961. Sacramento, CA: California Mosquito Control Association; 1963.
27. Mehringer PJ, Wood SF. A resampling of wood rat houses and human habitations in Griffith Park, Los Angeles, for *Triatoma protracta* and *Trypanosoma cruzi*. Bull South Calif Acad Sci. 1958;57:39–46.
28. Walsh JD, Jones JP. Public Health Significance of the cone-nosed bug, *Triatoma protracta* (Uhler) in the Sierra Nevada foothills of California. Calif Vector Views. 1962;9:33–7.
29. Sjogren RD, Ryckman RE. Epizootiology of Trypanosoma cruzi in southwestern North America. 8. Nocturnal flights of Triatoma protracta (Uhler) as indicated by collections at black light traps (Hemiptera: Reduviidae: Triatominae). J Med Entomol. 1966;3(1):81–92.
30. Georgieva AY, Gordon ERL, Weirauch C. Sylvatic host associations of Triatominae and implications for Chagas disease reservoirs: a review and new host records based on archival specimens. PeerJ. 2017;5:e3826.
31. Wood SF. Importance of feeding and defecation times of insect vectors in transmission of Chagas' disease. J Econ Entomol. 1951;44:52–4.
32. Klotz SA, Dorn PL, Klotz JH, Pinnas JL, Weirauch C, Kurtz JR, et al. Feeding behavior of triatomines from the southwestern United States: an update on potential risk for transmission of Chagas disease. Acta Trop. 2009;111(2):114–8.
33. Ikenga JO, Richerson JV. Trypanosoma cruzi (Chagas) (protozoa: Kinetoplastida: Trypanosomatidae) in invertebrate and vertebrate hosts from Brewster County in Trans-Pecos Texas. J Econ Entomol. 1984;77(1):126–9.

34. Pfeiler E, Bitler BG, Ramsey JM, Palacios-Cardiel C, Markow TA. Genetic variation, population structure, and phylogenetic relationships of Triatoma rubida and T. recurva (Hemiptera: Reduviidae: Triatominae) from the Sonoran Desert, insect vectors of the Chagas' disease parasite Trypanosoma cruzi. Mol Phylogenet Evol. 2006;41(1):209–21.
35. Wood SF. Additional observations on Trypanosoma cruzi Chagas, from Arizona in insects, rodents, and experimentally infected animals. Am J Trop Med Hyg. 1949;29(1):43–55.
36. Klotz JH, Dorn PL, Logan JL, Stevens L, Pinnas JL, Schmidt JO, et al. "Kissing bugs": potential disease vectors and cause of anaphylaxis. Clin Infect Dis. 2010;50(12):1629–34.
37. Reisenman CE, Lawrence G, Guerenstein PG, Gregory T, Dotson E, Hildebrand JG. Infection of kissing bugs with Trypanosoma cruzi, Tucson, Arizona, USA. Emerg Infect Dis. 2010;16(3):400–5.
38. Wood SF. Notes on the feeding of the cone-nosed bugs (Hemiptera: Reduviidae). J Parasitol. 1944;30:197–8.
39. Wood SF. Body weight and blood meal size in conenose bugs, Triatoma and Paratriatoma. Bull South Calif Acad Sci. 1959;58:116–8.
40. Elkens D. Nocturnal Flights of the *Triatoma (Hemiptera: Reduviidae)* in Sabino Cayon, Arizona, II. Neotoma lodge studies. J Med Entomol. 1984;21:140–4.
41. Pinnas JL, Lindberg RE, Chen TM, Meinke GC. Studies of kissing bug-sensitive patients: evidence for the lack of cross-reactivity between Triatoma protracta and Triatoma rubida salivary gland extracts. J Allergy Clin Immunol. 1986;77(2):364–70.
42. Reisenman CE, Gregory T, Guerenstein PG, Hildebrand JG. Feeding and defecation behavior of Triatoma rubida (Uhler, 1894) (Hemiptera: Reduviidae) under laboratory conditions, and its potential role as a vector of Chagas disease in Arizona, USA. Am J Trop Med Hyg. 2011;85(4):648–56.
43. Usinger RL, United States. Public Health Service. States Relations Division. The Triatominae of North and Central America and the West Indies and their public health significance. Washington, DC: Govt. Print. Off.; 1944. p. iv. 83.
44. Ryckman RE. The genus Paratriatoma in western North America. J Med Entomol. 1971;8(1):87–97.
45. Grundemann AW. Studies on the biology of the Triatoma sanguisuga (Leconte) in Kansas (Reduviidae: Triatominae). Kansas Entomol Soc. 1947;20:77–85.
46. Olsen PF, Shoemaker JP, Turner HF, Hays KL. Incidence of Trypanosoma cruzi (Chagas) in wild vectors and reservoirs in East-Central Alabama. J Parasitol. 1964;50:599–603.
47. Yaeger RG. The prevalence of Trypanosoma cruzi infection in armadillos collected at a site near New Orleans, Louisiana. Am J Trop Med Hyg. 1988;38(2):323–6.
48. Kjos SA, Snowden KF, Craig TM, Lewis B, Ronald N, Olson JK. Distribution and characterization of canine Chagas disease in Texas. Vet Parasitol. 2008;152(3-4):249–56.
49. Kimball BM. Conorhinus sanguisuga, its habits and life history. Kansas Acad Sci. 1894;14:128–31.
50. Waleckx E, Suarez J, Richards B, Dorn PL. Triatoma sanguisuga blood meals and potential for Chagas disease, Louisiana, USA. Emerg Infect Dis. 2014;20(12):2141–3.
51. Usinger RL. Descriptions of new Triatominae, with a key to genera (Hemiptera, Reduviidae). Berkeley, CA: University of California Press; 1939. p. 33–5. cover-title, 1 p. l.
52. Ryckman RE. The triatominae of North hand Central America and the West Indies: a checklist with synonymy (Hemiptera: Reduviidae: Triatominae). Bull Soc Vector Ecol. 1984;9:71–83.
53. Wood SF. The occurrence of Trypanosoma conorhini Donovan in the reduviid bug, Triatoma rubrofasciata (Degeer) from Oahu, TH. Proc Hawaii Entomol Soc. 1946;29:43–55.
54. Arnold HL, Bell DB. Kissing bug bites. J Clin Immunol. 1944;74:436–42.
55. Dias JC, Silveira AC, Schofield CJ. The impact of Chagas disease control in Latin America: a review. Mem Inst Oswaldo Cruz. 2002;97(5):603–12.
56. Garcia MN, Woc-Colburn L, Aguilar D, Hotez PJ, Murray KO. Historical perspectives on the epidemiology of human Chagas disease in Texas and recommendations for enhanced

understanding of clinical Chagas disease in the Southern United States. PLoS Negl Trop Dis. 2015;9(11):e0003981.
57. Woody NC, Woody HB. American trypanosomiasis (Chagas' disease); first indigenous case in the United States. JAMA. 1955;159(7):676–7.
58. Greer DA. Found: two cases of Chagas disease. Texas Health Bull. 1955;9:11–3.
59. Cantey PT, Stramer SL, Townsend RL, Kamel H, Ofafa K, Todd CW, et al. The United States Trypanosoma cruzi Infection Study: evidence for vector-borne transmission of the parasite that causes Chagas disease among United States blood donors. Transfusion. 2012;52(9):1922–30.
60. Garcia MN, Woc-Colburn L, Rossmann SN, Townsend RL, Stramer SL, Bravo M, et al. Trypanosoma cruzi screening in Texas blood donors, 2008-2012. Epidemiol Infect. 2016;144(5):1010–3.
61. Manne-Goehler J, Umeh CA, Montgomery SP, Wirtz VJ. Estimating the burden of Chagas Disease in the United States. PLoS Negl Trop Dis. 2016;10(11):e0005033.
62. Bern C, Montgomery SP, Katz L, Caglioti S, Stramer SL. Chagas disease and the US blood supply. Curr Opin Infect Dis. 2008;21(5):476–82.
63. Buekens P, Almendares O, Carlier Y, Dumonteil E, Eberhard M, Gamboa-Leon R, et al. Mother-to-child transmission of Chagas' disease in North America: why don't we do more? Matern Child Health J. 2008;12(3):283–6.
64. Centers for Disease. C, Prevention. Congenital transmission of Chagas disease - Virginia, 2010. MMWR Morb Mortal Wkly Rep. 2012;61(26):477–9.
65. Murillo J, Bofill LM, Bolivar H, Torres-Viera C, Urbina JA, Benhayon D, et al. Congenital Chagas' disease transmission in the United States: diagnosis in adulthood. IDCases. 2016;5:72–5.
66. Di Pentima MC, Hwang LY, Skeeter CM, Edwards MS. Prevalence of antibody to Trypanosoma cruzi in pregnant Hispanic women in Houston. Clin Infect Dis. 1999;28(6):1281–5.
67. Edwards MS, Rench MA, Todd CW, Czaicki N, Steurer FJ, Bern C, et al. Perinatal screening for Chagas Disease in Southern Texas. J Pediatric Infect Dis Soc. 2015;4(1):67–70.
68. Stimpert KK, Montgomery SP. Physician awareness of Chagas disease, USA. Emerg Infect Dis. 2010;16(5):871–2.
69. Schmunis GA, Yadon ZE. Chagas disease: a Latin American health problem becoming a world health problem. Acta Trop. 2010;115(1-2):14–21.
70. Bern C, Montgomery SP. An estimate of the burden of Chagas disease in the United States. Clin Infect Dis. 2009;49(5):e52–4.
71. Kapelusznik L, Varela D, Montgomery SP, Shah AN, Steurer FJ, Rubinstein D, et al. Chagas disease in Latin American immigrants with dilated cardiomyopathy in New York City. Clin Infect Dis. 2013;57(1):e7.
72. Traina MI, Sanchez DR, Hernandez S, Bradfield JS, Labedi MR, Ngab TA, et al. Prevalence and impact of Chagas Disease among Latin American immigrants with nonischemic cardiomyopathy in Los Angeles, California. Circ Heart Fail. 2015;8(5):938–43.
73. Hsia HH, Marchlinski FE. Electrophysiology studies in patients with dilated cardiomyopathies. Card Electrophysiol Rev. 2002;6(4):472–81.
74. Benaim G, Sanders JM, Garcia-Marchan Y, Colina C, Lira R, Caldera AR, et al. Amiodarone has intrinsic anti-Trypanosoma cruzi activity and acts synergistically with posaconazole. J Med Chem. 2006;49(3):892–9.
75. Kransdorf EP, Czer LS, Luthringer DJ, Patel JK, Montgomery SP, Velleca A, et al. Heart transplantation for Chagas cardiomyopathy in the United States. Am J Transplant. 2013;13(12):3262–8.
76. Fiorelli AI, Santos RH, Oliveira JL Jr, Lourenco-Filho DD, Dias RR, Oliveira AS, et al. Heart transplantation in 107 cases of Chagas' disease. Transplant Proc. 2011;43(1):220–4.
77. Godoy HL, Guerra CM, Viegas RF, Dinis RZ, Branco JN, Neto VA, et al. Infections in heart transplant recipients in Brazil: the challenge of Chagas' disease. J Heart Lung Transplant. 2010;29(3):286–90.

78. Campos SV, Strabelli TM, Amato Neto V, Silva CP, Bacal F, Bocchi EA, et al. Risk factors for Chagas' disease reactivation after heart transplantation. J Heart Lung Transplant. 2008;27(6):597–602.
79. Schwartz BS, Paster M, Ison MG, Chin-Hong PV. Organ donor screening practices for Trypanosoma cruzi infection among US Organ Procurement Organizations. Am J Transplant. 2011;11(4):848–51.
80. Kun H, Moore A, Mascola L, Steurer F, Lawrence G, Kubak B, et al. Transmission of Trypanosoma cruzi by heart transplantation. Clin Infect Dis. 2009;48(11):1534–40.
81. Bern C, Montgomery SP, Herwaldt BL, Rassi A Jr, Marin-Neto JA, Dantas RO, et al. Evaluation and treatment of Chagas disease in the United States: a systematic review. JAMA. 2007;298(18):2171–81.
82. Afonso AM, Ebell MH, Tarleton RL. A systematic review of high quality diagnostic tests for Chagas disease. PLoS Negl Trop Dis. 2012;6(11):e1881.
83. Diez M, Favaloro L, Bertolotti A, Burgos JM, Vigliano C, Lastra MP, et al. Usefulness of PCR strategies for early diagnosis of Chagas' disease reactivation and treatment follow-up in heart transplantation. Am J Transplant. 2007;7(6):1633–40.
84. Qvarnstrom Y, Schijman AG, Veron V, Aznar C, Steurer F, da Silva AJ. Sensitive and specific detection of Trypanosoma cruzi DNA in clinical specimens using a multi-target real-time PCR approach. PLoS Negl Trop Dis. 2012;6(7):e1689.
85. Centers for Disease Control and Prevention. Spotlight on CDC's parasitic disease work. Atlanta, GA: Centers for Disease Control; 2015.
86. Manne J, Snively CS, Levy MZ, Reich MR. Supply chain problems for Chagas disease treatment. Lancet Infect Dis. 2012;12(3):173–5.
87. Alpern JD, Lopez-Velez R, Stauffer WM. Access to benznidazole for Chagas disease in the United States-Cautious optimism? PLoS Negl Trop Dis. 2017;11(9):e0005794.
88. Forsyth CJ, Hernandez S, Olmedo W, Abuhamidah A, Traina MI, Sanchez DR, et al. Safety profile of nifurtimox for treatment of Chagas Disease in the United States. Clin Infect Dis. 2016;63(8):1056–62.

Part IV
Diagnosis

Diagnosis of Chagas Disease

Alejandro O. Luquetti and Alejandro G. Schijman

Abstract Diagnosis of Chagas disease is related to the phase of this protozoan infection. For acute phase, parasitological methods are preferred and for the chronic phase, serological ones. Parasitological methods comprise from the simplest wet smear, going through alternatives as concentration methods and tests that involve the multiplication of the parasite in media (hemoculture), triatomine insects (xenodiagnosis), or animals (inoculation of susceptible mammals), to more sophisticated molecular tests as polymerase chain reaction (PCR) and loop-mediated isothermal amplification (LAMP). All have indications, advantages, and disadvantages, as well as costs, all of which will be detailed in this chapter. Among serological tests, a vast repertoire has been also developed and standardized, from the simplest indirect hemagglutination to the sophisticated CMIA, including indirect immunofluorescence, ELISA, and rapid tests. As all are indirect tests, it is recommended to use at least two of them for a concluding laboratory result, in order to confirm or exclude the infection by *Trypanosoma cruzi*. Diagnosis may be used in different contexts as confirmation of the infection, exclusion of blood donors, epidemiological survey, congenital infection, or follow-up after specific treatment. In order to have a precise diagnosis, it is necessary to have commercial kits of proved performance and good laboratory practices, for which permanent training of laboratory personnel is mandatory. An external quality control will prove that these conditions have been fulfilled.

A. O. Luquetti
Núcleo de Estudos da doença de Chagas (NEDoC), Hospital das Clínicas, Federal University of Goias, Goiania, Brazil
e-mail: luquetti@ufg.br

A. G. Schijman (✉)
Laboratorio de Biología Molecular de la Enfermedad de Chagas, Instituto de Investigaciones en Ingeniería Genética y Biología Molecular "Dr. Hector Torres" (INGEBI-CONICET), Ciudad de Buenos Aires, Argentina
e-mail: schijman@dna.uba.ar

J. M. Altcheh, H. Freilij (eds.), *Chagas Disease*, Birkhäuser Advances in Infectious Diseases, https://doi.org/10.1007/978-3-030-00054-7_7

1 Introduction

Diagnosis of an infectious disease involves mainly three aspects: epidemiological, clinical, and laboratory tests. All of them should fit, and this applies also to Chagas disease [1]. The epidemiology is important as infected people come from endemic areas, and depending on the prevalence of the infection in each area, the probability to obtain a positive result changes. Also it is important to question the patient if he/she has relatives with the infection under suspicion. In endemic areas, up to 2/3 of them recall the mother or grandfather or siblings infected. For non-endemic areas, the fact to traveling to an endemic area or of have being born in areas of endemicity is also helpful. Some characteristic clinical findings that are frequent in *T. cruzi*-infected people are of utmost importance, such as the complete right bundle branch block at an electrocardiogram and the occurrence of megaesophagus or megacolon. In endemic areas, more than 90% of the patients with one of these manifestations are infected. Finally, the laboratory analyses will close the diagnosis, showing that parasitological and/or serological tests are positive.

The natural history of the infection is worth to describe in order to understand some characteristics related to diagnosis. The acute phase lasts for 60 days after symptoms start and is characterized by a high parasitemia, with easily detected parasites in any drop of blood. Nevertheless, symptoms are usually scarce, and more than 90% of the infections are not detected, because the physicians do not suspect them or because the symptoms may be only fever, which subsides in few weeks. Recovery of the acute phase is the rule, and a decrease in the number of parasites easily detected is observed after the first 4 weeks. Lethality is below 5%, mainly in young children or after a transfusion or by oral route, conditions in which parasitemia is exceedingly high. After the acute phase, those infected start a silent chronic phase, with few or no parasites detected in blood, a reason for asking indirect methods for diagnosis. This is the indeterminate or asymptomatic form of the chronic phase, in which only antibody detection is sensitive enough for accurate diagnosis. An estimated 2% per year goes through symptomatic forms as the cardiac or mega-esophagus and/or megacolon. Nearly half of all infected individuals never will present overt disease and will be dying of other causes. In many cases, diagnosis occurs only when individuals donate blood or when they submit themselves for routine checkups [2]. The acute phase comprises infection acquired through the vector (kissing bug) which is the more frequent route in endemic countries, as well as infection acquired through blood transfusion or organ transplantation. The oral route may be another way, by consuming drinks or food in whose preparation contaminated insects or their fecal samples were included; these occur as micro-epidemics when a number of people have been contaminated by the same preparation. Another mechanism is the vertical transmission from an infected mother to her newborn, causing congenital Chagas disease; nearly 5% of chagasic women may transmit *Trypanosoma cruzi* to their offspring, and the medical importance is that all newborns diagnosed may be cured by specific treatment [3]. Another possibility is the reactivation of an infected individual submitted to immunosuppression by corticoids or antineoplastic drugs or after acquiring HIV infection with low CD4 counts [4]. The suspicion of these epidemiological circumstances, together with clinical signs as fever, allows the physician to ask for parasitological laboratory tests, which will be detailed in the following sections. Nowadays, the more frequent situation is for chronic-phase infected individuals, when serological tests should be requested for confirmation.

In non-endemic regions, due mainly to migration of infected people of Latin-America to any part of the world, physicians may be aware of Chagas disease and be able to ask for proper laboratory tests to confirm or exclude the clinical suspicion [5].

Trypanosoma cruzi is not homogeneous, and there are at least six different lineages of this protozoan, named discrete typing units (DTUs TcI to TcVI) [6] related to different geographical areas, to the frequency in which any one of these DTUs is present in human infection, to clinical manifestations, and to the response to chemotherapeutic agents [6]. Other differences are under investigation as the frequency of congenital transmission and the capacity to be transmitted from blood to a receptor. The more frequent in humans are TcI, TcII, and TcV. TcI is distributed mainly to the North of Amazonas River, and children submitted to specific treatment with benznidazole underwent cure in short periods of time [7]. TcII is found mainly in Central Brazil, an area in which megaesophagus is more frequent than in other areas [8]. TcV is found in humans mainly in countries of the Southern Cone of South America (Chile, Argentina, Bolivia, Paraguay, Uruguay, and South of Brazil) and is associated with a higher congenital transmission [9]. Emigration to non-endemic countries (mainly Europe) has been mainly from people originated from the Southern Cone, and hence this is the type of Tc more frequently seen in Spain and other European countries. Migration to the United States is mainly from México [5], where TcI predominates.

2 Parasitological Tests

2.1 Direct Tests, Wet Smear; Concentration Methods, Strout and Microhematocrit

The easiest test is the wet smear, by a drop of blood (10 μL) delivered onto a slide and covered with a coverslip. The preparation should be pressed (with any object) to obtain a thin layer of red blood cells separated among them that allows to detect the quick movements of the parasite. It is necessary to examine on a microscope (10 × 40) at least 100 fields, because parasitemia may be low, mainly when symptoms started several weeks before. When this method is negative and the clinical suspicion persists, a concentration method may be applied [2].

The method of Strout is also easy, cheap, and sensitive but requires 2–5 mL of venous blood without anticoagulants [10]. Once the clot is formed, the liquid phase is transferred to a test tube and spun down for 5 min at 50–100 g. The supernatant is transferred to another tube and spun down at 400 g to allow parasites go to the bottom. This supernatant (serum) is discarded and the last drop mounted on a slide, with the same procedure as the wet smear, already described.

When few blood is available (neonates), the method of microhematocrit should be used, by filling up to four capillaries and, after spinning, looking on the interface between red blood cells and plasma in a microscope. The capillary tube may be broken at the interface and proceed as with the wet smear [11]. Special care should be taken to avoid accidental contamination, by use of personal protection equipment (PPE).

Stained smears are less sensitive and only appropriate with high parasitemias, as may be observed in reactivation (immunosuppression) or transfusional transmission. Nevertheless the thick smear (stained) used for malaria diagnosis may be useful in the field, when, instead of plasmodium, a flagellate is found. Health personnel working with malaria has been trained in some areas to be able to diagnose *T. cruzi* as well.

2.2 Multiplication Methods: Hemoculture, Xenodiagnosis, and Animal Inoculation

During the chronic phase, in some circumstances, it may be necessary to isolate the parasite, as on chemotherapeutic trials. The low parasitemia may be detected only by multiplication methods, i.e., from few parasites at the sample, offering them the proper conditions to multiply. As a consequence, all these methods require a time to allow *T. cruzi* to increase the original low numbers. All these methods are not routine and need to be performed in research institutions. They also are not commercially available. The main ones are hemoculture, xenodiagnosis, and animal inoculation.

Hemoculture is based on harvesting heparinized blood in special media as liver infusion tryptose. It is essential to include a rather large amount of blood (i.e., 20 mL) and exclude the plasma that has antibodies and complement, which may kill the parasite. Culture is performed in several tubes, each one with 1 mL of packed red blood cells and 2–3 mL of medium. Observations should be performed monthly, for 6 months. Contamination is a risk, and one of the disadvantages is to run all the procedure in sterility [12].

Xenodiagnosis was the first procedure used by the time the disease was described. The rationale is to feed triatomine bugs with blood of the patient. It is necessary to culture colonies of these bugs, a rather difficult task. After feeding, bugs are examined, one by one, at 30 and 60 days after feeding, looking at their feces. Usually 40 bugs are used per procedure. Formerly, bugs were applied into a box onto the arms and legs of the patient, but nowadays heparinized blood collected from the patient is offered to bugs through a latex membrane (artificial xenodiagnosis). The advantage is that bugs may be transported to endemic areas and do not need sterile procedures [13].

Animal inoculation with blood from the patient or from feces of bugs is another procedure, seldom used. Susceptible mice are employed, i.e., Balb C. Tail blood of inoculated mice should be examined daily for 1–2 months [14].

With all these methods, the positivity is low and variable (around 20%), being highly dependent on the operator skills and expertise. If the method is repeated, the positivity increases (up to 60%), but for some patients with very low parasitemia, even repeated examinations will be always negative.

2.3 *Molecular Methods: Polymerase Chain Reaction and Loop-Mediated Isothermal Amplification*

2.3.1 Polymerase Chain Reaction

PCR has been used for sensitive detection of *T. cruzi* DNA in human blood, firstly as qualitative tool [15–17] and later on as quantitative method to estimate parasitic load, using real-time PCR technology [18–20]. There are a few procedures already standardized and validated that employ whole blood treated with guanidine hydrochloride as a chaotropic agent [19, 21]; most of them use nuclear satellite DNA (satDNA) or minicircle molecule (kDNA) as parasite molecular targets plus an internal amplification standard [19, 20, 22]. High concordance was observed between real-time PCR targeted to the abovementioned sequences [22]. Analytical sensitivity is more uniform among different DTUs for kDNA qPCR than for SatDNA qPCR, being the latter less sensitive for some TcI and TcIV strains, due to a lower gene dosage, but recent characterization of satellite sequences from a higher number of strains allowed improvement of primer/probe design and consequently sensitivity [22]. In regions where *T. rangeli* infections concur with Chagas disease [22–24], satDNA is recommended for *T. cruzi*-specific detection. An external quality control program for evaluation of *T. cruzi*-qPCR performance has been recently implemented [25].

2.3.2 Loop-Mediated Isothermal Amplification (LAMP)

LAMP is able to amplify large amounts of DNA within 30–60 min of incubation at 60–65 °C, employing a complex design of primer sequences and strand displacement Bst DNA polymerase. LAMP reagents are stable at room temperatures up to 37 °C, avoiding the need of a cold chain [26, 27]. No thermocycler is needed for the reaction, and product visualization can be done by the naked eye or followed in real time by turbidity or fluorescence using intercalating dyes. In-tube visualization may be achieved using manganese loaded calcein. A first LAMP procedure targeting 18s rDNA gene that has been evaluated in triatomine feces showed a sensitivity of 100 fg of DNA per test but was cross-reactive with *Leishmania* sp. DNA [28], and in human blood detection level, sensitivity was 50 parasites/mL [29]. A recent prototype kit for detection of *T. cruzi* satDNA in human blood samples was developed by the Eiken Company [30]; it contains dried reagents on the inside of the microtube caps. It detected 1×10^{-2} parasite equivalents/mL in blood samples anticoagulated with EDTA and spiked with known concentrations of culture parasites, when DNA extraction was done using commercial columns or rapid boil and spin method and did not amplify *Leishmania* sp. or *T. rangeli* DNAs. The method appears highly sensitive for congenital Chagas disease and immunosuppressed patients with Chagas reactivation [30].

3 Serological Tests

These are employed for the diagnosis of all infected individuals during the chronic phase. They may be divided in those conventional, which are routinely used in the last 40 years and are all commercially available, and the more recent nonconventional, some of them not commercially available. Rapid tests are also employed in special circumstances. All these tests are designed to find IgG antibodies, which are present in large concentrations and show high affinity. Other immunoglobulins may be present as well, mainly anti-*T. cruzi* IgM, which may be useful in those acute cases when parasites are not easily found. Antibodies of IgM class may be present also in some chronic patients, so they are useful only in some cases during the acute phase [1].

3.1 Conventional Methods

These tests include indirect hemagglutination (IHA), indirect immunofluorescence (IIF), and the immunoenzymatic tests of ELISA (enzyme-linked immunosorbent assay). A large experience with all of them has been built in all endemic countries, and results are comparable in different centers. Several studies and publications allowed developing better products. Good laboratory practices should be followed, which include personnel training. Kits should be of good quality and retested with each new lot of reagents. Internal and external quality controls should be employed [1].

The World Health Organization recommends to employ at least two of these serological tests in order to avoid false results. The titer of each reaction should be included, and the possibility of errors with high titers is minimized. Each laboratory should include a table with the negative values, those on the gray region, and the positive values, which may differ from laboratory to laboratory [31].

IHA is the simplest and cheaper method, with few steps, which avoids errors. Sensitized red blood cells and serum from the patient are in contact for 1–2 h; after this time if antibodies are present, the red blood cells make a net on the bottom of the tube or well, which is visually read. If the red blood cells sediment on the bottom as a point, the reaction is negative. Serial dilutions permit to get the titer of the reaction, i.e., if reaction still occurs when the serum is diluted 1/100, this is the titer, indicating for sure the presence of antibodies. This test has a good specificity (>98%) and a reasonable sensitivity (>96%) [1].

IIF is used for serological diagnosis in many infectious diseases. It has several steps and incubations, it needs fluorescent microscopy, and the reading may be subjective and time-consuming. The main advantage is the sensitivity (>99%), but the specificity is lower (>96%) mainly at lower titers. Many diseases may yield a positive result at titers below 1/160, mainly leishmaniasis. The serum is placed in contact with the epimastigote form of the parasite, for 30 min at 37 °C. After washing, a further incubation with an anti-*T. cruzi* antiserum (mainly goat) conjugated with a fluorochrome is performed. The preparation is observed on the fluorescence microscope to look for fluorescent parasites, indicating a positive reaction. Serial dilutions are performed to obtain the final titer. Infected individuals show reactivity with dilutions of sera at the order of 1/2560. Again, good laboratory practices and reagents of recognized quality are necessary to obtain a confident result [1].

ELISA is rather similar to IIF and needs several incubations; it has a high sensitivity but lower specificity. The rationale includes the contact of serum with antigens of the parasite stick to the plastic material of a well of a microplate. After this incubation antigens are in contact with an antibody anti-*T. cruzi* conjugated with an enzyme. After washing the complex, "antigen-serum-enzyme" will react with a colorful substrate if the enzyme was not washed. The colored reaction is measured by a photocolorimeter giving a reading in optical density (OD). A scale of controls builds a figure which is the cutoff value. The OD of the sample divided by the OD of the cutoff gives a figure (index) which is considered positive if higher than 1.1. This test is more objective than IIF, and results are presented as OD or the index obtained.

For all these conventional tests, results obtained may be negative, positive, or borderline (gray zone), and two of them concordantly positive or negative assure the confidence of the results [31].

3.2 Nonconventional Methods

There are a number of recent tests based on different methodologies that have been employed for serological diagnosis of Chagas disease. The most employed one is chemiluminescence, which is commercially available and used in many blood banks (Chemiluminescent Microparticle Immuno Assay, CMIA) [32]. The sensitivity is circa 100%, but specificity may be lower, so it is essential to use this type of test together with a conventional one, mainly for ascertain diagnosis of a case. It may be used as a single test for exclusion purposes as blood banks.

Other nonconventional tests are RIPA (radioimmunoassay), which is no commercially available and used only in the United States, Western blot tests (TESA-blot), and lytic assays including flow cytometry (noncommercially available) [reviewed in [1]].

3.3 Rapid Tests

These are quick tests on the same basis than those available for other conditions as diagnosis of pregnancy, kalazar, HIV, and others. For Chagas disease a membrane is sensitized with several recombinant antigens, and a drop of serum or blood is placed in contact. After few minutes, a reaction may be seen as a band, if the test is positive. These rapid tests have a number of advantages: they may be used by any individual, may be transported to the field, and do not need special temperature conditions, and the result may be stored together with the file of the patient. Several research works have been published [33, 34] showing reasonable specificity and sensitivity of some of them. Again, they have precise indications and should not label an individual as infected unless a second, conventional test is used in parallel.

4 Other Tests

A number of tests not based in antibodies have been described, such as skin tests (delayed hypersensitivity) and the detection of circulating antigens in serum and in urine, but for different reasons they are not employed as routine tests. The search for anti-*T. cruzi* antibodies of the IgM class may be performed by IIF, when an acute case is suspected and parasites are not found. This is an "in-house" test, not commercially available that has some pitfalls as false positives when rheumatoid factor is present. Another difficulty is to have proper controls, such as sera from acute phase patients [1].

A nanoparticle assay [Chunap] has been developed for diagnosis of congenital Chagas disease in a single urine specimen at 1 month of life with more than 90% sensitivity and more than 95% specificity. The study demonstrated that poly[NIPAm] particles coupled with trypan blue dye capture and concentrate *T. cruzi* antigens in urine, and under experimental conditions these particles protect *T. cruzi* antigens in urine from enzymatic degradation [35].

5 Different Contexts for Diagnosis and Handling

5.1 *Diagnosis of Acute Phase*

Acute phase is defined by the presence of easily detected parasites. This includes concentration methods, already explained, but excludes multiplication techniques, as hemocultures, xenodiagnosis, and animal inoculation, because these may be also positive in some chronic-phase infected individuals.

After an incubation period, often of some days, symptoms may appear, and by this time, parasites are present in the circulation, where may be picked up for examination. Nevertheless, a number of cases may have only slight fever and not diagnosed. The physicians in endemic areas may suspect if another case was diagnosed at the same locality some time before.

The easiest test is the wet smear, but laboratory personnel should have some training before to recognize alive parasites. If a negative result comes out, concentration techniques already described may apply. Better results are obtained when fever is present. After some weeks, parasitemia declines and will be more difficult to find patent parasitemia [1].

A special situation is the acute phase that may emerge in seronegative recipients of organs from seropositive donors. In general the Strout method is used; molecular diagnosis may be useful to detect infection earlier; indeed PCR enabled detection of bloodstream *T. cruzi* DNA between 28 and 47 days earlier than "Strout" [36]. LAMP technology also has shown potential usefulness to follow-up these cases [30].

5.1.1 Vector Transmission

Vector transmission has been the usual mechanism of infection, with an incubation period of 7–10 days; a portal of entry may suggest the diagnosis, mainly Romaña sign, which nevertheless occurs in a minority of cases. Consequently, a large proportion of patients remain undiagnosed evolving to the chronic phase. If the parasite is difficult to find during the acute phase, these cases have frequently IgM anti-*T. cruzi*, which may help.

Direct or indirect detection of circulating parasites is the method of choice. Microscopical observation of fresh blood can reveal motile trypomastigotes. Stained blood smears allow the identification of morphological characteristics of *T. cruzi*; however these methods have only 70% sensitivity in acute infections. Concentration alternatives are employed to enhance sensitivity. Few serological methods have been developed for diagnosis of acute Chagas. An IgM-type humoral response against shed acute phase antigen (SAPA), a member of trans-sialidase family, was mostly investigated [37].

5.1.2 Oral Route and Outbreaks

Outbreaks by oral route have been recognized recently as a frequent mechanism, mainly in regions where vector transmission is under control [38–41]; a number of cases with fever, usually within the same family/school or after a social event, may indicate the presence of food-transmitted infections. As this route is very effective (is the usual way by which reservoir animals get infected), high numbers of parasites are found, and a different setup of clinical manifestations are observed, as severe digestive involvement, abdominal pain, and jaundice in some cases. Mortality is higher than by the vector route, probably due to the high numbers of parasites ingested and delay for diagnosis. Outbreaks may involve many individuals, as with sugarcane in Brazil and with guava at a school in Caracas, with more than 100 infected children [38].

Oral transmission is the most important route of infection in Brazilian Amazon and Venezuela, and reports exist from oral outbreaks in Colombia, Bolivian Amazonas, and French Guiana [38–43]. In most outbreaks molecular methods have been crucial for specific and rapid diagnosis and also for identification of the parasite genotype involved.

5.1.3 Transfusional Transmission

Transfusionally acquired infection may have a large period of incubation, whose reasons are not clearly understood and should be suspected in any case with fever after a transfusion of blood or their components. Very high parasitemias are observed, and often the diagnosis is suspected after finding flagellates in a stained smear for differential count of leukocytes. As patients have another disease that needed a transfusion, the prognosis is poor, and they may die without recognition of the superimposed infection with *T. cruzi*. Nevertheless, it should be emphasized that only 20% of the donors transmit the parasite, probably due to the paucity of bloodstream parasites, especially in chronically infected ones [1].

5.1.4 Transmission by Organ Transplantation

Transplantation involves the transmission of organs from an infected donor and the reactivation of the infected recipient. A difference between transfusion of blood from infected donors and organ transplantation from an infected donor is that parasites are always present in organs and the chance to acquire the infection increases. Furthermore, recipients of an infected organ are usually immunosuppressed, increasing the chances of severe acute phase [44].

5.1.5 Congenital Transmission

Congenital transmission is not usual (2–10% of infected mothers), but the importance of a correct diagnosis is remarkable, since all neonates detected can be easily cured after specific treatment if diagnosis is made. The presence of anti-*T. cruzi* maternal IgGs at the first months of age raises a different approach: it is recommended to search for IgGs after the ninth month of age, when maternal IgG disappears or has such low titers that assure the lack of transmission. Conversely the presence of antibodies against *T. cruzi*, at sizeable titers, at that age is a formal indication to start specific treatment. Parasitological diagnosis may be performed at birth, but a laboratory experienced personnel is necessary and available 24 h a day, which is far from the conditions usually present in endemic areas. As mentioned before, parasitological diagnosis is operator-dependent and needs a fresh sample to enable detection of motile trypomastigote forms. Furthermore, transmission may have placed during labor, and parasites will be demonstrable only 7–10 days after, when mother and baby are far away from the hospital. In these circumstances it is more feasible to look for IgG around 9 months of age, looking for antibodies with at least two conventional serological techniques [4]. At this end, it is recommended to avoid nonconventional techniques, which may yield false-positive results (mainly CMIA), as soon as more accurate strategies become validated. As explained before, at least a conventional test should be performed together with nonconventional ones to avoid misdiagnosis that may lead to treat a noninfected baby [3, 32].

Molecular methods in newborns/neonates could allow early diagnosis and bypass loss to follow-up [30, 45–48]. Noteworthy, those newborns with clinical signs present higher parasitic loads [49]. A kit prototype based on duplex TaqMan real-time PCR (qPCR) that starts from 1 mL of peripheral blood mixed with a DNA stabilizer solution has been built and validated in binomials of Chagas disease pregnant women and their newborns residing in endemic regions [50]. The accuracy of molecular detection in cord blood is still under study [50, 51].

5.1.6 Reactivation

The presence of easily detected parasites is seen in chronic-phase infected individuals that acquire AIDS or are submitted to immunosuppression for cancer, transplantation, autoimmune diseases, or other reasons [52].

As the presence of parasites is the hallmark of acute phase, this is called a reactivation of the infection and should be handled in the same way. The difference with the other forms of transmission is that the patient is already infected, so large concentrations of anti-*T. cruzi* IgGs are detectable, as well as the parasite. A proper exclusion of *T. cruzi* infection by serology, as already explained, should be mandatory in all these cases. Low CD4 cell counts in AIDS and immunosuppression with drugs for a long time and large doses, favor increase of parasitemias, and complications (unusual in other mechanisms of transmission) may occur, like panniculitis, meningoencephalitis by *T. cruzi*, and acute myocarditis, the latter two with higher mortality.

In heart transplantation, reactivation has been earlier detected and followed up by PCR methods carried out in peripheral blood and endomyocardial biopsies [44, 53, 54]. In HIV-coinfected Chagas disease patients, molecular methods are useful for differential diagnosis of meningoencephalitis due to *T. cruzi* or toxoplasmosis allowing prompt therapy decisions [55–57].

In conclusion in all these contexts of acute infection, direct parasitological tests should be performed, which may include concentration methods. The use of serology (search of antibodies of the IgM class against the parasite) may be only complimentary in those vectorial cases in which parasites were not found. The use of IgM in congenital cases has been withdrawn because most of them lack IgM. The same applies to other modes of transmission [1]. Upon validation, molecular methods will enable to close the gap of parasitological methods sensitivity and allow early and more sensitive diagnosis.

5.2 *Diagnosis of Chronic Phase*

Laboratorial diagnosis in this phase is performed with indirect tests (serological) because parasites are usually absent or in such low numbers dispersed on the blood that a chance to get some of them by venipuncture in one arm is remote. Nucleic acid amplification-based techniques are more sensitive than parasitological ones but due to the low and intermittent burden of bloodstream parasites are not accurate enough for diagnosis. On the other side, serological tests may have some pitfalls, and the use of two tests in parallel (on the same collection of blood), as WHO recommends, avoids most of the problems [31]. In the evaluation of serological results, two variables should be considered: specificity and sensitivity. There are tests with high specificity and others with high sensitivity, and the purpose of the diagnosis should be established. For diagnosis of a patient, specificity should be as higher as possible, to avoid mislabeling. When the exclusion of an infected sample is the final goal, a very sensitive test should be employed, even if some will be erroneously labeled as infected. The security of the blood is more important. What is not possible is to have nowadays a single test with 100% sensitivity and 100% specificity [1].

Cross-reactions, mainly with leishmaniasis, may be seen. The concentration of antibodies present (the titer or optical density) usually helps, because infected individuals usually have high titers and indexes. As emphasized before, the epidemiology and clinical context are very important for the interpretation of the laboratory result obtained. This is particularly important in cases of visceral leishmaniasis (kalazar) where patients have severe compromise of several organs and fevers. Because in kalazar there is a strong B-cell response (polyclonal activation), with a sizeable increase in gamma globulin, antibodies of different specificities favor cross-reactions with many infections, among them, Chagas disease. A chronic case of *T. cruzi* infection will not have fever, nor hepatic or spleen enlargement, or blood alterations, often seen in kalazar. Conventional serology usually gives false-positive results in kalazar cases, and the clinician should have this in mind [1].

Serological diagnosis may be applied with different aims and contexts. This implies a selection of tests and procedures. Some situations will be briefly described, as follows.

5.2.1 Clinical Diagnosis in a Patient

This is the common situation: the physician suspects of Chagas disease in a patient and needs to confirm the suspicion, by laboratory confirmation. According to WHO recommendation, diagnosis should be based on the positivity of at least two of the tests mentioned above [30]. Tests used should be of good specificity and ideally with high titers. To diagnose a patient as infected based on a single recombinant test may lead to a false-positive result with even legal consequences.

Demonstration of the parasite in the blood may be performed by xenodiagnosis, with the classic method (four boxes with ten parasites in each) or the artificial method; the latter has several advantages. Hemoculture shows a positivity of no more than 50% of cases. Positivity of these techniques may be increased when the examination is performed two or more times.

Molecular methods lack sensitivity at the chronic phase; different PCR strategies were evaluated, and a clinical sensitivity of around 70% was achieved when only one peripheral blood sample was analyzed. Serial sampling allowed increment of PCR clinical sensitivity [58]. These approaches could be useful in those cases with dubious results on serology if they give detectable results.

5.2.2 Exclusion on a Blood Bank

The final purpose of a blood bank is to offer a safe product, without infectious material. In this case, a test with 100% sensitivity is desirable. Such a test may be ELISA or even nonconventional commercially available, as CMIA. Provided that an external quality control exists, it is possible to use a single test, because the purpose is to exclude any donor whose serum gives signals in the system, even at a low titer or in the gray zone [1].

5.2.3 Epidemiological Survey

The purpose in surveys is to know if the infection is present in a depicted area. For operational reasons, hundreds of samples may be obtained in a single day. The rationale is to use filter paper or rapid tests, avoiding the time consumed between blood extraction, separation of sera, and labeling. Tests to be used should have high sensitivity. Those few that gave a positive result may be sorted out in a second visit which will involve only a low number of collections of blood [59].

5.2.4 Vertical Transmission

The first step is to confirm the positive serology of the mother, as for clinical diagnosis. Once confirmed, the offspring may be tested with parasitological tests at birth as already described. Wet smear, microhematocrit, or PCR may detect the parasite in the blood of the newborn, who should be immediately submitted to specific treatment. Nevertheless, for reasons already explained, this approach may not be available. If this is the case, all newborns from confirmed infected mothers should be investigated later. A good possibility is to perform PCR any time after delivery. If this is not feasible, all of them should have a collection of blood after their 8 month of life, looking for IgGs, as for clinical diagnosis.

5.2.5 Follow-Up After Specific Treatment

This is a different situation. Infected individuals that were submitted to trypanocidal drugs (benznidazole or nifurtimox; see corresponding chapter) need to be followed up for a period of time, in order to know if antibodies disappear or titers are coming down. This is attained in months for neonates or patients treated during the acute phase but demands some years for those children treated during the recent chronic phase. For those in late chronic phase, a switch may be observed after some decades. Cure is obtained when no more antibodies are present, and failure is an outcome better investigated through parasitological and/or molecular tests, when they persist or become positive after treatment completion. As complete disappearance of antibodies may take time, these cases should be investigated by as many tests as possible, of the conventional type, and three are desirable [60].

Parasitological tests have limited sensitivity, and accordingly a negative result does not necessarily mean the absence of parasites, but in contrast, a positive result indicates treatment failure. A highly sensitive and reliable method to assure cure is urgently needed. Molecular methods are useful tools for treatment monitoring [61–65]. Blood-based qPCR techniques are being consistently used to detect therapeutic response or failure in clinical trials with traditional and novel drugs, which were administered with different regimens and combinations [59, 65–69]. Most trials in chronic CD have shown a lower efficacy of ravuconazole and posaconazole in comparison to benznidazole [59, 66, 69]. However, clearance of parasitic loads exerted by drugs can be transient and lead to misleading conclusions when follow-up is performed at the short term. Ideally, molecular methods used for monitoring chronic patients should be performed for several years after treatment to confirm or discard available data. Recent findings of dormant amastigote subpopulations, refractory to benznidazole action, may represent a key factor leading to treatment failure, which deserves further investigation [70, 71].

6 Concluding Remarks

Laboratorial diagnosis of Chagas disease is well established. For acute phase, direct parasitological tests should be employed, which are rather simple but need skilled technicians to perform them. For chronic phase, search of IgG anti-*T. cruzi* is easily performed through several conventional tests. The purpose of the diagnosis may delimitate the type of tests to be used. The use of two techniques in parallel is necessary to ascertain a positive result or exclude the infection in a patient. In order to obtain a good performance, kits of recognized quality and good laboratory practices are necessary. To fulfill these needs, internal and external quality control are imperative.

Standardized and validated, in-house and commercial PCR and LAMP methodologies have been a research priority to improve current Chagas disease diagnosis [71], especially in the following scenarios: early diagnosis of congenital infection, oral outbreaks, reactivation follow-up due to immunosuppression, and treatment response monitoring [21, 25, 29, 72]. Target product profiles (TPPs) of molecular strategies for diagnosis of *T. cruzi* infection have been addressed, pointing to the need of developing point-of-care assays [61, 73]. Their evaluation in field studies is needed to predict their usefulness in the clinical practice and for public health applications.

References

1. Luquetti AO, Schmuñis GA. Diagnosis of *Trypanosoma cruzi* infection. Chapter 29. In: Telleria J, Tibayrenc M, editors. American Trypanosomiasis. Chagas Disease. One hundred years of research. 2nd ed. Amsterdam: Elsevier; 2017.
2. Luquetti AO, Rassi A. Diagnóstico Laboratorial da Infecção pelo *Trypanosoma cruzi*. In: Brener Z, Andrade AZ, Barral-Neto M, editors. *Trypanosoma cruzi* e Doença de Chagas. 2nd ed. Rio de Janeiro: Guanabara Koogan; 2000. p. 344–78.
3. Carlier Y, Torrico F, Sosa-Estani S, Russomando G, Luquetti AO, Freilij H, Vinas PA. Congenital Chagas disease: recommendations for diagnosis, treatment and control of newborns, siblings and pregnant women. PLoS Negl Trop Dis. 2011;5:e1250.
4. Luquetti AO, Ferreira MS. Diagnóstico da doença de Chagas na coinfecção *T. cruzi*/HIV. In: Almeida EA, editor. Epidemiologia e clínica da coinfecção *Trypanosoma cruzi*/HIV. Campinas: Editora Universidade Estadual de Campinas; 2015.
5. Schmunis GA, Yadon ZE. Chagas disease: a Latin American health problem becoming a world health problem. Acta Trop. 2010;115:14–21.
6. Zingales B, Miles MA, Campbell DA, Tibayrenc M, Macedo AM, Teixeira MM, Schijman AG, Llewellyn MS, Lages-Silva E, Machado CR, Andrade SG, Sturm NR. The revised *Trypanosoma cruzi* subspecific nomenclature: rationale, epidemiological relevance and research applications. Infect Genet Evol. 2012;12(2):240–53. https://doi.org/10.1016/j.meegid.2011.12.009. Review.

7. Yun O, Lima MA, Ellman T, Chambi W, Castilho S, Flevaud L, Roddy P, Parreño F, Viñas PA, Palma PP. Feasibility, drug safety, and effectiveness of etiological treatment programs for Chagas disease in Honduras, Guatemala, and Bolívia: 10-year experience of Mèdecins Sans Frontières. PLoS Negl Trop Dis. 2009;3(e):488. https://doi.org/10.1371/journal.pntd.0000488.
8. Rassi A, Rezende JM, Luquetti AO, Rassi A Jr. Clinical phases and forms of chagas disease. Chapter 28. In: Telleria J, Tibayrenc M, editors. American Trypanosomiasis. Chagas Disease. One hundred years of research. 2nd ed. Amsterdam: Elsevier; 2017.
9. Luquetti AO, Tavares SB, Siriano L da R, de Oliveira RA, Campos DE, de Morais CA, de Oliveira EC. Congenital transmission of *Trypanosoma cruzi* in central Brazil. A study of 1,211 individuals born to infected mothers. Mem Inst Oswaldo Cruz. 2015;110:369–76.
10. Strout RG. A method for concentrating hemoflagellates. J. Parasit. 1962;48:100.
11. Freilij H, Altcheh J. Chagas congénito. In: Storino R, Milei J, editors. Enfermedad de Chagas. Buenos Aires: Edit. Doyma; 1994.
12. Castro AM, Luquetti AO, Rassi A, Chiari E, Galvão LMC. Detection of parasitemia profiles by blood culture after treatment of human chronic *Trypanosoma cruzi* infection. Parasitology Research. 2006;99:379–83.
13. Santos AH, Silva IG, Rassi A. Estudo comparativo entre o xenodiagnóstico natural e o artificial em chagásicos crônicos. Rev Soc Bras Med Trop. 1995;28:367–73.
14. Oliveira EC, Stefani MMA, Luquetti AO, Vencio EF, Moreira MAR, Souza C, Rezende JM. *Trypanosoma cruzi* and experimental Chagas disease: characterization of a stock isolated from a patient with associated digestive and cardiac form. Rev Soc Bras Med Trop. 1993;26: 25–33.
15. Avila HA, Pereira JB, Thiemann O, et al. Detection of *Trypanosoma cruzi* in blood specimens of chronic chagasic patients by polymerase chain reaction amplification of kinetoplast minicircle DNA: comparison with serology and xenodiagnosis. J Clin Microbiol. 1993;31:2421–6.
16. Britto C, Cardoso MA, Vanni CM, et al. Polymerase chain reaction detection of *Trypanosoma cruzi* in human blood samples as a tool for diagnosis and treatment evaluation. Parasitology. 1995;110(Pt 3):241–7.
17. Moser DR, Kirchhoff LV, Donelson JE. Detection of *Trypanosoma cruzi* by DNA amplification using the polymerase chain reaction. J Clin Microbiol. 1989;27(7):1477–82.
18. Piron M, Fisa R, Casamitjana N, et al. Development of a real-time PCR assay for Trypanosoma cruzi detection in blood samples. Acta Tropica. 2007;103:195–200.
19. Duffy T, Cura CI, Ramirez JC, et al. Analytical performance of a multiplex real-time PCR assay using TaqMan probes for quantification of *Trypanosoma cruzi* satellite DNA in blood samples. PLoS Negl Trop Dis. 2013;7(1):e2000.
20. Ramírez JC, Cura CI, da Cruz Moreira O, et al. Analytical validation of quantitative real-time PCR methods for quantification of Trypanosoma cruzi DNA in blood samples from Chagas disease patients. J Mol Diagn. 2015;17(5):605–15.
21. Schijman AG, Bisio M, Orellana L, et al. International study to evaluate PCR methods for detection of *Trypanosoma cruzi* DNA in blood samples from Chagas disease patients. PLoS Negl Trop Dis. 2011;5(1):e93.
22. Ramírez JC, Torres C, Curto MLA, Schijman AG. New insights into *Trypanosoma cruzi* evolution, genotyping and molecular diagnostics from satellite DNA sequence analysis. PLoS Negl Trop Dis. 2017;11(12):e0006139.
23. Guhl F, Vallejo GA. *Trypanosoma* (Herpetosoma) *rangeli* Tejera, 1920: an updated review. Mem Inst Oswaldo Cruz. 2003;98:435–42.
24. Saldana A, Samudio F, Miranda A, et al. Predominance of *Trypanosoma rangeli* infection in children from a Chagas disease endemic area in the west-shore of the Panama canal. Mem Inst Oswaldo Cruz. 2005;100:729–31.
25. Ramírez JC, Parrado R, Sulleiro E, de la Barra A, Rodriguez M, Villarroel S, Irazu L, Alonso-Vega C, Alves F, Curto M, Garcia L, Ortiz L, Torrico F, Gascon J, Flevaud L, Molina I, Ribeiro I, Schijman AG. First External quality assurance program for real-time PCR monitoring of treatment response in clinical trials of Chagas disease. PLoS One. 2017;12(11):e0188550. https://doi.org/10.1371/journal.pone.0188550.

26. Mori Y, Nagamine K, Tomita N, et al. Detection of loop-mediated isothermal amplification reaction by turbidity derived from magnesium pyrophosphate formation. Biochem Biophys Res Commun. 2001;289:150–4.
27. Notomi T, Okayama H, Masubuchi H, et al. Loop-mediated isothermal amplification of DNA. Nucleic Acids Res 28: p. 2000;E63:266.
28. Thekisoe OM, Rodriguez CV, Rivas F, Coronel-Servian AM, Fukumoto S, Sugimoto C, et al. Detection of *Trypanosoma cruzi* and *T. rangeli* infections from *Rhodnius pallescens* bugs by loop-mediated isothermal amplification (LAMP). Am J Trop Med Hyg. 2010;82:855–60. https://doi.org/10.4269/ajtmh.2010.09-0533.
29. Rivero R, Bisio M, Velázquez EB, Esteva MI, Scollo K, González NL, Altcheh J, Ruiz AM. Rapid detection of *Trypanosoma cruzi* by colorimetric loop-mediated isothermal amplification (LAMP): a potential novel tool for the detection of congenital Chagas infection. Diagn Microbiol Infect Dis. 2017;89:26. https://doi.org/10.1016/j.diagmicrobio.2017.06.012. pii: S0732-8893(17)30189-X.
30. Besuschio SA, Llano Murcia M, Benatar AF, Monnerat S, Cruz I, Picado A, Curto MLÁ, Kubota Y, Wehrendt DP, Pavia P, Mori Y, Puerta C, Ndung'u JM, Schijman AG. Analytical sensitivity and specificity of a loop-mediated isothermal amplification (LAMP) kit prototype for detection of *Trypanosoma cruzi* DNA in human blood samples. PLoS Negl Trop Dis. 2017;11:e0005779.
31. WHO. WHO technical report series no. 905. Control of Chagas Disease. Second report of the WHO Expert Committee. Geneva: World Health Organization; 2002.
32. Abras A, Gállego M, Llovet T, Tebar S, Herrero M, et al. Serological diagnosis of chronic Chagas disease: is it time for a change? J Clin Microbiol. 2016;54:1566–72.
33. Luquetti AO, Ponce C, Ponce E, Esfandiari J, Schijman A, Revollo S, Añez N, Zingales B, Aldao RR, Gonzalez A, Levin M, Umezawa E, Franco da Silveira J. Chagas disease diagnosis: a multicentric evaluation of Chagas Stat-Pak, a rapid immunochromatographic assay with recombinant proteins of *Trypanosoma cruzi*. Journal of Diagnostic Microbiology and Infectious disease. 2003;46:265–71.
34. Sanchez-Camargo CL, Albajar-Vinas P, Wilkins PP, Nieto J, Leiby DA, Paris L, et al. Comparative evaluation of 11 commercialized rapid diagnostic tests for detecting *Trypanosoma cruzi* antibodies in serum banks in areas of endemicity and nonendemicity. J Clin Microbiol. 2014;52(7):2506–12.
35. Castro-Sesquen YE, Gilman RH, Galdos-Cardenas G, et al. Use of a novel Chagas urine nanoparticle test (chunap) for diagnosis of congenital Chagas disease. PLoS Negl Trop Dis. 2014;8(10):e3211.
36. Cura CI, Lattes R, Nagel C, et al. Early molecular diagnosis of acute Chagas disease after transplantation with organs from *Trypanosoma cruzi*-infected donors. Am J Transplant. 2013;13:3253–61.
37. Affranchino JL, Ibanez CF, Luquetti AO, Rassi A, Reyes MB, Macina RA, Aslund L, Pettersson U, ACC F. Identification of a *Trypanosoma cruzi* antigen that is shed during the acute phase of Chagas' disease. Mol. Bioch. Parasitol. 1989;34:221–8.
38. Alarcon de Noya B, Diaz-Bello Z, Colmenares C, Ruiz-Guevara R, Mauriello L, Zavala-Jaspe R, et al. Large urban outbreak of orally acquired acute Chagas disease at a school in Caracas, Venezuela. J Infect Dis. 2010;201:1308 1315.
39. Shikanai-Yasuda MA, Carvalho NB. Oral transmission of Chagas disease. Clin Infect Dis. 2012;54:845–52. https://doi.org/10.1093/cid/cir956.
40. Silva-Dos-Santos D, Barreto-de-Albuquerque J, Guerra B, Moreira OC, Berbert LR, Ramos MT, Mascarenhas BAS, Britto C, Morrot A, Serra Villa-Verde DM, Garzoni LR, Savino W, Cotta-de-Almeida V, Meis J. Unraveling Chagas disease transmission through the oral route: gateways to *T. cruzi* infection and target tissues. PLoS Negl Trop Dis. 2017;11(4):e0005507. https://doi.org/10.1371/journal.pntd.0005507.
41. Noya BA, Díaz-Bello Z, Colmenares C, Ruiz-Guevara R, Mauriello L, Muñoz-Calderón A, Noya O. Update on oral Chagas disease outbreaks in Venezuela: epidemiological, clinical and diagnostic approaches. Mem Inst Oswaldo Cruz. 2015;110:3786. https://doi.org/10.1590/0074-02760140285. Review.

42. Ramírez JD, Montilla M, Cucunubá ZM, Floréz AC, Zambrano P, Guhl F. Molecular epidemiology of human oral Chagas disease outbreaks in Colombia. PLoS Negl Trop Dis. 2013;7:e2041. https://doi.org/10.1371/journal.pntd.0002041.
43. Blanchet D, Breniere SF, Schijman AG, et al. First report of a family outbreak of Chagas disease in French Guiana and posttreatment follow-up. Infect Genet Evol. 2014;28:245–50.
44. Diez M, Favaloro L, Bertolotti A, Burgos JM, Vigliano C, Lastra MP, Levin MJ, Arnedo A, Nagel C, Schijman AG, Favaloro RR. Usefulness of PCR strategies for early diagnosis of Chagas' disease reactivation and treatment follow-up in heart transplantation. Am J Transplant. 2007;7:1633–40.
45. Cura CI, Ramírez JC, Rodríguez M, Lopez-Albízu C, Irazu L, Scollo K, Sosa-Estani S. Comparative study and analytical verification of PCR methods for the diagnosis of congenital Chagas Disease. J Mol Diagn. 2017;19:673. pii: S1525-1578(17)30108-3.
46. Bua J, Volta BJ, Perrone AE, et al. How to improve the early diagnosis of *Trypanosoma cruzi* infection: relationship between validated conventional diagnosis and quantitative DNA amplification in congenitally infected children. PLoS Negl Trop Dis. 2013;7(10):e2476.
47. Schijman AG, Altcheh J, Burgos JM, et al. Aetiological treatment of congenital Chagas' disease diagnosed and monitored by the polymerase chain reaction. J Antimicrob Chemother. 2003;52(3):441–9.
48. Mora MC, Sanchez-Negrette O, Marco D, et al. Early diagnosis of congenital *Trypanosoma cruzi* infection using PCR, hemoculture, and capillary concentration, as compared with delayed serology. J Parasitol. 2005;91:1468–73.
49. Messenger LA, Gilman RH, Verastegui M, Galdos-Cardenas G, Sanchez G, Valencia E, Sanchez L, Malaga E, Rendell VR, Jois M, Shah V, Santos N, Del Carmen Abastoflor M, LaFuente C, Colanzi R, Bozo R, Bern C, Working Group on Chagas disease in Bolivia and Peru. Towards improving early diagnosis of congenital Chagas disease in an endemic setting. Clin Infect Dis. 2017;65:268. https://doi.org/10.1093/cid/cix277.
50. Benatar AF, Besuschio SA, Bortolotti S, Ramirez JC, Cafferata ML, Danesi E, Lopez Albizu C, Ciganda A, Lara L, Agolti G, Seu S, Uequìn V, Curet L, Adamo EL, Black F, Lucero H, Esteva M, Bua J, Longhi C, MdeA S, Poeylaut-Palena A, Scollo K, Althabe F, Capriotti G, Rojkin F, Sosa Estani S, Schijman AG. Validation of a real time PCR kit prototype for early diagnosis of congenital Chagas disease in a multicenter field study. Medicina. 2017;77(Suppl I):400.
51. Basombrío MA, Nasser J, Segura MA, Marco D, Sánchez Negrette O, Padilla M, Mora MC. The transmission of Chagas disease in Salta and the detection of congenital cases. Medicina (B Aires). 1999;59(Suppl 2):143–6. Spanish.
52. Bern C. Chagas disease in the immunosuppressed host. Curr Opin Infect Dis. 2012;25:450–7.
53. Burgos JM, Diez M, Vigliano C, et al. Molecular identification of *Trypanosoma cruzi* discrete typing units in end-stage chronic Chagas heart disease and reactivation after heart transplantation. Clin Infect Dis. 2010;51:485–95.
54. da Costa PA, Segatto M, Durso DF, de Carvalho Moreira WJ, Junqueira LL, de Castilho FM, de Andrade SA, Gelape CL, Chiari E, Teixeira-Carvalho A, Junho Pena SD, Machado CR, Franco GR, Filho GB, Vieira Moreira MDC, Mara Macedo A. Early polymerase chain reaction detection of Chagas disease reactivation in heart transplant patients. J Heart Lung Transplant. 2017;36:797–805.
55. Burgos JM, Begher SB, Freitas JM, Bisio M, Duffy T, Altcheh J, Teijeiro R, Lopez Alcoba H, Deccarlini F, Freilij H, Levin MJ, Levalle J, Macedo AM, Schijman AG. Molecular diagnosis and typing of Trypanosoma cruzi populations and lineages in cerebral Chagas disease in a patient with AIDS. Am J Trop Med Hyg. 2005;73:1016–8.
56. Perez-Molina JA, Rodriguez-Guardado A, Soriano A, et al. Guidelines on the treatment of chronic coinfection by *Trypanosoma cruzi* and HIV outside endemic areas. HIV Clin Trials. 2011;12:287–98.
57. Almeida EA, Ramos-Junior AN, Correia D, et al. Co-infection *Trypanosoma cruzi*/HIV: systematic review (1980-2010). Rev Soc Bras Med Trop. 2011;44:762–70.
58. Torrico F, Gascon J, Lourdes O, Cristina A-V, María-Jesús P, Alejandro S, Almeida Igor C, Fabiana A, Nathalie S-W, Isabela R, on behalf of the E1224 Study Group. Treatment of adult

chronic indeterminate Chagas disease with benznidazole and three E1224 dosing regimens: a proof-of-concept, randomised, placebo-controlled trial. The Lancet Infectious Diseases. 2018;18:419.
59. Luquetti AO, Passos ADC, Silveira AC, Ferreira AW, Macedo V, Prata AR. O inquérito nacional de soroprevalência de avaliação do controle da doença de Chagas no Brasil (2001-2008). Rev. Soc Brasileira Medicina Tropical. 2011;44(Suppl 2):108–21.
60. Rassi A, Luquetti AO. Capítulo 53: Critérios de Cura da Infecção pelo *Trypanosoma cruzi* na Espécie Humana. In: Coura JR, editor. Dinâmica das doenças infecciosas e parasitárias, vol. 1. 2nd ed. Rio de Janeiro: Guanabra Koogan; 2013. p. 729–35.
61. Pinazo MJ, Thomas MC, Bua J, et al. Biological markers for evaluating therapeutic efficacy in Chagas disease, a systematic review. Expert Rev Anti Infect Ther. 2014;12:479–96.
62. Murcia L, Carrilero B, Muñoz MJ, et al. Usefulness of PCR for monitoring benznidazole response in patients with chronic Chagas' disease: a prospective study in a non-disease-endemic country. J Antimicrob Chemother. 2010;65:1759–64.
63. Viotti R, Alarcon de Noya B, Araujo-Jorge T, et al. Towards a paradigm shift in the treatment of chronic Chagas disease. Antimicrob Agents Chemother. 2014;58(2):635–9.
64. Moreira OC, Ramírez JD, Velázquez E, Melo MF, Lima-Ferreira C, Guhl F, Sosa-Estani S, Marin-Neto JA, Morillo CA, Britto C. Towards the establishment of a consensus real-time qPCR to monitor Trypanosoma cruzi parasitemia in patients with chronic Chagas disease cardiomyopathy: a substudy from the BENEFIT trial. Acta Trop. 2013;125:23–31.
65. Alonso-Padilla J, Gallego M, Schijman AG, Gascon J. Molecular diagnostics for Chagas disease: up to date and novel methodologies. Expert Rev Mol Diagn. 2017;17:699–710.
66. Molina I, Gomez i Prat J, Salvador F, et al. Randomized trial of posaconazole and benznidazole for chronic Chagas' disease. N Engl J Med. 2014;370(20):1899–908.
67. Morillo CA, Marin-Neto JA, Avezum A, et al. Randomized trial of benznidazole for chronic Chagas' cardiomyopathy. N Engl J Med. 2015;373:1295–306.
68. Álvarez MG, Vigliano C, Lococo B, Bertocchi G, Viotti R. Prevention of congenital Chagas disease by benznidazole pre-treatment in reproductive-age women. An observational study. Acta Trop. 2017;174:149. https://doi.org/10.1016/j.actatropica.2017.07.004. pii: S0001-706X(16)30750-1.
69. Morillo CA, Waskin H, Sosa-Estani S, Del Carmen Bangher M, Cuneo C, Milesi R, Mallagray M, Apt W, Beloscar J, Gascon J, Molina I, Echeverria LE, Colombo H, Perez-Molina JA, Wyss F, Meeks B, Bonilla LR, Gao P, Wei B, McCarthy M, Yusuf S, STOP-CHAGAS Investigators. Benznidazole and posaconazole in eliminating parasites in asymptomatic T. cruzi carriers: the STOP-CHAGAS trial. J Am Coll Cardiol. 2017;69:939–47. https://doi.org/10.1016/j.jacc.2016.12.023.
70. Valdez F, Padilla A, Tarleton R. Identification of a rare dormant sub-population of *Trypanosoma cruzi* amastigotes able to reassume proliferation, infection and generation of new quiescent forms. Medicina. 2017;77(Suppl I):26.1.
71. WHO. Research priorities for Chagas disease, human African trypanosomiasis and Leishmaniasis. World Health Organization technical report series. Geneva: World Health Organization/Special Programme for Research and Training in Tropical Diseases (TDR); 2012. (975):v-xii, 1-100.
72. Schijman AG, Burgos JM, Marcet P. Molecular tools and strategies for diagnosis of Chagas Disease and leishmaniasis, Chapter 9. In: da Silva S, Cano MI, editors. Molecular and cellular biology of pathogenic trypanosomatids; Frontiers in parasitology, vol. 1. Sharjah: Bentham Science Publishers; 2016. p. 394–452.
73. Porras AI, Yadon ZE, Altcheh J, et al. Target product profile (TPP) for Chagas disease point-of-care diagnosis and assessment of response to treatment. PLoS Negl Trop Dis. 2015;9:e0003697.

Part V
Clinical Aspects

Acute Vector-Borne Chagas Disease

Guillermo Moscatelli and Samanta Moroni

Abstract It has been estimated that there are between six and seven million people in the world infected with *Trypanosoma cruzi*, the parasite that causes Chagas disease, most of them in Latin America. The vector-borne transmission is produced through insects of the subfamily Triatominae carrying the parasite. The most important species in the Southern Cone of the Americas is *Triatoma infestans*. Currently, this route of infection is observed in the Americas.

The treatment produces the cure if the infection is recent. During the chronic phase of the disease, an antiparasitic treatment can slow down or prevent the progression of the disease.

The most useful method to prevent Chagas disease in Latin America is the control of the vector insects.

1 Introduction

The Brazilian physician Dr. Carlos Ribeiro Justiniano das Chagas was the medical researcher to first discover the disease in 1909 when he described the sign-symptomatology of a girl, documenting in this way the first case of acute vector-borne Chagas disease.

In 1907 Dr. Carlos Chagas had been recommended to participate in an anti-malaria campaign with the railroad workers in Belo Horizonte, in the north of the state of Minas Gerais, Brazil. There, he observed hematophagous insects that the natives called "barbaeiros" or "chupoes" that hid within the cracks of adobe or mud

G. Moscatelli (✉) · S. Moroni
Servicio de Parasitología y Enfermedad de Chagas, Hospital de Niños "Ricardo Gutiérrez", Buenos Aires, Argentina

Instituto Multidisciplinario de Investigación en Patologías Pediátricas (IMIPP-GCBA), Consejo Nacional de Investigaciones Científicas y Técnicas (CONICET), Buenos Aires, Argentina

J. M. Altcheh, H. Freilij (eds.), *Chagas Disease*, Birkhäuser Advances in Infectious Diseases, https://doi.org/10.1007/978-3-030-00054-7_8

huts. These bugs fed on the blood of humans and domestic animals. Driven by his curiosity, Dr. Chagas analyzed the intestinal content of these hematophagous insects and found flagellated parasites. Later he sent these to Dr. Osvaldo Cruz who let the insects fed on monkeys of the genus *Callitrix*. Twenty to 30 days after the biting, large quantities of protozoans of the genus *Trypanosoma* were found, which Dr. Carlos Chagas named *Trypanosoma cruzi* in honor to his mentor Dr. Osvaldo Cruz. This was the first of a series of discoveries: the disease on humans, its causal agent, and the transmission vector [1].

Following this discovery, the search for this parasite in blood tests among individuals living in the houses where the hematophagous bugs were found was initiated. During this search, Dr. Chagas was called to analyze the blood of a girl in serious condition with fever that was previously healthy. The main symptoms in this girl, besides the fever of up to 40 °C, were spleen and liver enlargement and groups of peripheral lymph nodes with augmented size. Her face was swollen, with edema similar to myxedema. The microscopic examination revealed the presence of a large number of parasites, identified as *Trypanosoma cruzi*. In this way, the existence of a new trypanosomiasis in humans was proved [2].

During the time of its discovery, Chagas disease was wrongly identified as the cause of goiter and endemic cretinism. This error led to many years of indifference from politicians, physicians, and researchers. Later, when the disease could be studied in areas free of other parasitosis and deficiencies, it was recognized as its own entity and entered the knowledge of physicians working on endemic regions, with them being able to diagnose Chagas disease without the need of lab tests. It also became recognized among populations from areas of Argentina and Brazil, which were familiarized with some ocular symptoms particularly visible [2].

This revolution of new knowledge led to an intense activity around this new parasitosis: Vianna discovered fundamental lesions found in the pathological anatomy; Guerreiro and Machado successfully essayed the serological technique of complement fixation for the diagnosis of chronic cases; and in 1924 Petrocchi and Zuccarini discovered in northern Argentina, in Tucumán and Salta provinces, the first two acute cases when testing blood samples from individuals suspected of suffering paludism. In 1932, Romaña point out the first two cases of chronic chagasic myocarditis in Chaco, Argentina. Also in Argentina, Dr. Salvador Mazza dedicated his life to the study of the disease providing care for many patients and completing the research initiated by Dr. Chagas [1].

2 Introduction: Worldwide Magnitude of the Problem

Chagas disease is produced by the flagellated parasite *Trypanosoma cruzi*. The means of transmission are:

- Vector-borne (through the contact with the vector's feces)
- Transplacental

- Through blood transfusion
- Organ transplantation
- Oral, through the ingestion of food contaminated with the vector's feces

The last two being the least frequent.

The increasing number of migrants from rural to urban areas occurring in Latin America in the last years changed the traditional epidemiological pattern; the Chagas disease became urban. It is for this reason that it affects the entire American continent, spreading to countries from Europe and Asia [3].

The vector-borne disease is transmitted by insects of the order Hemiptera, subfamily Triatominae (kissing bugs or assassin bugs, also named vinchucas, chinches, or barbaeiros in Latin America), which are capable of colonizing rural, suburban, or urban substandard housings. The most important species in the Southern Cone of South America is *Triatoma infestans*. The *T. cruzi* parasite enters the digestive tube of the insect when it bites an infected person or mammal. The parasite actively divides on the intestine of the insect, giving origin to the infective stages that are transmitted through the insect's feces that are deposited on the person or mammal's skin, while the insect sucks blood a few millimeters away from the bite. The parasite eliminated with the insect's feces can invade the definitive host tissues multiplying within its cells and latter being released to its blood flow [4]. When feeding, the triatomine bugs ingest circulating parasites perpetuating in this manner their evolu-

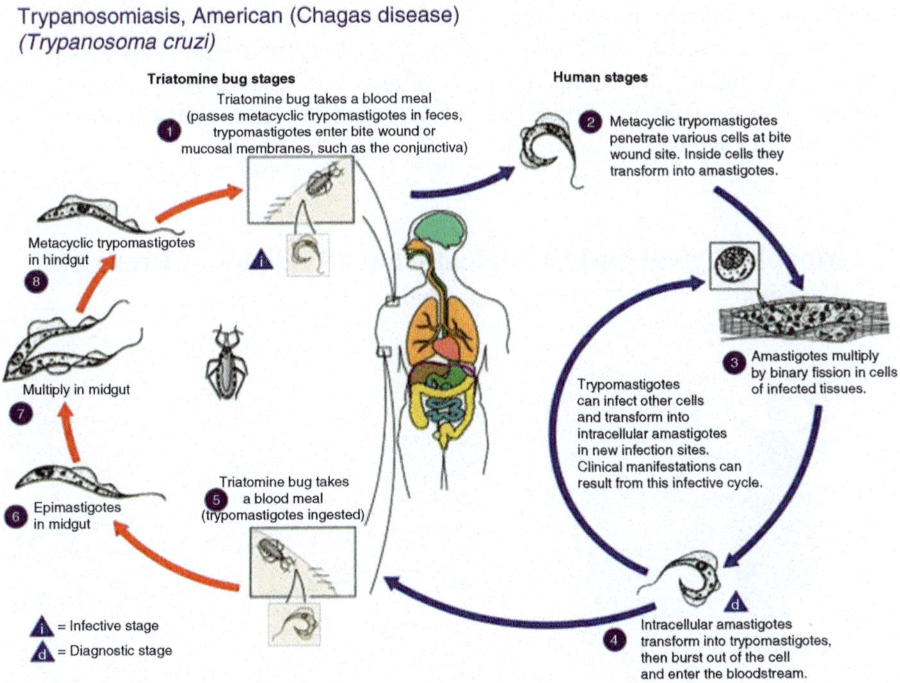

Fig. 1 *Trypanosoma cruzi*'s life cycle. Source: Centers for Disease Control and Prevention (CDC)

tionary cycle (Fig. 1). This route of transmission, called vector-borne, can occur in the distribution area of triatomine bugs within the Americas comprehended from the limit between Mexico and the United States south to Chile and Argentina.

Even though the infection acquired through vector-borne transmission can present itself at any age, the greatest risk occurs within children less than 10 years old. In areas non-treated with insecticides, the largest incidence of the disease is registered before the age of 14 years old. The acute phase of the infection acquired through vector-borne transmission can last between 2 and 4 months [4]. The triatomine bugs can infect rodents, marsupials, and other wild mammals. These triatomine bugs can also infect domestic animals such as dogs and cats and carry the *T. cruzi* parasite (causal agent of the disease) into human housing.

Generally, there is a greater incidence of acute cases during summer season coinciding with an increase biological activity in the kissing bugs. It is of great value to investigate the presence of triatomine bugs within the patient's house or peridomiciliary area.

3 Vector Insects

Even though there are around 130 species of triatomine bugs and more than half of them has been shown to be either naturally or artificially infected with *T. cruzi*, the epidemiologically important species for humans are less than ten since these are capable of colonizing households and they tend to feed from people. The most important insect vectors are *Triatoma infestans* in Argentina, Bolivia, Brazil, Chile, Paraguay, Uruguay, and Peru; *Rhodnius prolixus* in Colombia, Venezuela, and Central America; *Triatoma dimidiata* in Ecuador and Central America; and *Rhodnius pallescens* in Panama [5].

4 Morphological and Biological Characteristics of Prevalent Vectors

Triatoma infestans.

Triatoma infestans: Adults are 25–30 mm long, colored brown to blackish, dark abdominal connexivum with transversal light yellowish spots, and dark legs, except for the trochanter and the adjacent basal femur area which are yellow. It is distributed mostly in warm and dry regions where they can produce up to two generations per year if they dispose of a continuous food source, such as those of households and peridomestic areas.

It is a mostly domestic species; its habitat is restricted to ecotypes either created or modified by humans. It hides in thatch roofs and cracks on mud walls or unplastered or poorly plastered brick walls, behind furniture, beds, boxes, etc. Within peridomestic areas it is found in henhouses, corrals, pigeon lofts, warehouses, firewood, etc.

Rhodnius prolixus

Rhodnius prolixus: They possess elongated heads with the antennae inserted in the anterior part of it, near the clypeus, and predominant eyes. They are pale brown to yellowish spotted dark brown. It is a first-order Chagas disease transmitter given its high vector-borne capacity and its population dynamics that led it to develop numerous colonies that invade human households. This species was accidentally introduced to Central America in the first decades of the XX century from Northern South America (Venezuela and Colombia); it can potentially be eliminated through its control or as a consequence of positive transformations in the quality of life and in the structure of rural and suburban housings [5].

Triatoma dimidiata

Triatoma dimidiata: Its antennae are inserted in the central area of the head; it possesses dark banded connexivum and black scutellum. It is a jungle species that lives in birds' nests, mammal's dens, caves, and holes and also lives under rocks, tree roots, and leaves from different palm trees [6]. In domestic and peridomestic areas, it can be found in mud walls' cracks, within firewood or wood storage, in areas for breeding or resting of domestic or farm animals, and behind objects on the walls, roofs, furniture, beds and wherever clothes accumulate, boxes, or sacs. They are attracted by light.

Rhodnius pallescens

Rhodnius pallescens: It presents a great capacity of adaptation to different habitats and environmental conditions near or inside households, and therefore its presence constitutes a contact risk factor with *T. cruzi* for humans, either directly or through contaminated food. In general, its coloration is yellowish-brown with dark brown spotting. Its head is elongated, with the area before the eyes being three times larger than the postocular area when seen from a dorsal perspective. The antennae are inserted in the frontal area near the clypeus.

5 Epidemiological Situation in Latin America

This disease is considered within the group of neglected diseases or diseases of poverty. It is endemic in 21 countries of the region. About 65 million people in the Americas region live in areas of exposure and are at risk of contracting Chagas disease. It is estimated that between six and seven million persons in the world, a great proportion of these in Latin America, suffer from the infection with *Trypanosoma cruzi*. The PAHO/WHO estimates an incidence in the Americas of 30,000 annual cases [7].

The control plan against the vector and the improvement in the controls over blood products encouraged by the Pan American Health Organization/World Health Organization (PAHO/WHO) in 1991 allowed the interruption of the domiciliary vector-borne transmission in 17 affected countries, the elimination of indigenous vector species (*T. infestans*, *T. brasiliensis*, *T. sordida*), the implementation of universal screening on blood donors in the 21 endemic countries, the detection and treatment of infested subjects, the increment of cover for the diagnosis, access to treatment, and clinical attention of patients. Later, the Andean Initiative was organized in 1997 (for *R. prolixus*, *T. dimidiata*, *T. maculata*, *R. ecuadoriensis*), together with the Central American Initiative (for *R. prolixus*, *T. dimidiata*, *T. barberi*, *R. pallescens*), and finally in the Amazonian Initiative in 2004 (for *R. prolixus*, *R. robustus*, *R. geniculatus*, *R. brethesi*).

In El Salvador, Costa Rica, and Mexico, *Rhodnius prolixus* was eliminated as the main vector between 2009 and 2010. In South America the elimination of the *Triatoma infestans* vector was achieved in Brazil and Uruguay in 2012 and 2014, respectively (Fig. 2) [7].

It is important to consider that when we discuss about Chagas disease epidemiology, we include infected subjects and not only the areas where the vector inhabits (Table 1).

Fig. 2 Vector's geographic distribution—PAHO 2014

Table 1 Estimated demographic and epidemiological parameters of Chagas disease in Latin America, WHO 2015

Data	Year 2010
Population	543,877,115
Number of infected people	5,742,167
Annual new cases due to vector-borne transmission	29,925
Exposed population in endemic areas	70,199,360

Source: WHO, Weekly Epidemiological Record, 2015(6), 90:33–44

In Argentina, the PAHO estimates the number of infected people to be of 1,505,235. There are isolated rural populations at the north of our country with a higher prevalence, around 15%. In neighboring countries like Bolivia, prevalence rates of up to 30% have been reported. This percentage is smaller in Paraguay and Peru [7].

6 Clinical Presentation

Even though the infection through vector-borne transmission can present itself at any age, the greatest risk occurs in children less than 10 years old. The acute phase of the acquired infection through this means of transmission can last between 2 and 4 months and is, in most cases, asymptomatic. Only 10% of the patients present clinical compromise during the acute period. The acute period is the moment in which the pathology offers the most evident symptomatology, with pathognomonic signs, with other very characteristic and many symptoms or systemic manifestations common to many other diseases. Romaña's classification (Romaña [8]) has been generalized as "forms with signs of apparent entry of the parasite" and "forms with no signs of apparent entry of the parasite" [8]. The findings on 339 acute cases, following this classification, are described in Table 2.

These signs and symptoms can show between 4 and 15 days after the contact with the vector has occurred (incubation period).

6.1 Forms with Signs of Apparent Entry of the Parasite

Ophthalmo-lymphonodal complex (Romaña's sign): This clinical presentation, which Cecilio Romaña called "unilateral schizotrypanosomic ophthalmia" (Fig. 3), is not pathognomonic and acquires true diagnostic value with the presence of *T. cruzi* in blood samples.

- Characteristics: Sudden appearance of elastic and slightly painful edema on one eye's eyelids, erythema, homolateral satellite lymphadenopathy, conjunctival hyperemia, and dacryoadenitis; symptoms such as exophthalmia, dacryocystitis, keratitis, and hemifacial edema are less frequently observed. The region acquires a quite characteristic red-purplish coloration. The edema extends to neighboring

Table 2 Clinical findings on 339 acute cases from Santiago del Estero, Argentina [9]

With signs of apparent entry of the parasite	Ophthalmo-lymphonodal complex	82%
	Inoculation chagoma (in limbs, face) single or multiple	5.6%
Without signs of apparent entry of the parasite		
Typical forms	Hematogenous chagoma	1.7%
	Lipochagoma	0.8%
	Generalized edema	3.5%
Atypical forms	Prolonged fever	0.5%
	Anemic	0.2%
	Visceral (hepatosplenomegaly)	1.1%
	Cardiac (heart failure)	0.2%
	Neurological (meningoencephalitis)	3.5%

Source: Ledesma O. Aspectos clínicos de la enfermedad de Chagas aguda. Congreso Argentino de Protozoología y Reunión sobre enfermedad de Chagas. Huerta Grande, Córdoba, 1984

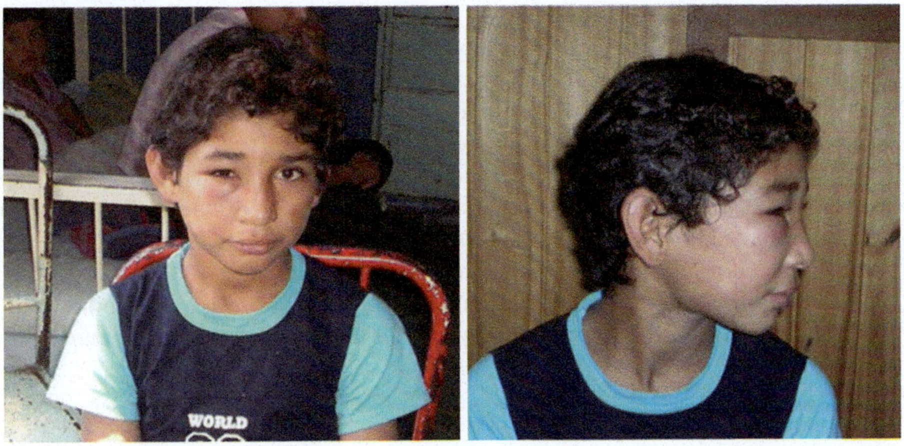

Fig. 3 Ophthalmo-lymphonodal complex (Romaña's sign). Bipalpebral edema with mild edema in the buccal region. *Own source*

areas of the face on the same side, the latter generalizing to the other side of the face and the rest of the body. There is scarce conjunctival secretion that is found on the eyelashes when the child awakes and which microscopic examination shows large amounts of degenerating polymorphonuclear cells. When the edema in the eyes is very intense, it could be overlapping of the upper eyelid over the inferior one with complete occlusion of the palpebral fissure.

- The satellite lymphadenopathy is practically never absent, especially in the preauricular region, although it is also frequent in the cervical, submandibular, and parotid ones. The size is variable, usually being free and non-painful, and in the cervical region, there is one larger than the rest, which Mazza named "the perfect lymph node" [10].

- Duration: Without treatment it spontaneously disappears within 1 or 2 months from its beginning, although it may present periodic exacerbations that would be related to new ruptures of parasite conglomerates or with an allergic phenomenon.
- Laboratory: Unlike the pyogenic processes that are accompanied with neutrophilia, in Chagas disease there is a marked lymphomonocytosis. Most patients usually have a mild anemia, generally hypochromic; in some cases it can be important and show a decrease of 30% on hemoglobin levels. The values of total and relative serum proteins show modifications during the acute phase, generally showing total hypoproteinemia falling back especially over albumin. On the other hand, there is hypergammaglobulinemia that depends on the increase of the fractions alpha$_2$ and gamma; alpha$_1$ and beta remain normal [11, 12].

Inoculation chagoma (cutaneous-lymphonodal complex or inoculation chancre): It may present in any body part, but it is more frequent in uncovered areas such as the face, arms, and legs. The cheeks are the preferred sites within the face, and the forearm and thighs are preferred within the limbs. It is only slightly painful or non-painful. The lesion appears as a red-purplish zone, warm, and with edema. There is permanent underlying infiltration that includes both the skin and the subcutaneous cellular tissue which is easily perceived by touch. Its evolution is torpid and long, almost always accompanied with fever and other general symptoms. When the chagoma is near complete healing, the skin peels with small scales, while the affected area darkens. Later, the phenomena of reparation at the superficial and deep tissue level lead to a retraction of the subcutaneous and muscular cellular tissues provoking face deformations that are particularly visible. The different forms are furunculoid, erysipeloid-like, inflammatory tumor, and lupoid.

Inoculation chagoma, and particularly the ophthalmo-lymphonodal complex, is indicative of the parasite entry way in the vector-borne transmission.

6.2 *Forms Without Signs of Apparent Entry of the Parasite*

The first general symptoms of the acute phase appear simultaneously with the signs of entry way of the parasite.

Among children these manifest as a feverish state accompanied of prostration, restlessness, hyporexia, exaggerated irritability, or accentuated somnolence. Sometimes, children may present vomits or diarrhea or signs of bronchitis. Older children and adults may present headache, especially of the frontal type, ocular pain, and arthralgia.

Hematogenous chagoma: Mazza and Freire [10] have described the formation of numerous subcutaneous tumors that set primarily on the adipose cellular tissue during the period of major generalization of the acute phase, which have been named hematogenous chagoma. These are flat tumors that grow dermis and subcutaneous cellular tissue, generally not attached to deep planes, single or multiple

[10]. Their size can vary from very small to large plaques. The most frequent location is the inferior abdomen, buttocks, and thighs. In general, these are not painful and can be sensitive to pressure. Usually, these do not alter the color of the upper skin.

Buccal lipochagoma: This chagoma takes the Bichat's fat pad, and it was considered as pathognomonic by Freire. It can have a lipomatous or a hard consistency. Generally, it is painful and breastfeeding in nurslings can be difficult [13].

Fever: This symptom is present in all or at least the great majority of acute cases from the beginning of the disease [14, 15]. Fever is usually high, reaching up to 41 °C, continuous, and with afternoon peaks (with double or triple daily peak, similar to leishmaniasis or kala-azar). These high records generally coincide with the presence of large numbers of trypanosomes in the blood. Hypothermia can exist in severe clinical forms that are accompanied by meningoencephalitis. There are also cases in which the acute phase of the infection goes by without fever.

Hepatosplenomegaly: It can be present at the beginning of the clinical picture being, together with the fever, the only manifestations; or it could appear later during the advanced evolution [16]. It may be accompanied of mild alterations of the hepatic enzyme values.

Circulatory system and electrocardiographic alterations: It is clinically expressed as tachycardia and low blood pressure. Chagas described the heartbeat as "frequent, small and filiform in the gravest cases, not being related, in most cases, the number of heartbeats with the thermal reaction." The electrocardiographic abnormalities appear in around 30–45% of the patients with prolongation of the PR interval and repolarization disorder being the most frequent ones (Fig. 4). The presence of blockage of the right branch in the acute myocarditis has bad prognosis, unlike what happens with chronic myocarditis [16–19].

Edema: It usually shows up 10–15 days after the beginning of the feverish state; it is generalized, white, and most evident in the face, limbs, and testicles. When it is very accentuated, it may get the aspect of a real anasarca; in these cases the face shows a particular infiltration that altogether give the face a particular aspect named "bouffi" by Chagas, which physiognomy can also be found among patients with African trypanosomiasis. It presents variable intensity and duration, and generally the edema retrogrades as other symptoms cease.

Digestive system: Lack of appetite, vomit, and diarrhea are common in small children [20, 21]. The abdomen is usually tense and tympanitic, and in certain grave cases, diarrhea dominates the clinical picture leading patients to a state of toxicosis.

Infrequent symptoms: Chagas described necrotic plaques in the skin following the formation of blisters and the swelling of surrounding tissues. The necrosis develops more or less circularly and might reach the deepest layers of the dermis.

Lesions in the testicles and the epididymis seem to occur with certain frequency. Romaña pointed out an orchiepididymitis in a boy of a little more than 1 year old [22]. Freire also observed sick patients with orchiepididymitis [23].

Romaña, Mazza, and Benitez have pointed out an increase in the parotid gland in the same side as the ocular lesion of entry of the parasite [22].

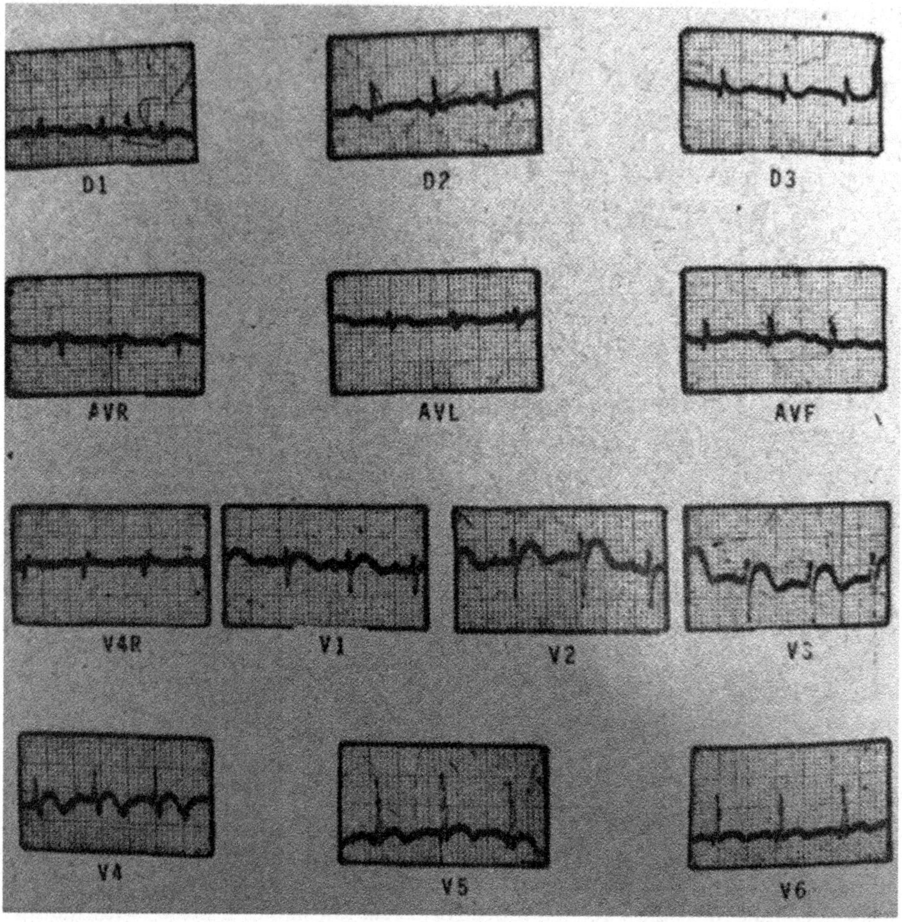

Fig. 4 Myopericarditis in an acute Chagas disease patient. Diffuse disorders of ventricular repolarization can be observed characterized by supra-elevation of the ST segment in the anterolateral face suggestive of subepicardial ischemia of the left ventricle with acute myopericarditis, right branch complete blockage

Generally, the evolution of the vector-borne acute phase is favorable and benign. However, there is a minority of grave forms that affect preferentially small and malnourished children and present high mortality due to meningoencephalitis and myocarditis with cardiac insufficiency.

The vector-borne acute infection constitutes a real epidemiological urgency given that it indicates the presence of the vector and the active transmission in the region for which the implementation of evaluation and entomological control measures in the area is required. The vector-borne acute infection is one of the events under clinical surveillance.

7 Differential Diagnosis

The differential diagnosis of the vector-borne form of the Chagas disease varies according to its clinical presentation.

Chagas disease differential diagnosis should be considered in every patient with prolonged feverish picture and compatible epidemiology.

The ocular entry syndrome can be confused with a sty or a chalazion accompanied with eyelid edema, but the localized and painful inflammation in the infections of the palpebral gland and the lack of general symptoms clarify the diagnosis. Also, the pyogenic processes in the eyes are accompanied by an increase of neutrophils in the leukocyte count, while in trypanosomiasis there is an important lymphomonocitary reaction.

The sting of insects near the eyes can produce an edema that leads us to think of a vector-borne form. The bite of triatomine insects can cause a similar picture to that of the ocular syndrome, but it is not accompanied by general symptoms and rapidly disappears after 2–3 days. Bees and wasps' sting can also produce some confusion, but the bite precedent and the generalization of the edema to both eyes resolve the diagnosis.

The cutaneous parasite entry syndrome leads to the differential diagnosis with pyogenic furunculoid or erysipeloid-like infections according to the aspect of the lesion. The evaluation of general symptoms solves the clinical problem.

The patient's clinical history, the possibility of contact with triatomine insects, and the presence of these insects in the household orient the diagnosis toward vector-borne Chagas disease.

8 Diagnostic Methods

The diagnosis of the acute vector-borne infection is based on direct parasitological methods. The direct parasitological method **microhematocrit (MH)** possesses several advantages: it uses small blood volumes (0.3 mL), its fulfillment takes little time (30 min), and it has high sensitivity. For all of this, we consider that MH is the method of choice for the study of this route of infection [4, 24]. This is the suggested method by public health organisms. It is of fundamental importance to observe the blood with anticoagulants.

The dependency of this technique on the operator is a drawback of the method; a sensibility varying between 80% and 93% has been reported in specialized centers during the perinatal period.

The **indirect** parasitological methods, **xenodiagnosis**, **inoculation of lactating mice**, and **blood culture**, are high-sensitivity techniques, but these require a complex infrastructure, and the results are obtained between 15 and 60 days after the extraction of the sample.

New techniques such as PCR are currently on standardization phase for their posterior clinical use. Currently their employment is in experimental phase.

9 Serological Diagnosis

The seroconversion between two analyses separated with and interval of 30–90 days can also serve as confirmatory diagnosis of the acute phase if the parasitemia cannot be obtained. However, it is necessary to remember that seroconversion has less acute-phase diagnostic value for patients under treatments or suffering diseases that generate immunosuppression. The serological tests are used to detect circulating antibodies (immunoglobulin G (IgG)) against the parasite. The IgG can be detected after 30 days from the occurrence of the acute infection, reaching their maximum level at the third month. The most common techniques are ELISA, indirect hemagglutination, immunofluorescence, and particle agglutination [24].

10 Treatment

Because of the singular evolution of the disease, it is very hard to establish a clinical parameter for the evaluation of the effectiveness of a specific treatment since the signs and symptoms of the acute phase spontaneously disappear within weeks [25].

The first studies and publications about the treatment were performed on the acute form of the infection. The therapeutic response was evaluated through the improvement of the inoculation chagoma and the thermal curve. This criterion led to the mistaken interpretation of the efficacy of certain medications that only improved the symptoms but had no parasiticide effect [26, 27].

The objectives of the etiological treatment are to cure the infection and prevent lesions on the organs ensuring in this way the avoidance of long-term complications. To achieve this, the employed drugs need to be capable of destroying the circulating and intracellular parasites. The current treatment, employed for over 40 years, is based on the use of nitroheterocyclic compounds nifurtimox (Bayer 2502; Bayer Leverkusen, Germany) and benznidazole (Laboratório Farmacéutico do Estado de Pernambuco (LAFEPE), Recife, Brazil; and Laboratorio ELEA, Autonomous City of Buenos Aires, Argentina). Both nifurtimox and benznidazole fulfill with this criterion because they have a trypanocide activity against all parasitic forms and have largely demonstrated their efficacy in acute as well as in chronic forms of the disease [28, 29]:

- **Nifurtimox**: It was the first drug employed for the treatment of Chagas disease. The first clinical essays with this drug are from 1965 in South America obtaining the best results for the acute phase and the treatment of children. Different therapeutic schemes have been proposed, using 8–10 mg/kg/day in three doses during 60–90 days. Currently, 30-day schemes have been shown to have adequate effectiveness. The parasitological cure is obtained for 88–100% of the vector-borne acute Chagas disease patients. It comes in presentations of 120 mg pills. A pill of

30 mg is being developed. The adverse effects more commonly found are anorexia, weight loss, paresthesia, somnolence, psychomotor excitation, and gastrointestinal symptoms such as vomit, nausea, and abdominal pain. Sometimes these symptoms impose the need to interrupt the treatment.

- **Benznidazole**: It acts inhibiting the *Trypanosoma cruzi*'s protein and RNA synthesis. Since the start of its employment in the 1970s, different doses were tested; it was even used with increasing doses at the beginning of treatment until reaching the currently employed dose of 5–8 mg/kg/day in two takes for 60 days. Schemes of 30 days with a maximum of 300 mg/day were shown to have adequate efficacy for vector-borne acute cases. It comes in presentations of 50 and 100 mg pills. A pill of 12.5 mg is currently under development.

The response to treatment is near 100% in the vector-borne acute phase [25, 27, 29, 30]. In the following table, the studies that show the effectiveness of the treatment with benznidazole and nifurtimox in the acute phase of the vector-borne disease are described (Table 3).

The collateral effects of these drugs are similar: lack of appetite, irritability, sleep disorders, leukopenia, thrombocytopenia, cutaneous erythema, and digestive disorders. Of 65 patients treated for vector-borne transmission at the Hospital de Niños Ricardo Gutiérrez, 31% presented adverse events to the treatment, but in no case it was necessary to abandon it. In children older than 7 years old, we observed exanthema with benznidazole, the suspension for 2–3 days plus the incorporation of an antihistaminic allowed completing the therapeutic scheme.

In two cases we observed leukopenia and thrombocytopenia at the beginning of medication [28].

11 Post-therapeutic Controls

In patients that initiate treatment during the acute phase with detectable parasitemia, direct parasitological control (micromethod) is advised every 7 days from the beginning of treatment [4]. With an adequate therapeutic response and at the end of the treatment, the parasitemia must be negative. In case the parasitemia persists, the administration of the treatment needs to be evaluated, and, especially, the dose must be verified, before therapeutic failure is considered. In case of persistent parasitemia, another available drug should be used following the recommended scheme. Once treatment is complete, it is recommended to conduct serologic tests to detect IgG for the control of its efficacy every 3 months during the first year and to continue controls every 6 months afterward (Fig. 5). Serological controls are realized until seroconversion is achieved, which is the current criterion of cure. The less time from the acute infection, the more precocious is the serological negativization [27].

Table 3 Efficacy of nifurtimox and benznidazole in acute vector-borne Chagas disease

References	Age (years)	Benznidazole			Nifurtimox			Design[a]	Follow-up (months) lost to FU (%)	Efficacy endpoints[b]	
		n	Dose (mg/kg/day)	Length (days)	*n*	Dose (mg/kg/day)	Length (days)			Serological tests (% neg)	Parasitological tests (% pos)
Ferreira 1967; 1969	5–10	–	–	–	1 3 1 1	30+20 25+15 30+15 30+15	30+30 15+75 5+60 30+30	C nR nB	20 months – – –	1/1 (MGR) – – –	0/1 (Xeno) – – –
Fernandez 1969	1–55	–	–	–	24	30	60	nC nR nB	24 months 50%	(CFT) 42%	(Xeno) 50%
Lugones 1969; Cerisola 1969; Cerisola 1970	Children 92% <15	–	–	–	43 40 367	NT[c] 15 25 + 15	– 90 15 + 75	C nR B (sero)	24 months 46% }43%	(IHA, IFA, CFT) 0% 81%	(Xeno) 61% Neg. Sero: 0% Pos. Sero: 44%
Barclay 1978	Children ?	107 32	7.5–10 5	30 30	–	–	–	C nR nB	18 months } 30%	87% (CFT) 91% (IFA)	12% (Xeno) 14% (Strout)
Ferreira 1988	6–13 2–18	– 17	– 5	– 60	21 –	15 –	90 –	C nR nB	15 years: 28% 9 years: 41%	(IHA, IFA, CFT) 100% 100%	(Xeno) 0% 0%
Solari 2001	Children	–	–	–	66	–	–	nC	3 years	(IHA, EIA) 94.4%	(Xeno, PCR) 0%
Cancado								C	13 years	(EIA, IHA, IFA)	Not used
2002	<10 11–60	6 15	10–20 5–10	40–60 30–60	–	–	–	nR nB	0%	76%	

[a]Design: C (controlled, control or comparative group), nC (not controlled), R (randomized), nR (not randomized), B (blinded), nB (not blinded)
[b]Efficacy endpoints: MGR (Machado Guerreiro reaction), CFT (complement fixation test), IHA (indirect hemagglutination assay), IFA (immunofluorescence assay), EIA (enzyme immunoassay), IC (immunochromatography), Xeno (xenodiagnosis), Strout (Strout technique), MH (microhematocrit)
[c]*NT* Not treated

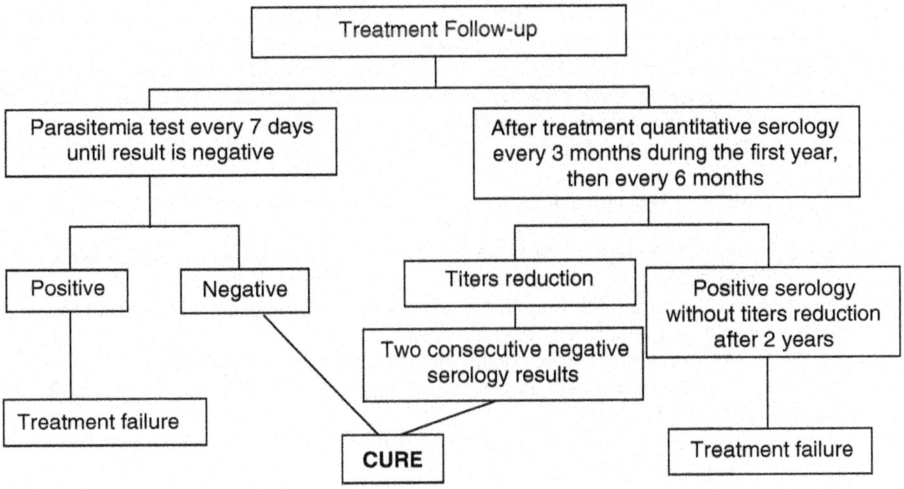

Fig. 5 Follow-up protocol for children treated with benznidazole or nifurtimox

References

1. Posse RA. A 75 años del descubrimiento de la enfermedad de Chagas. Rev Argent Cardiol. 1984;52:175.
2. Antecedentes históricos. En Enfermedad de Chagas. Editorial Doyma, Argentina. 1994.
3. Moscatelli G, García Bournissen F, Freilij H, et al. Impact of migration on the occurrence of new cases of Chagas disease in Buenos Aires city, Argentina. J Infect Dev Ctries. 2013;7(8):635–7. https://doi.org/10.3855/jidc.2930.
4. Guías para la atención al paciente infectado con *Trypanosoma cruzi* (Enfermedad de Chagas). Buenos Aires: Ministerio de Salud de la Nación. 2012.
5. Carcavallo RU, GalindezGiron I, Jurberg J, Lent H. Atlas of Chagas' disease vectors in the Americas, vol. I, II, III. Rio de Janeiro: FioCruz; 1998.
6. Calderón-Arguedas O, Chinchilla M, García F, Vargas M. Preferencias alimentarias de Triatoma dimidiata (Hemiptera: Reduvíidae) procedente de la meseta central de Costa Rica a finales del siglo XX. Parasitol Día. 2001;25(3-4):78–81. https://doi.org/10.4067/S0716-07202001000300002.
7. WHO. Wkly Epidemiol Rec. 2015;90(6):33–44. http://www.who.int/wer.
8. Romaña C. Acerca de un síntoma inicial de valor para el diagnóstico de forma aguda de la enfermedad de Chagas, la conjuntivitis esquizotripanósica unilateral (Hipótesis sobre puerta de entrada conjuntival de la enfermedad). MEPRA. 1935;22(3):16–28.
9. Ledesma O. Aspectos clínicos de la enfermedad de Chagas aguda. In: Congreso Argentino de Protozoología y Reunión sobre enfermedad de Chagas. Hurta Grande, Córdoba. 1984.
10. Mazza S, Freire R. Manifestaciones cutáneas de inoculación, mestastásicas y hematógenas en enfermedad de Chagas, chagomas de inoculación, chagomas metastásicos y chagomas hematógenos. MEPRA. 1940;46(1):3–38.
11. Lugones H, Peralta F, Canal Feijoo D, de Marteleur A. Evolución de la sintomatología clínica y la función hepática en la enfermedad de Chagas aguda tratada con Bay 2502. Bol Chil Parasitol. 1969;24:19–24.
12. Salum J, et al. Electroforesis de las proteínas séricas en la fase aguda de la enfermedad de Chagas. Rev Goiana Med. 1959;5(1):13.

13. Freire R. Sobre el valor diagnóstico de las diversas manifestaciones cutáneas de la Enfermedad de Chagas. In: Primera Conferencia Nacional sobre Enfermedad de Chagas, pp. 54–64. 1954.
14. Pellegrino J, Lobo Rezende C. A Doenca de Chagas nainfancia. Rev Bras Med. 1953;9:765.
15. Ferreira H. Fase aguda da Doenca da Chagas na infância, aspectos clínicos observados 57 casos. Jornada Brasileira de Pediatria e Puericultura, pp. 393–397. 1960.
16. Chagas C. Tripanosomiase Americana, forma aguda da molestia. Mem Inst Oswaldo Cruz. 1916;8(2):153.
17. Storino R, Milei J. Miocardiopatía Chagásica Crónica. Buenos Aires: Ed. Club de Estudio; 1986.
18. Laranja F, Dias E, Nobrega G, Miranda A. Chagas' disease. A clinical epidemiologic and pathologic study. Circulation. 1956;14:1035.
19. Freilij H, Leon L, Grinstein S, Bodino J. Chagas agudo: edema generalizado como forma de presentación. Rev Hospital Niños. 1981;XXII(94)
20. Lugones H. Consideraciones acerca de los síntomas del período agudo de la enfermedad de Chagas en la infancia. In: Anais do Congreso Internacional da Doenca de Chagas, Río de Janeiro (Brasil), vol. 3, pp. 861–870. 1959.
21. Bordin C, Cibeira de Totera ME, Abraham de Kablan I, Barros F. Enfermedad de Chagas aguda, formas atípicas. Arch Argent Pediatr. 1974;72:148.
22. Romaña C. Algunos síntomas poco observados en la Enfermedad de Chagas. Anal Instit Med Region. 1953;3:251.
23. Freire R. Hidrocele agudo chagásico, observaciones personales. Anal Instit Med Region. 1957;4:256.
24. Freilij H, Altcheh J. Congenital Chagas disease: diagnostic and clinical aspects. Clin Infect Dis. 1995;21:551–5.
25. Altcheh J, Moscatelli G, Mastrantonio G, Moroni S, Giglio N, Koren G, Freilij H, García Bournissen F. Drugs for half the world: pediatric clinical pharmacology population pharmacokinetics study of benznidazole in children with Chagas disease. 16th World Congress of Basic and Clinical Pharmacology, 17–23 July 2010, Copenhagen, Denmark. Basic Clin Pharmacol Toxicol. 2010;107(Suppl. 1):162–692.
26. Barclay C, et al. Aspectos farmacológicos y resulta dos terapéuticos del benznidazol en el tratamiento de la infección chagásica. Prensa Med. 1978;65:239.
27. Cancado JR. Long term evaluation of etiological treatment of Chagas disease with benznidazole. Rev Inst Med Trop. 2002;44(1):29–37.
28. Altcheh J, Moscatelli G, Moroni S, Garcia-Bournissen F, Freilij H. Adverse events associated with benznidazole in infants and children with Chagas disease: a cohort study. Pediatrics. 2011;127:e212–8.
29. García-Bournissen F, Altcheh J, Giglio N, Mastrantonio G, Della Védova CO, Koren G. Pediatric clinical pharmacology studies in Chagas disease: focus on Argentina. Paediatr Drugs. 2009;11:33–7.
30. Cerisola JA. Evolución serológica de pacientes con enfermedad de Chagas aguda tratados con Bay 2502. Bol Chil Parasitol. 1969;24:54–9.

Congenital Chagas Disease

Jaime Marcelo Altcheh

Abstract Congenital infection with *Trypanosoma cruzi* is a global problem, occurring on average in 5% of children born from chronically infected mothers, with variations depending on the region. In endemic areas with inadequate vector control, Chagas disease, also known as American trypanosomiasis, is usually a vector-borne disease. In areas with effective vector control and in urban areas, the main route of new cases of CD is through mother-to-child transmission. The estimated number of new cases of congenital *T. cruzi* infection is 8668 cases/year [2]. This is also the case of non-endemic countries where, due to population migration, CD is increasingly becoming a public health problem in non-endemic settings. The majority of infected infants are asymptomatic, and the diagnosis is based on the microscopic observation of the parasite in the blood of newborns and by serology in infants older than 8 months of age when maternal transplacental antibodies have waned. Two drugs are available for treatment, benznidazole and nifurtimox. Treatment should be implemented immediately after diagnosis since the sooner the infection is treated, the better the prognosis. The treatment efficacy is around 95% in different cohorts with a good drug safety profile. Recent studies have shown that its transmission can be prevented through treatment of infected women before they become pregnant. Considering that etiological treatment of the child is highly effective if performed before 1 year of age, the diagnosis of infection in pregnant women and their newborns has to become the standard of care and integrated into the surveillance programs of syphilis and human immunodeficiency virus.

J. M. Altcheh (✉)
Servicio de Parasitología y Enfermedad de Chagas, Hospital de Niños "Ricardo Gutiérrez", Buenos Aires, Argentina

Instituto Multidisciplinario de Investigación en Patologías Pediátricas (IMIPP-GCBA), Consejo Nacional de Investigaciones Científicas y Técnicas (CONICET), Buenos Aires, Argentina

J. M. Altcheh, H. Freilij (eds.), *Chagas Disease*, Birkhäuser Advances in Infectious Diseases, https://doi.org/10.1007/978-3-030-00054-7_9

1 Introduction

Chagas disease or American trypanosomiasis is caused by the hemoflagellate parasite *Trypanosoma cruzi*. Infection most often occurs in Latin America, the endemic area of Chagas disease, via vector transmission. In the past few decades, a growing migratory current has brought large numbers of patients from rural to urban areas, changing the epidemiology of Chagas disease. Currently, the largest number of infected people in Latin America dwells in cities with no vector transmission, making the mother-to-child transmission the main route of infection of the disease in these areas and the main route of transmission overall. Also, migration of Latin Americans to non-endemic countries has brought the disease, and congenital cases, to areas where Chagas disease had never been present before, putting newborns at risk of misdiagnosis or no diagnosis at all in the case of asymptomatic patients [1].

T. cruzi infection initially presents with an acute phase of variable duration (2–8 weeks) characterized by high parasitemia and non-specific symptoms (e.g., fever, hepatosplenomegaly), followed by a phase known as the chronic indeterminate or latent asymptomatic phase that can last for decades before developing cardiac and/or digestive disorders in about 30% of the patients.

Endemic countries, with the support of the Pan American Health Organization (PAHO), have established regional programs for the prevention and control of CD. These programs mostly focus on the elimination of the triatomine vector by insecticide spraying, prevention of transfusion of contaminated blood, and screening for maternal infection during pregnancy. In large part due to these programs, the current burden of CD in Latin America has been reduced from 2.8 million disability-adjusted life years to less than 500,000 in 2015 [2].

Unfortunately, strategies to prevent vector transmission have no impact on vertical transmission, which leads to a continuous stream of new congenital cases from mothers with the disease.

The PAHO has stressed the need to consider congenital Chagas disease as a public health problem and recommends that each country develops protocols for universal pregnancy screening and treatment of infected newborns. Most pregnant women with the disease are in the asymptomatic chronic phase and have vertical transmission rates of approximately 5% [3].

Universal screening of pregnant women for detection of Chagas disease is extremely important given the asymptomatic nature of the disease in the vast majority of these women and the high effectiveness and safety of pharmacological treatment in infected newborns. Implementation of universal screening has also led to the realization that most infected newborns are asymptomatic and would hardly be identified if screening is not performed.

2 Epidemiology

It is important to keep in mind that the epidemiology of Chagas disease should refer to the epidemiology of the patients with the disease and not to the areas where the vector can be found. This is a common confusion when discussing Chagas disease that may leave most of the patients (i.e., those who live in urban areas without exposure to the vector) out of the picture.

Chagas disease patients can be found in all of the Americas, including North America, and, due to migratory currents, also in most continents of the world. Cases have been reported in countries as far from the Americas as Japan and Australia and in virtually all of Europe. The surge in cases in non-endemic countries has prompted the implementation of diagnostic and treatment guides in these areas [4].

Recently the PAHO has estimated that there are 5,700,000 infected people in Latin America, with Argentina having the largest absolute number of patients and Bolivia the highest prevalence of the infection. Approximately 8700 babies are born with congenital Chagas disease each year in Latin America [2].

Chagas disease prevalence in pregnant women ranges from 3% to 40% depending on the region and increases significantly after 20 years of age, which suggests that vector and transfusion control programs have been highly effective in thwarting disease transmission [5]. However, certain areas of Argentina, Bolivia, and Paraguay still have high infection rates in children under 10 years old due to uncontrolled vector presence. The same is true for some areas in Central America, which has an important impact on the epidemiology of North American countries such as Mexico and the United States due to migration of Latin American patients to these areas.

Migration has also led to a prevalent urban picture of the disease, with most new cases occurring due to congenital transmission [6]. The highest incidence of congenital cases can be found in cities, where most infected patients live, in spite of being an area with no vector transmission [5, 7]. The explanation for this situation is that large migrations took place from rural areas to the cities, which brought a large proportion of infected patients from areas with vector presence to cities without the vector.

The number of infected women of childbearing age has been estimated [2] to be approximately 1,250,000 in Latin America, and the overall incidence of congenital Chagas disease infection has been estimated to be 0.9 per 1000 live newborns. This would lead to approximately 8700 infected newborns every year. However, the actual number of congenital transmissions diagnosed and treated is much lower than this estimate, most likely due to a lack in implementation of universal screening programs and to a reluctance to treat infected newborns.

In the United States, approximately 300,000 cases are believed to be present [8], and 40,000 pregnant women may be infected, resulting in 2000 congenital cases through mother-to-child transmission [9]. Clearly, there is a need for active surveillance data in order to better refine these prevalence data.

3 Infection in Pregnancy

Chagas disease does not seem to affect fertility [10, 11]. Reports from the 1970s suggested a high incidence of prematurity and high morbidity in infected newborns. However, these observations were based on populations of symptomatic women in the acute phase of the infection. Maternal chronic infection has no effect on the outcome of pregnancy or on the health of newborns as long as there is no maternal transmission of parasites to the unborn child [12]. Fetal infection in early gestation increases the risk of premature delivery, low birth weight, and premature rupture of amniotic membranes, effects that may be related to placental inflammation observed in these cases [13].

Most pregnant women are in the chronic phase of infection and lack any symptoms or signs related to the infection [14]. Absence of clinical and electrocardiogram findings in pregnant women may be explained by the fact that the peak of fertility in the areas with high prevalence of Chagas disease is approximately 20–30 years of age, whereas cardiac or gastrointestinal tract involvement does not appear in infected patients until well past 40 years of age.

Fetal infection requires the presence of circulating parasites in the pregnant mother's bloodstream. *T. cruzi* persistently infects the host and can be found in the bloodstream in both acute and chronic phases, meaning that a woman can transmit the infection in any of her pregnancies [15, 16]. However, acute infection exacerbates the risk of transplacental transmission of the infection, possibly due to the high parasitemia observed in the acute phase. In the published case series, infection was observed in 71% of newborns of infected mothers in the acute phase, 57% of whom were born prematurely [17, 18].

Most women of fertile age are in the chronic asymptomatic phase, which implies a low level of circulating parasites. However, parasite replication and circulation have been shown to increase during pregnancy, particularly in the last trimester. In fact, a direct relationship between maternal parasitemia and risk of congenital infection has been identified [19–21]. This increased risk provides strong support for pharmacological treatment (i.e., with benznidazole or nifurtimox) of women before they become pregnant, as children or young adults, to abolish parasitemia and prevent congenital transmission [22, 23].

4 Factors Associated to Congenital Transmission of Chagas Disease

Reported rates of congenital transmission show significant geographic variability that ranges from 1% to 18.2% [3]. The reasons for this variability remain unexplained to date.

Congenital transmission of *T. cruzi* requires an interaction between the parasite, the maternal immune system, and the placenta.

1. *The parasite*: *T. cruzi* circulates in the bloodstream as trypomastigotes, which can disseminate to tissues and invade the placenta. After accessing the placental cells, the parasite transforms into the intracellular form, the amastigote, and starts dividing. After bursting out of the host cell, amastigotes become trypomastigotes again, and, if the parasite manages to reach the fetal circulation, it can infect the fetus. *T. cruzi* has been classified into six groups (called discrete typing units (DTU)). However, no relationship has been observed between DTUs and the risk of congenital transmission of the infection [24]. Large gaps still exist in

the understanding of the role of the genetic diversity of *T. cruzi*, in the infection, the risk of disease progression, and the risk of congenital transmission and of the reactivation in immunosuppressed subjects [25].

2. *Maternal immune system*: Most pregnant women can be found with low-level parasitemia in the chronic stage. Maternal IgG antibodies are thought to control the infection and limit placental invasion, thus limiting placental transfer of the parasite to the fetus. In acute infections, where the immune response to the parasite is still immature and fails to control parasitemia, congenital transmission rates are significantly higher [21, 26].
3. *Placenta*: The placenta plays a vital role as a barrier that protects the fetus from infectious agents. *T. cruzi* infection is believed to take place toward the end of the pregnancy since the majority of infected newborns are asymptomatic.

Histological examination of placentas from infected mothers shows necrotic areas with inflammation and high numbers of amastigote nests. These areas can be found mainly in the areas of placental membrane insertions, which lead to the suggestion that the main infection route of the organ is via the marginal sinus to the chorion, infecting macrophages and fibroblasts, eventually reaching the fetal vessels [27].

Placentas from women in the chronic stage commonly lack significant signs of inflammation and have low numbers or no amastigote nests. These observations suggest the existence of antiparasitic mechanisms protecting the placenta of women at this stage of the infection.

Transmission of the infection seems to be lower during early pregnancy possibly due to the intervillous space being close to the arterial spaces. Placental blood flow is continuous and diffuses starting on week 12 of gestation, which has been proposed as a reason for the observed infections at a higher gestational age [28].

5 Chagas Disease and Breastfeeding

Chagas disease can be acquired by ingestion of foods or juice contaminated with vector feces containing the parasite. This fact suggests the possibility that the disease can also be transmitted by breastmilk from an infected woman. However, oral infections due to contaminated food have been linked to a heavy load of parasites in the food, something that is unlikely in breastmilk.

There are very few epidemiological data suggesting infant exposure through breastmilk [29]. A few studies have observed trypomastigotes in breastmilk from women in the acute phase of the infection, but these observations have not been confirmed by other authors [30]. One case of possible human infection linked to contaminated breastmilk has been documented, in a breastfed infant who became infected and whose mother had bleeding nipples, which raises the possibility of an oral route of infection, most likely due to the ingestion of blood and not breastmilk [31].

Experimental mouse models have shown a low rate of infection via breastmilk in breastfeeding mice fed from mice in the acute infection stage [29].

Considering the data previously mentioned, and given that most infected women nowadays would be in the chronic stage (i.e., low parasitemia) at the time of breastfeeding, the risk of transmission of Chagas disease to their breastfed babies would be close to negligible. *Therefore, maternal Chagas disease is not considered a contraindication to breastfeeding*. The only possible exception is a woman in the acute phase of the infection who also has bleeding nipples.

6 Clinical Findings in the Newborn

Most infected newborns are asymptomatic. Clinical manifestations of Chagas disease in newborns vary widely, from premature babies with important multiorgan involvement and high risk of mortality to completely asymptomatic full-term babies [15, 32–36]. This wide range of clinical presentations has been observed in several geographical areas and may be explained to some extent by variations in parasite strains, maternal nutritional state, and variations in maternal immune response to the infection.

Symptomatic newborns can have varying degrees of morbidity. Clinical signs may be related to the timing of fetal infection. Infections in the first and/or second trimester of pregnancy lead to prematurity and significant clinical manifestations such as low birth weight, hepatosplenomegaly, anemia, thrombocytopenia, and, more rarely, meningoencephalitis, myocarditis, and pneumonitis [32, 35, 37, 38].

Widespread maternal screening for Chagas disease has brought about the diagnosis of a larger number of infected infants, most of whom turned out to be completely asymptomatic. This observation provides support for the idea that most of the infections take place during the third trimester or near the end of the pregnancy [15, 39].

Symptomatic cases show clinical manifestations that resemble those observed in other intrauterine infections such as toxoplasmosis and *Cytomegalovirus* infection. The most common clinical sign is hepatomegaly and anemia and thrombocytopenia. Much less commonly, eye and/or digestive tract involvement has been described [40, 41]. Mortality has been reported to be approximately 5%, mainly in patients with acute-disseminated infection with myocarditis and/or meningoencephalitis [12].

Several case series of congenital cases can be found in the literature. In Cordoba City, Argentina, 52 cases were described (62% term babies), with an incidence of 30% abnormal weight for gestational age [42]. Approximately 30% of cases were asymptomatic, 31% had hepatosplenomegaly and tachycardia, and the remaining 38% had delayed clinical signs, mostly hepatosplenomegaly. Only four infants had signs of heart failure and none had meningoencephalitis. No deaths were observed in this series, and no cardiovascular sequelae were found after 10 years of follow-up [43].

In a case series of 102 infants collected over 18 years in the Province of Salta, Argentina, 33.3% of patients were asymptomatic, 28% were premature babies, and 59% had hepatomegaly, 42% splenomegaly, 40% jaundice, 39% anemia, 5% hydrops fetalis, and 4% meningoencephalitis [44].

Another case series, from Bolivia, described infants with congenital Chagas infection; 54% were asymptomatic, but there was also a high rate of prematurity, and 25% had respiratory distress, 16% hepatomegaly, 14% splenomegaly, 13% cardiomegaly, 11% neurological involvement, 8% anasarca, and 8% petechia [12].

Finally, a case series from the Parasitology and Chagas Service of the Buenos Aires Children's Hospital "Ricardo Gutierrez," Argentina, described 168 patients, 76% of whom were asymptomatic, and 9% had hepatosplenomegaly, 1% sepsis, 2% myocarditis, and 2% hepatitis [35].

Some cases of coinfection of *T. cruzi* and HIV have been reported; most of them are severe with profound central nervous system involvement and marked parasitemia observed by microhematocrit (MH) method at the time of diagnosis [45].

7 Diagnosis

Diagnosis of Chagas disease in pregnant women is made by serological testing. In areas without vector transmission, a single serological test (with two different serological methods) should suffice to confirm or rule out maternal infection. The most commonly used tests are ELISA, indirect hemagglutination (IHA), indirect immunofluorescence (IFI), and direct agglutination. No test is sensitive or specific enough to allow use of any single test for diagnosis. Current guidelines recommend the use of two serological tests based on a different set of antigens (e.g., whole-parasite lysate and recombinant antigens) to ensure high enough specificity and sensitivity in the diagnosis of Chagas disease [46]. In areas with active vector transmission, serological testing should be performed at the beginning and near the end of the pregnancy period and considered for any pregnant mother with a febrile illness without a clear explanation to rule out acute infection.

Diagnostic tests for Chagas disease have to be carried out for any newborn with an infected mother. Infected newborns are considered to be in the acute phase of the infection. Thus, initial diagnostic procedures involve searching for parasites in blood, commonly using direct parasitological methods with a concentration step. These methods are based on direct visualization of the parasite in the white blood cell phase after centrifugation of a very small volume of blood obtained in a heparinized capillary tube [15, 47]. Direct detection of the parasite in blood is facilitated by the fact that the trypanosome has a characteristic movement pattern and a relatively large size that allows its visualization by optical microscopy. In addition, the high numbers of parasites present in the blood of infected newborns make detection a relatively simple task for adequately trained laboratory personnel. The peak of parasitemia is observed at 1 month of age.

Diagnosis of congenital infection during the first weeks of life should be based on the examination of the infant's blood using a direct parasitological method (i.e., microhematocrit) to look for the presence of T. cruzi.

Serological tests are not useful during the first months of life due to passive transfer of IgG from the mother during gestation. Serological tests for specific IgM would be potentially useful but have not been fully developed to date.

If the infant has a negative parasitemia during the first few weeks of life, a serological test at 8 months of age should be performed to rule out or confirm the infec-

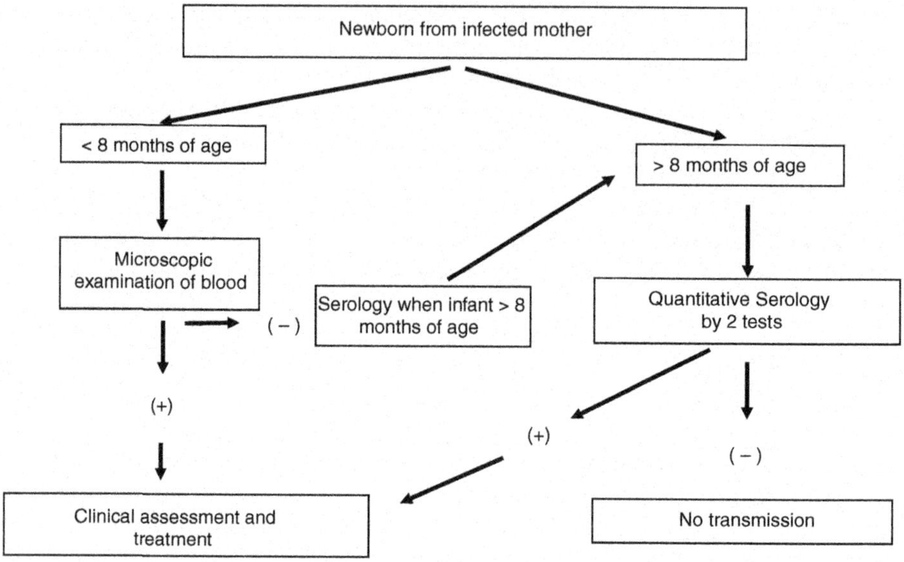

Fig. 1 Algorithm for diagnosis of congenital Chagas disease

tion. It is assumed that maternal antibodies disappear after 8 months of age, meaning that if the serological tests are positive at this time, the detected IgG reflects an immune response of the infant against the parasite, and therefore there is an active infection. If the patient has a negative serological test at 8 months, the congenital infection is ruled out [15, 47] (Fig. 1). In the few cases that have quantitative serological values close to the cutoff level at 8 months, the tests should be repeated after 1 month to confirm the diagnosis.

7.1 Diagnosis in Newborns

Direct parasitological detection methods such as the microhematocrit (MH) method are the diagnostic methods of choice for newborns. The MH method is based on the observation of parasites (trypomastigotes) in the white blood cell layer after centrifuging blood from the newborn in a heparinized capillary tube. The MH method has numerous advantages, including a low volume of blood requirement (0.3 ml), fast turnover (under 30 min), and high sensitivity in the hands of adequately trained personnel. The MH method is the suggested procedure by the Pan American Health Organization and public health Organization for diagnosis of newborns [47].

A drawback of this method is the requirement for a trained operator, which may lead to variability in diagnostic sensitivity across centers depending on the level of training and experience [32]. Another consideration is that the age of the newborn at the time of blood sampling is important as parasitemia increases during the first days of life, which has led to the suggestion that repeated blood sampling could increase sensitivity of the method. However, repeated sampling is difficult to implement in screening programs aimed at mostly asymptomatic newborns [32, 37]. Infants with clinical signs had higher parasite loads and were significantly more likely to be detected by direct parasitological test.

Indirect parasitological methods (i.e., xenodiagnosis, blood culture, and inoculation of lactating mice) have high sensitivity but require access to highly trained personnel and complex infrastructure and take weeks to months to yield results. None of these indirect methods are in routine use currently [15].

Histological examination of the placenta has limited sensitivity, and placental involvement does not closely correlate to fetal infection [48].

Modern PCR techniques are being developed and standardized for clinical use and have shown higher sensitivity than conventional parasitological diagnostic techniques in some series [49–51]. However, false positives have been reported, possibly due to the presence of free parasite DNA from the mother in fetal circulation [52, 53]. Currently, PCR methods should be considered experimental since the PCR sensitivity varies widely depending on the DNA extraction methods, primers and population tested.

The use of isothermal amplification molecular methods, such as loop-mediated isothermal amplification and nucleic acid sequence-based amplification, is becoming increasingly popular for the detection of trypanosomes as they offer simple, rapid, and cheap alternatives to traditional PCR-based methods [54, 55]. Isothermal tests involve a single reaction in a single tube incubated at a constant temperature; therefore, these techniques do not require the expensive thermocycling equipment that is necessary for PCR.

Widespread clinical use of these molecular tests still requires further testing in larger series to accurately evaluate sensitivity and specificity. Alternative serological methods, such as the detection of *T. cruzi*-specific IgM in the infant, have unfortunately never been adequately investigated, mainly due to the perception of low sensitivity of these methods and the consequent lack of interest in the laboratory diagnostics industry. Nontraditional antigens, such as SAPA [56–58], have been proposed as good markers of recent infection in newborns but have not been shown to be sufficiently sensitive in other studies [32].

Development of immune complexes (i.e., antibody-antigen complexes) due to high levels of parasitemia has been shown to produce false negatives in antibody-based tests due to *T. cruzi*-specific antibody sequestration in those complexes, particularly using hemagglutination test [15].

7.2 *Diagnosis for Children Over 8 Months of Age*

Infants have circulation antibodies which originated in the mother and which were transferred through the placenta during the last months of gestation. Due to this, identification of *T. cruzi*-specific IgG is not useful for diagnosis of infant infection as these antibodies may reflect maternal *T. cruzi*-specific IgG.

The age at which maternal antibodies disappear has been reported from 6 to 9 months. The more sensitive the test used, the longer the amount of time the transferred maternal antibodies will be detected. Using indirect hemagglutination and immunofluorescence, which are less sensitive tests, the maternal antibodies are not detected at 6 months. When a highly sensitive ELISA test is employed, maternal antibodies can be detected in a small number of infants at 8 months. Children over 8 months of age are expected to have eliminated these maternal antibodies that could have produced false-positive results in antibody-based tests for the diagnosis of Chagas disease. After this age, the methods of choice for the diagnosis of *T. cruzi* infection, or to rule out the infection, are serological tests. No single serological test has sufficient sensitivity and specificity to be relied on alone. Therefore, Chagas disease is diagnosed in children over 8 months of age if two distinct serological IgG tests, principally ELISA using whole-parasite lysates or recombinant antigens, are positive [47].

Diagnosis in children over 8 months of age is confirmed if two serological tests are positive.

8 Therapy

Only two drugs are available and have been shown to be efficacious in the treatment of congenital and pediatric Chagas disease: nifurtimox and benznidazole. Both drugs lead to a high response (close to 100% in children under 3 years of age), as measured by the decrease of antibody titers and conversion to negative serology at follow-up. Several clinical studies have shown that the earlier the treatment is administered, the higher the chance of complete response paired with a significantly lower risk of adverse events to the drugs. Successful treatment of infected infants will prevent the development of later cardiological and gastrointestinal complications in adulthood. These facts highlight the need for early diagnosis and treatment [35, 43, 59–62].

Benznidazole was developed in 1971. It was initially being developed as a chemotherapeutic agent, and hence this may explain the mg/kg dosing approach. Usual doses can vary between 5 and 10 mg/kg/day BID. Benznidazole is formulated in scored tablets of 50 and 100 mg. A pediatric formulation (a 12.5 mg dispersible tablet) is currently in development.

Nifurtimox was developed by Bayer in 1970; usual doses are 10–15 mg/kg/day TID. Only a 120 mg scored tablet is currently available. A pediatric 30 mg dispersible tablet is under development.

Both drugs are well absorbed with good tissue distribution.

Treatment duration with any of these two drugs has been empirically set at 60 days, based on studies performed on older children [60, 63].

Newborns and infants have high response rates (over 95%) [35, 50, 59], after 60 days of treatment. An excellent response was also observed in a cohort of newborns treated for 30 days with once-daily (OD) benznidazole dosing [61]. These results suggest that further clinical studies are urgently needed to define the optimal dosing schedule of these drugs in order to improve compliance and decrease adverse drug reaction rates, given that efficacy rates seem to be close to optimal.

Table 1 summarizes the published clinical trials of congenital cases treated with benznidazole or nifurtimox with sufficient therapeutic data to provide meaningful conclusions. In the summarized studies, a decrease in antibody titers until negative seroconversion was observed.

Clinical studies have consistently shown a better tolerance of benznidazole or nifurtimox in young children and infants compared to older children (i.e., above 7 years of age) and adults, with treatment dropout rates close to zero in newborns [35, 43, 65, 70] and around 10% in infants and young children [63]. These rates increase up to 20–40% in adults [71].

The main adverse drug reactions (ADRs) associated with benznidazole and nifurtimox are similar and include anorexia, headache, irritability, sleep disturbances, leukopenia, thrombocytopenia, rash, gastrointestinal upset, increased liver enzymes, and neuropathy. Most of these ADRs are rare, except for mild rashes (more common with benznidazole) and anorexia, irritability, and headache (more common with nifurtimox) [42]. Gastrointestinal upset is also common, but problems associated with the administration of formulations which are not specifically developed for children may play a role in this ADR since tablets need to be crushed for administration.

In a large cohort of infants and children followed at the Parasitology and Chagas Service, Buenos Aires Children's Hospital "Ricardo Gutierrez," 69% of patients had no ADRs [72]. Most of the ADRs observed in this cohort were mild and did not require treatment interruption. Children older than 7 years of age had a higher frequency of mild rash that in some cases temporary treatment interruption and administration of antihistamines before completing treatment are required.

Premature and low-weight newborns seem to have higher rates of leukopenia and thrombocytopenia associated with the treatment. This observation led to the development of a treatment protocol in our institution that starts treatment with half the dose (e.g., 2.5 mg/kg/day instead of 5 mg/kg/day) until there is a normal blood count at 1 week of treatment after checking for ADRs. If no hematological ADRs are observed, full dose is then instituted until the full 60 days of treatment are completed [15].

Recent pediatric pharmacokinetic studies of benznidazole have shown that children have lower drug levels in blood compared to older children and adults, possibly due to a higher clearance rate. However, the therapeutic response was excellent, and ADRs were rare in the study population, suggesting that adults and older children may be receiving higher-than-necessary benznidazole doses and possibly dose reduction in this older population may lead to decreased ADR rates without affecting response rates [68]. This hypothesis remains to be formally proven in a clinical study to date.

In our experience, in a series of 62 congenitally infected children treated with nifurtimox, adverse effects were common, but most were mild (24% poor feeding, 14.5% irritability, and 6.5% vomiting). Three newborns had reversible leukopenia and thrombocytopenia [15]. Pediatric pharmacokinetic studies of nifurtimox in

Table 1 Efficacy of nifurtimox and benznidazole in congenital Chagas disease

Author	Age (years)	Benznidazole			Nifurtimox			Follow-up (months) lost to FU (%)	Efficacy endpoints[a]	
		n	Dose (mg/kg/day)	Length (days)	*n*	Dose (mg/kg/day)	Length (days)		Serological tests (% neg)	Parasitological tests (% pos)
Rubio and Donoso [64]	Congenital <1 year	–	–	–	4	25 + 12.5	15 + 90	18 m 25%	IHA 75% CFT 100%	Xeno 0%
Moya [42]	Congenital <1.5 years	–	–	–	40	15	60 = 90	6 years 0%	IHA, IFA, CFT 92%	Xeno 2%
Russomando [65]	Congenital <2 years	6	7–10	60	–	–	–	24 months 0%	EIA, IFA 100%	MH, PCR 0%
Blanco et al. [66]	Congenital <1 years	3	5	30	29	10	60	24 months 6%	EIA, IHA, IFA 94%	MH 0%
Schijman [50]	Congenital <2 years	16	5–8	60		10–15	60	36 months 0%	EIA, IHA 87%	MH, PCR 0%
Altcheh [35]	Congenital 15 days to 10 years	–	–	–	126	10–15	60	3 years 22%	EIA, IHA 87%	MH 0%
Chipaux [67]	Congenital <1 year	59 52	5 7.5	60 30	–	–	–	10 months 2% 4%	EIA 91% 90%	MH 0% 0%
Altcheh [68]	Congenital 2–12 years	107	5–8	60	–	–	–	36 months 0%	EIA, IHA 40%, and titers decrease	qPCR 0%
Balouz [69]	Congenital 8 days to 143 days 8 months to 9 years	12 6	5–8	60	–	–	–	36 months 0%	EIA 100% 38%	MH, qPCR 0% 0%

[a]Efficacy endpoints: CFT (complement fixation test), IHA (indirect hemagglutination assay), IFA (immunofluorescence assay), EIA (enzyme immune assay), Xeno (xenodiagnosis), MH (microhematocrit)

children are scarce. Currently a large clinical trial of a nifurtimox pediatric formulation is underway, and it includes a pharmacokinetic arm to try to address this gap [Clinicaltrials.gov#NCT02625974].

8.1 Follow-Up of Treatment Response

Treatment protocols of children with Chagas disease should include evaluation of potential ADRs, including blood counts and liver function tests, at 7, 30, and 60 days of treatment. If a newborn has a positive parasitemia at the start of the treatment, a weekly evaluation of parasitemia by a direct parasitological test like the microhematocrit test is recommended until it becomes negative usually between the second and third weeks of treatment [15, 35, 61]. The main criterion for therapeutic failure is the identification of the presence of the parasite by direct parasitological methods. If parasitemia persists, inappropriate compliance with the treatment should be considered, as well as (less likely) medication intolerance (e.g., vomiting) and parasite resistance. However, it should be mentioned that parasite resistance to benznidazole or nifurtimox seems to be extremely rare and well-documented examples of this event are extremely scarce. In case of treatment failure or parasite resistance, change of medication (benznidazole to nifurtimox or vice versa) is advisable.

After treatment is completed, patients need to be monitored with quantitative serological tests every 3 months during the first year and then every 6 months until serology becomes negative. After two consecutive negative serological tests, the patient is considered cured (Fig. 2). Cure rates, as evaluated by conventional serology, are over 90% in infants treated before 1 year of age [15, 35, 50, 59, 61, 65]. Newborns with positive parasitemia are considered cured when their serological tests become negative, around 8–12 months of age.

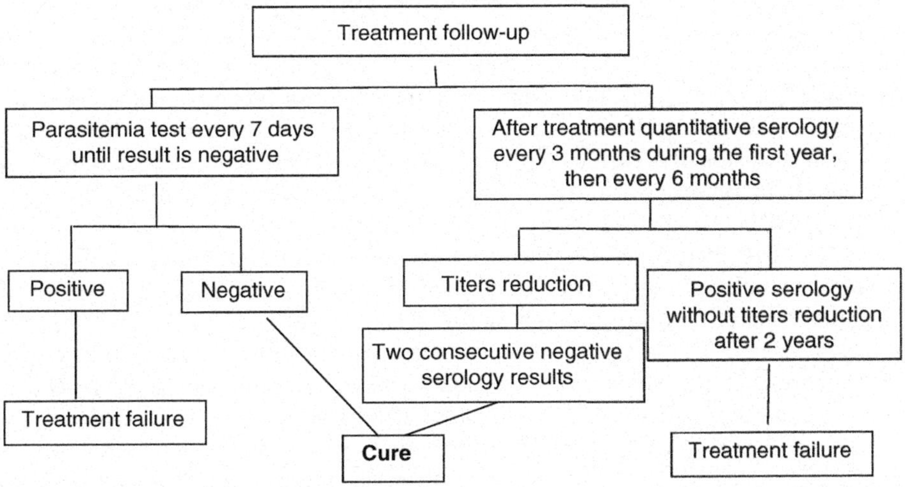

Fig. 2 Follow-up protocol for children treated with benznidazole or nifurtimox

The shorter the duration of the infection, the faster the serological tests become negative. Infants treated during the first year of life commonly become negative within 1 year of treatment. Older children show a slower rate of decrease in antibody titers that can take years to become negative [60, 63].

In patients with persistent high serological titers during posttreatment follow-up, a direct parasitological test such as PCR should be considered, and if positive, a new round of treatment should be evaluated. In such a case, evaluation should focus on compliance with the treatment, as the most common reason for treatment failure is, by far, non-compliance.

It is important to keep in mind that the longer the follow-up, the higher the probability of negative seroconversion.

9 Prevention of Congenital Infection

Primary prevention of congenital infection can be achieved by controlling vector transmission, which will avoid maternal infection. This strategy has been successful in many areas in Latin America. However several regions with high rates of vector transmission remain. These are linked to poor housing and house infestation with the vectors.

Secondary prevention by screening and treatment of pregnant women is currently not viable due to the lack of information on fetal safety for nifurtimox and benznidazole. An alternative way of secondary prevention, namely, pharmacological treatment of girls and women before pregnancy, has been shown to be effective in preventing vertical transmission of the disease [22, 23, 73].

Effective tertiary prevention by newborn screening, diagnosis, and treatment of infected infants requires universal screening strategies that identify infected pregnant women and allow evaluation of every newborn to these women. Unfortunately, these strategies are hampered by the high rate of migration of infected subjects and poor compliance with screening strategies in some areas by governments and health systems.

According to the recommendations of the Technical Group on "Prevention and Control of Congenital Transmission and Case Management of Congenital Infections," [47] serological testing is recommended for pregnant women (1) who are living in disease-endemic areas, (2) who are living in disease non-endemic areas and have occasionally received blood transfusion in disease-endemic areas, and (3) who are living in disease non-endemic areas and are born or have lived previously in disease-endemic areas or whose mothers were born in such areas.

Since 2010, Pan American Health Organization (PAHO) member states have committed to the elimination of mother-to-child transmission (EMTCT) of HIV and syphilis in Latin America. In addition, the PAHO Strategy and Plan of Action for Chagas Disease Prevention, Control, and Care [74] includes the specific objective of supporting implementation of secondary prevention of congenital Chagas disease. It recognizes that reducing mother-to-child transmission of Chagas disease requires *T. cruzi* screening for pregnant women as part of universal prenatal care as well as monitoring, diagnosing, and treating all newborns of infected mothers.

These commitments were renewed and expanded in 2016 through the approved Plan of Action for the Prevention and Control of HIV and Sexually Transmitted Infections (2016–2021), contributing to the end of AIDS and sexually transmitted infections as a public health problem in the Americas. The plan of action expands the EMTCT initiative (labeled "EMTCT Plus"), leveraging the maternal and child health platform to include elimination of other preventable communicable diseases in the Americas, such as hepatitis B and Chagas disease. A PAHO document was published in 2017 [75] proposing several interventions at different levels of the health system listed below.

9.1 EMTCT Plus Interventions at Different Levels of the Health System

Adolescence and Pre-pregnancy

- Diagnosis and treatment of *T. cruzi*-infected girls and women of childbearing age.

Pregnancy

- Routine serological screening for HIV and syphilis as well as Chagas disease and hepatitis B (HBsAg) when part of national policy.
- Follow-up of Chagas-seropositive pregnant women.

Childbirth

- *T. cruzi* parasitological and serological screening of newborns with infected mothers.

Postnatal Period: Mothers

- Treatment of *T. cruzi*-seropositive mothers after pregnancy (benznidazole or nifurtimox).

Postnatal Period: Newborns

- *T. cruzi* serology of newborns with infected mothers (at 8 months).
- Treatment of *T. cruzi*-seropositive children before 1 year of age (benznidazole or nifurtimox) and clinical and serological monitoring until negative.
- Immediate treatment of all newborns with positive parasitology for *T. cruzi*.

Cross-Cutting Interventions

- Chagas disease: accelerate actions to interrupt domiciliary transmission by the principal vectors.
- Chagas disease: consider serological testing of siblings of infants infected with *T. cruzi* (cluster approach).

10 Conclusion

Patients with Chagas disease can be found throughout the Americas as well as in most countries in Europe. This is due to the migration of infected women who run the risk of giving birth to infected babies in non-endemic areas. There is a need for active surveillance data in order to better refine prevalence data. Also, reported rates of congenital transmission show significant geographic variability. The reasons for this variability remain unexplained to date and should be investigated. Large gaps still exist in the understanding of the role of the genetic diversity of *T. cruzi*, if any, on the infection, risk of disease progression, and risk of congenital transmission. The actual number of congenital cases diagnosed and treated is lower than estimated. A significant obstacle for the diagnosis and treatment of infected newborns is the lack of highly sensitive diagnostic tests applicable at an early age and the high failure rate in tracking and following up infected pregnant mothers and their offspring. The treatment with benznidazole or nifurtimox is highly effective if it is administered as early as possible. Further clinical studies are urgently needed to define the optimal dosing schedule of drugs for treatment in order to improve compliance and decrease rates of adverse events. Pharmacokinetic studies in infants have shown that lower concentrations of benznidazole have produced excellent therapeutic responses. These results suggest that older children and adults may be receiving higher doses than necessary. Dose reduction in this older population may lead to decreased ADR rates without affecting response rates, but this hypothesis still has to be formally proved in a clinical study.

It is imperative to establish strategies that will allow:

- To systematically screen pregnant women for Chagas disease infection.
- To systematically evaluate every child born to an infected mother.
- To treat every infected child, as early as possible.

References

1. Verani JR, Montgomery SP, Schulkin J, Anderson B, Jones JL. Survey of obstetrician-gynecologists in the United States about Chagas disease. Am J Trop Med Hyg. 2010;83(4):891–5.
2. WHO. Chagas disease in Latin America: an epidemiological update based on 2010 estimates. Wkly Epidemiol Rec. 2015;90(6):11.
3. Howard EJ, Xiong X, Carlier Y, Sosa-Estani S, Buekens P. Frequency of the congenital transmission of Trypanosoma cruzi: a systematic review and meta-analysis. BJOG. 2014;121(1):22–33.
4. Soriano-Arandes A, Angheben A, Serre-Delcor N, Trevino-Maruri B, Gomez IPJ, Jackson Y. Control and management of congenital Chagas disease in Europe and other non-endemic countries: current policies and practices. Trop Med Int Health. 2016;21(5):590–6.
5. Moscatelli G, Garcia Bournissen F, Freilij H, Berenstein A, Tarlovsky A, Moroni S, et al. Impact of migration on the occurrence of new cases of Chagas disease in Buenos Aires city, Argentina. J Infect Dev Ctries. 2013;7(8):635–7.

6. Moscatelli G, Berenstein A, Tarlovsky A, Siniawski S, Biancardi M, Ballering G, et al. Urban Chagas disease in children and women in primary care centres in Buenos Aires, Argentina. Mem Inst Oswaldo Cruz. 2015;110(5):644–8.
7. Ministerio de salud A. Boletin epidemiologico anual. 2010;2010:63.
8. Hotez PJ. Neglected infections of poverty in the United States of America. PLoS Negl Trop Dis. 2008;2(6):e256.
9. Buekens P, Almendares O, Carlier Y, Dumonteil E, Eberhard M, Gamboa-Leon R, et al. Mother-to-child transmission of Chagas' disease in North America: why don't we do more? Matern Child Health J. 2008;12(3):283–6.
10. Cevallos AM, Hernandez R. Chagas' disease: pregnancy and congenital transmission. Biomed Res Int. 2014;2014:401864.
11. Bittencourt AL, Barbosa HS, Rocha T, Sodre I, Sodre A. Incidence of congenital transmission of Chagas' disease in premature births in the Maternidade Tsylla Balbino (Salvador, Bahia). Rev Inst Med Trop Sao Paulo. 1972;14(2):131–4.
12. Torrico F, Alonso-Vega C, Suarez E, Rodriguez P, Torrico MC, Dramaix M, et al. Maternal Trypanosoma cruzi infection, pregnancy outcome, morbidity, and mortality of congenitally infected and non-infected newborns in Bolivia. Am J Trop Med Hyg. 2004;70(2):201–9.
13. Duaso J, Rojo G, Cabrera G, Galanti N, Bosco C, Maya JD, et al. Trypanosoma cruzi induces tissue disorganization and destruction of chorionic villi in an ex vivo infection model of human placenta. Placenta. 2010;31(8):705–11.
14. Brutus L, Schneider D, Postigo J, Romero M, Santalla J, Chippaux JP. Congenital Chagas disease: diagnostic and clinical aspects in an area without vectorial transmission, Bermejo, Bolivia. Acta Trop. 2008;106(3):195–9.
15. Freilij H, Altcheh J. Congenital Chagas' disease: diagnostic and clinical aspects. Clin Infect Dis. 1995;21(3):551–5.
16. Sanchez Negrette O, Mora MC, Basombrio MA. High prevalence of congenital Trypanosoma cruzi infection and family clustering in Salta, Argentina. Pediatrics. 2005;115(6):e668–72.
17. Moretti E, Basso B, Castro I, Carrizo Paez M, Chaul M, Barbieri G, et al. Chagas' disease: study of congenital transmission in cases of acute maternal infection. Rev Soc Bras Med Trop. 2005;38(1):53–5.
18. Salas NA, Cot M, Schneider D, Mendoza B, Santalla JA, Postigo J, et al. Risk factors and consequences of congenital Chagas disease in Yacuiba, south Bolivia. Trop Med Int Health. 2007;12(12):1498–505.
19. Hermann E, Truyens C, Alonso-Vega C, Rodriguez P, Berthe A, Torrico F, et al. Congenital transmission of Trypanosoma cruzi is associated with maternal enhanced parasitemia and decreased production of interferon-gamma in response to parasite antigens. J Infect Dis. 2004;189(7):1274–81.
20. Kaplinski M, Jois M, Galdos-Cardenas G, Rendell VR, Shah V, Do RQ, et al. Sustained domestic vector exposure is associated with increased Chagas cardiomyopathy risk but decreased parasitemia and congenital transmission risk among young women in Bolivia. Clin Infect Dis. 2015;61(6):918–26.
21. Brutus L, Castillo H, Bernal C, Salas NA, Schneider D, Santalla JA, et al. Detectable Trypanosoma cruzi parasitemia during pregnancy and delivery as a risk factor for congenital Chagas disease. Am J Trop Med Hyg. 2010;83(5):1044–7.
22. Fabbro DL, Danesi E, Olivera V, Codebo MO, Denner S, Heredia C, et al. Trypanocide treatment of women infected with Trypanosoma cruzi and its effect on preventing congenital Chagas. PLoS Negl Trop Dis. 2014;8(11):e3312.
23. Moscatelli G, Moroni S, Garcia-Bournissen F, Ballering G, Bisio M, Freilij H, et al. Prevention of congenital Chagas through treatment of girls and women of childbearing age. Mem Inst Oswaldo Cruz. 2015;110(4):507–9.
24. Burgos JM, Altcheh J, Bisio M, Duffy T, Valadares HM, Seidenstein ME, et al. Direct molecular profiling of minicircle signatures and lineages of Trypanosoma cruzi bloodstream populations causing congenital Chagas disease. Int J Parasitol. 2007;37(12):1319–27.

25. Messenger LA, Miles MA, Bern C. Between a bug and a hard place: Trypanosoma cruzi genetic diversity and the clinical outcomes of Chagas disease. Expert Rev Anti-Infect Ther. 2015;13(8):995–1029.
26. Torrico MC, Solano M, Guzman JM, Parrado R, Suarez E, Alonzo-Vega C, et al. Estimation of the parasitemia in Trypanosoma cruzi human infection: high parasitemias are associated with severe and fatal congenital Chagas disease. Rev Soc Brasil Med Trop. 2005;38(Suppl 2):58–61.
27. Kemmerling U, Bosco C, Galanti N. Infection and invasion mechanisms of Trypanosoma cruzi in the congenital transmission of Chagas' disease: a proposal. Biol Res. 2010;43(3):307–16.
28. Carlier Y, Truyens C. Congenital Chagas disease as an ecological model of interactions between Trypanosoma cruzi parasites, pregnant women, placenta and fetuses. Acta Trop. 2015;151:103–15.
29. Norman FF, Lopez-Velez R. Chagas disease and breast-feeding. Emerg Infect Dis. 2013;19(10):1561–6.
30. Amato Neto V, Matsubara L, Campos R, Moreira AA, Pinto PL, Faccioli R, et al. Trypanosoma cruzi in the milk of women with chronic Chagas disease. Rev Hosp Clin. 1992;47(1):10–1.
31. Bittencourt AL, Sadigursky M, Da Silva AA, Menezes CA, Marianetti MM, Guerra SC, et al. Evaluation of Chagas' disease transmission through breast-feeding. Mem Inst Oswaldo Cruz. 1988;83(1):37–9.
32. Bern C, Martin DL, Gilman RH. Acute and congenital Chagas disease. Adv Parasitol. 2011;75:19–47.
33. Moya P, Moretti E, Paolasso R, Basso B, Blanco S, Sanmartino C, et al. Neonatal Chagas disease: laboratory diagnosis during the first year of life. Medicina. 1989;49(6):595–9.
34. Torrico F, Vega CA, Suarez E, Tellez T, Brutus L, Rodriguez P, et al. Are maternal re-infections with Trypanosoma cruzi associated with higher morbidity and mortality of congenital Chagas disease? Trop Med Int Health. 2006;11(5):628–35.
35. Altcheh J, Biancardi M, Lapena A, Ballering G, Freilij H. Congenital Chagas disease: experience in the Hospital de Ninos, Ricardo Gutierrez, Buenos Aires, Argentina. Rev Soc Brasil Med Trop. 2005;38(Suppl 2):41–5.
36. Bittencourt AL. Congenital Chagas disease. Am J Dis Child. 1976;130(1):97–103.
37. Azogue E, Darras C. Congenital Chagas in Bolivia: comparative study of the effectiveness and cost of diagnostic methods. Rev Soc Bras Med Trop. 1995;28(1):39–43.
38. Carlier Y, Sosa-Estani S, Luquetti AO, Buekens P. Congenital Chagas disease: an update. Mem Inst Oswaldo Cruz. 2015;110(3):363–8.
39. Bayer AM, Hunter GC, Gilman RH, Cornejo Del Carpio JG, Naquira C, Bern C, et al. Chagas disease, migration and community settlement patterns in Arequipa, Peru. PLoS Negl Trop Dis. 2009;3(12):e567.
40. Bittencourt AL, Mota E, Ribeiro Filho R, Fernandes LG, de Almeida PR, Sherlock I, et al. Incidence of congenital Chagas' disease in Bahia, Brazil. J Trop Pediatr. 1985;31(5):242–8.
41. Bittencourt AL, Sadigursky M, Barbosa HS. Congenital Chagas' disease. Study of 29 cases. Rev Inst Med Trop Sao Paulo. 1975;17(3):146–59.
42. Moya PR, Paolasso RD, Blanco S, Lapasset M, Sanmartino C, Basso B, et al. Treatment of Chagas' disease with nifurtimox during the first months of life. Medicina. 1985;45(5):553–8.
43. Moya P, Basso B, Moretti E. Congenital Chagas disease in Cordoba, Argentina: epidemiological, clinical, diagnostic, and therapeutic aspects. Experience of 30 years of follow up. Rev Soc Brasil Med Trop. 2005;38(Suppl 2):33–40.
44. Zaidenberg M. Congenital Chagas' disease in the province of Salta, Argentina, from 1980 to 1997. Rev Soc Bras Med Trop. 1999;32(6):689–95.
45. Freilij H, Altcheh J, Muchinik G. Perinatal human immunodeficiency virus infection and congenital Chagas' disease. Pediatr Infect Dis J. 1995;14(2):161–2.
46. WHO EC. Control of Chagas disease. World Health Organization technical report series. Brussels: WHO EC; 2002. 905:i-vi, 1–109, back cover
47. Carlier Y, Torrico F, Sosa-Estani S, Russomando G, Luquetti A, Freilij H, et al. Congenital Chagas disease: recommendations for diagnosis, treatment and control of newborns, siblings and pregnant women. PLoS Negl Trop Dis. 2011;5(10):e1250.

48. Fernandez-Aguilar S, Lambot MA, Torrico F, Alonso-Vega C, Cordoba M, Suarez E, et al. Placental lesions in human Trypanosoma cruzi infection. Rev Soc Brasil Med Trop. 2005;38(Suppl 2):84–6.
49. Duffy T, Bisio M, Altcheh J, Burgos JM, Diez M, Levin MJ, et al. Accurate real-time PCR strategy for monitoring bloodstream parasitic loads in chagas disease patients. PLoS Negl Trop Dis. 2009;3(4):e419.
50. Schijman AG, Altcheh J, Burgos JM, Biancardi M, Bisio M, Levin MJ, et al. Aetiological treatment of congenital Chagas' disease diagnosed and monitored by the polymerase chain reaction. J Antimicrob Chemother. 2003;52(3):441–9.
51. Messenger LA, Gilman RH, Verastegui M, Galdos-Cardenas G, Sanchez G, Valencia E, et al. Toward improving early diagnosis of congenital Chagas disease in an endemic setting. Clin Infect Dis. 2017;65(2):268–75.
52. Bern C, Verastegui M, Gilman RH, Lafuente C, Galdos-Cardenas G, Calderon M, et al. Congenital Trypanosoma cruzi transmission in Santa Cruz, Bolivia. Clin Infect Dis. 2009;49(11):1667–74.
53. Oliveira I, Torrico F, Munoz J, Gascon J. Congenital transmission of Chagas disease: a clinical approach. Expert Rev Anti-Infect Ther. 2010;8(8):945–56.
54. Besuschio SA, Llano Murcia M, Benatar AF, Monnerat S, Cruz I, Picado A, et al. Analytical sensitivity and specificity of a loop-mediated isothermal amplification (LAMP) kit prototype for detection of Trypanosoma cruzi DNA in human blood samples. PLoS Negl Trop Dis. 2017;11(7):e0005779.
55. Rivero R, Bisio M, Velazquez EB, Esteva MI, Scollo K, Gonzalez NL, et al. Rapid detection of Trypanosoma cruzi by colorimetric loop-mediated isothermal amplification (LAMP): a potential novel tool for the detection of congenital Chagas infection. Diagn Microbiol Infect Dis. 2017;89(1):26–8.
56. Leguizamon MS, Campetella OE, Reyes MB, Ibanez CF, Basombrio MA, Rincon J, et al. Bloodstream Trypanosoma cruzi parasites from mice simultaneously express antigens that are markers of acute and chronic human Chagas disease. Parasitology. 1991;102(Pt 3):379–85.
57. Reyes MB, Lorca M, Munoz P, Frasch AC. Fetal IgG specificities against Trypanosoma cruzi antigens in infected newborns. Proc Natl Acad Sci U S A. 1990;87(7):2846–50.
58. Mallimaci MC, Sosa-Estani S, Russomando G, Sanchez Z, Sijvarger C, Alvarez IM, et al. Early diagnosis of congenital Trypanosoma cruzi infection, using shed acute phase antigen, in Ushuaia, Tierra del Fuego, Argentina. Am J Trop Med Hyg. 2010;82(1):55–9.
59. Altcheh J, Corral R, Biancardi MA, Freilij H. Anti-F2/3 antibodies as cure marker in children with congenital Trypanosoma cruzi infection. Medicina. 2003;63(1):37–40.
60. de Andrade AL, Zicker F, de Oliveira RM, Almeida Silva S, Luquetti A, Travassos LR, et al. Randomised trial of efficacy of benznidazole in treatment of early Trypanosoma cruzi infection. Lancet. 1996;348(9039):1407–13.
61. Chippaux JP, Clavijo AN, Santalla JA, Postigo JR, Schneider D, Brutus L. Antibody drop in newborns congenitally infected by Trypanosoma cruzi treated with benznidazole. Trop Med Int Health. 2010;15(1):87–93.
62. Sosa-Estani S, Segura EL. Etiological treatment in patients infected by Trypanosoma cruzi: experiences in Argentina. Curr Opin Infect Dis. 2006;19(6):583–7.
63. Sosa Estani S, Segura EL, Ruiz AM, Velazquez E, Porcel BM, Yampotis C. Efficacy of chemotherapy with benznidazole in children in the indeterminate phase of Chagas' disease. Am J Trop Med Hyg. 1998;59(4):526–9.
64. Rubio M, Donoso F. Enfermedad de Chagas en niños y tratamiento con Bay 2502. Bol Chile Parasitol. 1969;24:43–8.
65. Russomando G, de Tomassone MM, de Guillen I, Acosta N, Vera N, Almiron M, et al. Treatment of congenital Chagas' disease diagnosed and followed up by the polymerase chain reaction. Am J Trop Med Hyg. 1998;59(3):487–91.
66. Blanco SB, Segura EL, Cura EN, Chuit R, Tulián L, et al. Congenital transmisión of Trypanosoma cruzi: and operational outline for detecting and treating infected infants in northwestern Argentina. Trop Med Int Health. 2000;5:293–301.

67. Chippaux JP, Clavijo AN, Santalla JA, Postigo JR, Schneider D, Brutus L. Antibody drop in newborns congenitally infected by Trypanosoma cruzi treated with benznidazole. Trop Med Int Health. 2010;15:87–93.
68. Altcheh J, Moscatelli G, Mastrantonio G, Moroni S, Giglio N, Marson ME, et al. Population pharmacokinetic study of benznidazole in pediatric Chagas disease suggests efficacy despite lower plasma concentrations than in adults. PLoS Negl Trop Dis. 2014;8(5):e2907.
69. Balouz V, Melli LJ, Volcovich R, Moscatelli G, Moroni S, González N, Ballering G, Bisio M, Ciocchini AE, Buscaglia CA, Altcheh J. The trypomastigote small surface antigen from trypanosoma cruzi improves treatment evaluation and diagnosis in pediatric chagas disease. J Clin Microbiol. 2017;55:3444–53.
70. Chippaux JP, Salas-Clavijo AN, Postigo JR, Schneider D, Santalla JA, Brutus L. Evaluation of compliance to congenital Chagas disease treatment: results of a randomised trial in Bolivia. Trans R Soc Trop Med Hyg. 2013;107(1):1–7.
71. Lescure FX, Le Loup G, Freilij H, Develoux M, Paris L, Brutus L, et al. Chagas disease: changes in knowledge and management. Lancet Infect Dis. 2010;10(8):556–70.
72. Altcheh J, Moscatelli G, Moroni S, Garcia-Bournissen F, Freilij H. Adverse events after the use of benznidazole in infants and children with Chagas disease. Pediatrics. 2011;127(1):e212–8.
73. Murcia L, Simon M, Carrilero B, Roig M, Segovia M. Treatment of Infected Women of Childbearing Age Prevents Congenital Trypanosoma cruzi Infection by Eliminating the Parasitemia Detected by PCR. J Infect Dis. 2017;215(9):1452–8.
74. PAHO/WHO. Strategy and plan of action for Chagas disease prevention, control and care. Washington, DC: PAHO; 2010. p. CE146/14.
75. PAHO. EMTCT plus. Framework for elimination of mother-to-child transmission of HIV, syphilis, hepatitis B, and Chagas. Washington, DC: PAHO PAHO; 2017. p. 17–009.

Clinical Care for Individuals with Chronic *Trypanosoma cruzi* Infection: Decision-Making in the Midst of Uncertainty

Juan Carlos Villar and Pablo Andrés Bermudez

Abstract This chapter discusses the reasoning behind clinical decisions for most individuals with chronic *Trypanosoma cruzi* infection (i.e., seropositive adults). It covers the goals and suggested strategies for diagnosis and treatment across the clinical spectrum of chronic Chagas cardiomyopathy (CCC), as the most relevant outcome for this population.

Uncertainty surrounds most clinical decisions for individuals at risk or with diagnosis of CCC. For such complex, neglected tropical disease, it has been very challenging to produce rigorous, longitudinal studies testing modern prognosis and therapy tools and recording a sufficient number of clinical events.

As a condition, being seropositive increases the risk of mortality at any clinical stage. Since the emergence of CCC is relatively infrequent, but substantial clinically, ruling out disease is critical upon serology diagnosis. We therefore propose a high-sensitivity initial approach, combining B-type natriuretic peptide with an electrocardiography (with a conventional ECG, adding a continuous monitoring for at least 1 h when negative). For those diagnosed with CCC, we propose to make use of Rassi's (the only externally validated) scale for a risk-oriented clinical staging and disease monitoring.

As high-risk individuals, treatment should be considered always, paralleling complexity with CCC staging. Those with early or no disease will probably derive more benefit from trypanocidal therapy (TT) in order to clear parasitism and may not need another measure while CCC-free. A special, very important case is that of seropositive women in childbearing age, who should receive TT to prevent congeni-

J. C. Villar (✉)
Research Department, Fundación Cardioinfantil-Instituto de Cardiología, Bogotá, Colombia

Faculty of Health Sciences, Universidad Autónoma de Bucaramanga,
Bucaramanga, Colombia
e-mail: jvillar@unab.edu.co

P. A. Bermudez
Research Department, Fundación Cardioinfantil-Instituto de Cardiología, Bogotá, Colombia

J. M. Altcheh, H. Freilij (eds.), *Chagas Disease*, Birkhäuser Advances in
Infectious Diseases, https://doi.org/10.1007/978-3-030-00054-7_10

tal transmission in future pregnancies. Adjuvant, proven treatment for heart failure from other origins is increasingly important for CCC patients and should be offered aggressively from early stages as an attempt to prevent its devastating complications and mortality.

As described in other sections of this book, most humans infected by *Trypanosoma cruzi* recover from acute infection. Although this initial phase may be life-threatening (half of the cases of orally transmitted outbreaks show severe symptoms and about 5% are fatal) [1, 2], most cases pass without specific diagnosis [3]. That majority surviving the acute phase will go through to a long-term pathophysiological process, the chronic phase of this parasitic infection. Over their life span, the host's immune system battles to clear parasites and minimize the tissue damage through a complex interaction [4].

The scarcity of circulating forms of the parasite (*Trypomastigote*) and the concurrent emergence of antibodies distinguish the chronic phase. Once more, most individuals will keep biological equilibrium with the parasite over their life span. From the clinical point of view, that majority will have positive serology tests, but little or no symptoms. In 1985, an expert consensus labeled as *indeterminate phase* of chronic infection the status of asymptomatic seropositive individuals with no findings in the electrocardiogram (ECG) or X-rays [5]. Those individuals will have, for the most part, clinically uneventful lives [6].

It is still not fully understood why a relatively small proportion (10–30%) of individuals with the chronic phase will develop clinical disease and its complications [7]. This is the so-called determinate phase, which may express as cardiac, digestive, or both forms. On the basis of an initial parasitic infection as a trigger, a variety of mechanisms may lead to tissue damage (and clinical progression) in the chronic phase [8]. Since most chronically infected will not develop disease, parasitism is a necessary but not sufficient condition to develop complications. However, the continued presence of parasites may be a key (although perhaps not the only) factor to maintain pathophysiological mechanisms [9].

This chapter will discuss the reasoning behind clinical decisions for individuals with diagnosis of chronic *T. cruzi* infection. As starting assumption, we will equate having a positive serology (i.e., two positive reactions to at least two different techniques) as working definition for chronic *T. cruzi* infection [10, 11]. We will focus on chronic Chagas cardiomyopathy (CCC), which is by far the most important complication of Chagas disease.

1 Two Working Clinical Scenarios

As starting point, two common cases in the field will serve to set our clinical scenarios. The first, referring to primary prevention, is that of an individual coming to office because of a positive serology as an incidental laboratory finding. More often this has to do with young or middle-aged adults with a positive serology, but no symptoms suggesting CCC. Typical cases include referred blood donors from banks or, increasingly, future mothers with a positive screening tests in Latin America or abroad [12–15].

The second scenario refers to secondary prevention. Here, individuals seeking medical care because of symptoms or clinical abnormalities raise suspicion of CCC, where serology is required for confirmation. This is a broad scenario, ranging from a patient having incidental findings in a medical checkup to someone coming to emergencies with a pulmonary edema or life-threatening arrhythmias [16, 17]. These cases represent the range of CCC, mainly the most severe CCC cases, demanding more specialized care.

The main difference in the above situations is that in primary prevention, serology comes first, mostly as a screening policy or requirement (e.g., for blood banks), rather than the usual process of medical diagnosis. Seropositive blood donors (which in truth are false-negative cases of the screening questionnaires at the banks) should be the exemption, but yet remain the major source of identification of such cases [18, 19]. The rule should be identifying infected individuals as a result of awareness campaigns promoting the diagnosis of infection among those more likely (i.e., when analysis of a pretest probability makes a doctor order a test) [20, 21]. In the second scenario, with symptoms coming first, diagnosis of *T. cruzi* infection, the exposure, comes decades after the adverse outcome, CCC [22]. Going from the clinical event to investigate the risk factor would be, in modern medicine, like diagnosing hypertension in a patient admitted for a hemorrhagic stroke.

Being common, our two scenarios illustrate the shortcomings of the process of care (i.e., identifying, making a prognosis, and avoiding consequences) for individuals at risk, that minority of seropositive individuals who eventually develop CCC. These two situations, highly unusual in today's medicine, point out how neglected Chagas disease still is [23].

1.1 Management Strategies for Each Scenario

Patients with chronic *T. cruzi* infection facing individual situations require their physicians to share the best evidence available to assist them in making informed decisions [24]. However, as neglected disease, despite the burden of disease, medical care for patients at risk of CCC is surrounded by abundant knowledge gaps. One of the reasons is a vicious circle fed by the lack of access or dispersion of care and the relatively lack of clinical research.

The above scenarios represent somewhat extreme situations (the former with very little or no disease, the latter with very advanced, irreversible diseases), outside what should be the main target for care of seropositive population. Those who could derive more benefit (i.e., the cases of mild to moderate CCC) are somehow absent from these scenarios. Such population should be the bulk of the work of clinics specializing in Chagas disease, accruing patients, nesting clinical research, and delivering increasingly better care. In the absence of awareness and diagnostic campaigns for *T. cruzi* infection, such clinics remain an exemption, unfortunately. In the meantime, medical decisions will, for the most part, have to rely on scarce or low-quality evidence. That includes having a majority of studies at moderate or high risk of bias, inconsistent results, biased selection or citation of positive studies, and extrapolation from other heart diseases along with clinical judgment or mechanistic inferences.

So, what should be the focus for each of the above scenarios? The natural history of CCC indicates that, as disease progresses, fewer people demand increasingly more health-care resources. The condition of being seropositive confers, even from asymptomatic or early stages of CCC, two- to threefold higher risk of mortality [25, 26]. And once fully blown, CCC seems to behave as a more severe form of dilated cardiomyopathy, compared with other causes [27, 28]. As a condition, the annual risk of death or major cardiac events can be around 1% for unselected seropositive individuals in the community, mostly asymptomatic. But it can go up to around 5%, 15%, or even 30% a year, for population in outpatient clinics, or series of CCC cases with moderate to severe disease, respectively (Fig. 1).

This underscores the need for prevention, with specific goals along the continuum of CCC natural history, in terms of diagnosis and treatment. Since only a minority progress to advanced forms, the goal for the first (primary prevention) scenario should be to rule out subclinical, early CCC in primary care for that majority of infected individuals. It has been estimated that the rate of progress to nonfatal CCC among blood donors found seropositive (assuming they were all CCC-free at baseline) is 1.85% per year in a 10-year span [29]. The secondary prevention scenario requires, when CCC is still mild, ruling out advanced CCC and preventing deterioration. It will vary greatly according to individual clinical stages (discussed in a section below). For advanced CCC cases, medical decisions will be oriented toward increasing survival free of complications, maintaining quality of life while monitoring progression (Fig. 2).

Clinicians should also adapt their treatment tools along natural history. The dynamics of pathophysiological mechanisms may be simpler and slower in the first stages, with increasing complexity and speed when disease progresses. Following that line, treatment approach may be simpler, likely requiring trypanocidal therapy (TT) only, seeking parasite clearance. The importance of (or responsiveness to) TT may decline in more advanced stages, where ancillary therapy (AT) should be instated. But the center of therapy should shift gradually to using the therapeutic arsenal available to manage heart failure of other causes, including more complex therapies (i.e., electric or mechanical devices to support cardiac function) for those already suffering major cardiac events or at high risk of death.

In the same line, follow-up strategies should adapt to disease progression (bottom part of Fig. 2). This implies different clinical attitudes in ruled-in CCC cases and those without diagnostic findings. Infected individuals with no findings in the primary prevention scenario will probably not derive benefit from reassessments in the short or median term (i.e., less than 3 years). In contrast, patients with moderate or severe CCC will need more regular follow-up (i.e., within months), in order to monitor risk and start, tune up, or modify their therapy.

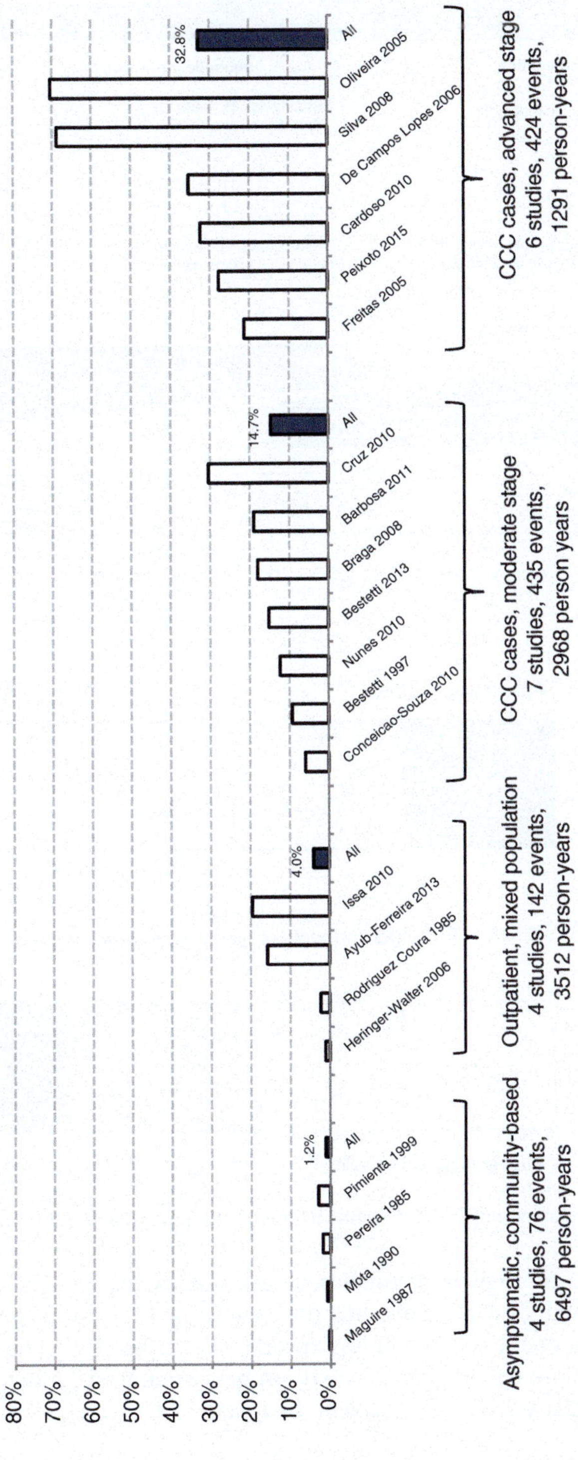

Fig. 1 Annual incidence of death or major cardiac events in 21 Brazilian cohort studies including different subsets of *T. cruzi* seropositive individuals. Data from individual studies as abstracted by Cucunuba and colleagues [25], excluding two studies originated outside Brazil and one Brazilian study focusing on participants over 60 years old

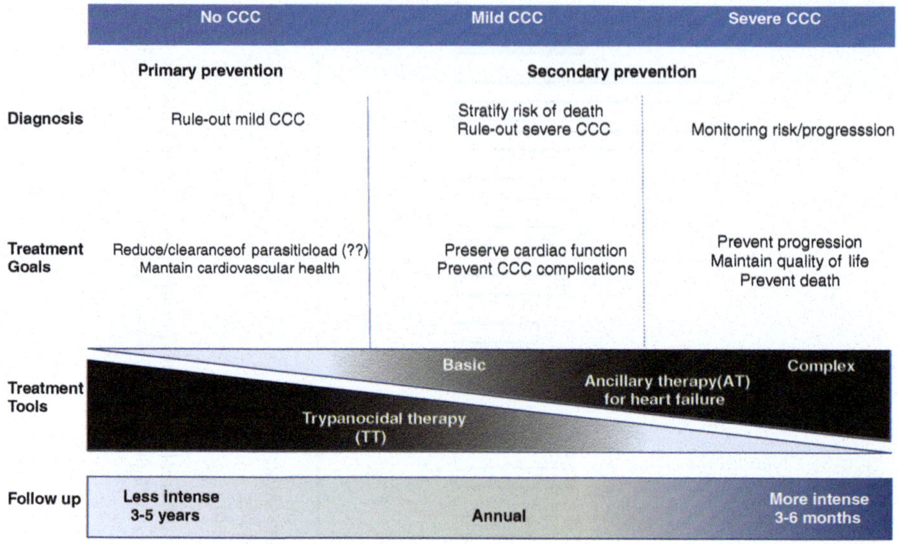

Fig. 2 Management strategy and goals across the natural history of chronic chagas cardiomyopthy

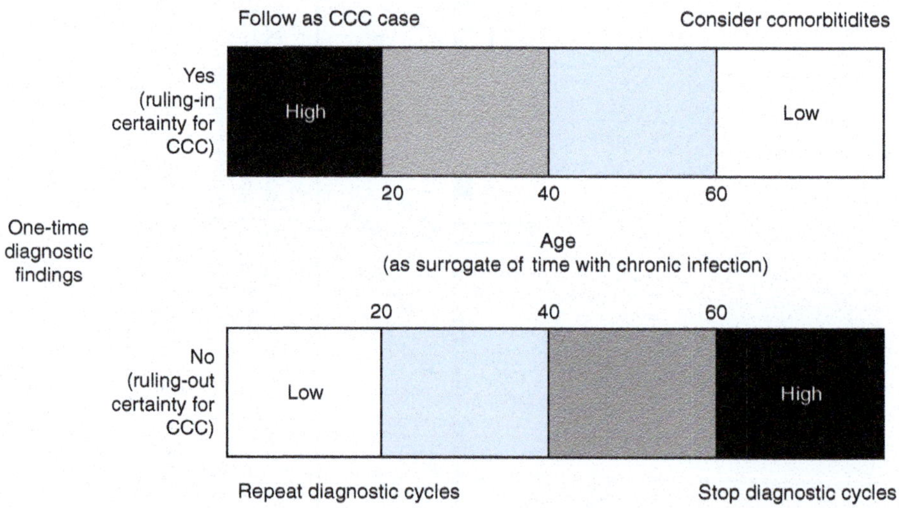

Fig. 3 Varying interpretation and use of diagnostic findings along the continuum of disease

Applying diagnostic tools in population at risk of CCC needs also to take into account the continuum of disease over the life span (Fig. 3). A positive cardiac test in the early decades of life may be highly specific (i.e., it had little chance to be a false positive). In those cases, clinicians will probably not need further diagnostic investigation to classify (rule in) a patient as a case of CCC. This would be the opposite in the elderly, where doctors may experience difficulty attributing diagnostic findings to *T. cruzi* infection rather than other causes (e.g., they may need to rule out an overlapping coronary artery disease). In the opposite situation, the absence of diagnostic findings, ruling-out criteria should also change across the life span. In the early decades of life, negative screening tests will demand future follow-up visits, as the situation may change over relatively few years. But having no such findings in middle age, especially older adults, may suffice to rule out disease.

1.2 Use of Clinical Tools for Disease Staging and Prognosis in CCC

Staging and prognosis are key considerations to establish treatment goals and make shared expectations with patients. This section discusses how clinicians can make an evidence-based rational use of the tools in hand to maintain health (by preventing the emergence of CCC and preserving cardiac function) among seropositive or survival and quality of life (by avoiding deterioration and medical complications) for those already diagnosed with CCC. We will first consider the challenges in classifying seropositive individuals by making disease staging and prognosis for those at risk of or already with CCC. Treatment options are covered in the following section of this chapter.

Traditional staging of those with chronic *T. cruzi* infection has relied, for the most part, on two classification schemes, both based on the presence of symptoms (in this case, those suggesting heart failure), X-rays and the 12-lead ECG. The most widely used form to classify both individuals and populations with a positive *T. cruzi* serology is as indeterminate and, consequently, determinate forms (cardiac, in this case) of Chagas disease. A second classification came originally from Kuschnir and colleagues who (in order to analyze the results in a diagnostic study) decided to group their seropositive population into a four-level ordinal scale [30]. In the first of these levels, there are no findings other than a positive serology. It equates to the indeterminate phase (called 0 in Kuschnir's work). The following (class I) level requires ECG abnormalities; the next will add cardiac enlargement (originally when seen at chest X-rays), and the final (class III) compiles ECG changes and cardiomegaly with symptoms suggesting heart failure.

Despite their popularity, practicality, and generally good risk discrimination capacity, those classifications have limitations, mainly in two fronts. The first criticism comes from the variety of abnormalities discovered when individuals in early stages are tested with diagnostic tools other than the ECG [31–37]. Since ECG status may be inconsistent over time in seropositive asymptomatic individuals [38], Kuschnir's classes 0 and I are almost interchangeable. That results in misclassification, especially in the early stages, given the reliance on the ECG as the only diagnostic tool. In fact, most of the discrimination ability of this classification comes from differences between early stages (classes 0 and I) with the more advanced [39–41]. Furthermore, for definitive CCC cases (classes II and III), Kuschnir's levels may not change much, while patients are changing in their functional status. That is, among those with established ECG changes, chest radiographs may not have the responsiveness to change (i.e., NYHA's functional class) that patients and clinicians will need. In short, classifying seropositive populations into the dichotomous (indeterminate or determinate) status or Kuschnir's ordinal scale may be helpful at a group level, but hardly for clinicians following individual cases.

We will now discuss the use of ECG and other clinical tools that may assist staging seropositive individuals in both the primary and secondary prevention scenarios.

Primary prevention scenario: As stated above, it is expected that only a minority of *T. cruzi* seropositive cases will develop CCC. Since the driver for decision-making in most cases should be to rule out disease accurately, sensitivity (i.e., having the lowest possible false-negative cases of CCC) is critical at this point. Ideally, such a test would be one that has little cost, is portable and accurate, and if applied once could rule out every possible case of CCC at least for a reasonable period of time.

12-lead surface electrocardiogram (ECG), *T. cruzi* chronic infection treatment options and primary prevention scenario has been widely used as ruling-out diagnostic field tool in the field. However, its sensitivity may be questionable, as indicated above. A few population-based cohort studies in Brazilian population recording mortality data after long-term follow-up provide the more solid evidence available in that regard [42, 43]. Fig. 4 shows such data for seropositive individuals, using study-specific definitions for ECG abnormalities (right bundle branch block

Mg Guire et al. 1987

a. BBBR and VE

		Death +	Death -	TOTAL		
ECG	+	5	4	9	SENS	29,4%
	-	12	182	194	SPEC	97,8%
	Total	17	186	203	PPV	55,5%
					NPV	93,8%
					+LR	13,7
					-LR	0,72

b. BBBR or VE

		Death +	Death -	TOTAL		
ECG	+	14	49	63	SENS	82,4%
	-	3	137	140	SPEC	73,6%
	Total	17	186	203	PPV	22,2%
					NPV	97,8%
					+LR	3,12
					-LR	0,23

Mota et al. 1990

c. Abnormal findings

		Death +	Death -	TOTAL		
ECG	+	20	130	150	SENS	62,5%
	-	12	350	362	SPEC	72,9%
	Total	32	480	512	PPV	13,3%
					NPV	96,6%
					+LR	2,30
					-LR	0,51

d. Borderline or abnormal findings

		Death +	Death -	TOTAL		
ECG	+	27	233	260	SENS	84,3%
	-	5	247	252	SPEC	51,5%
	Total	32	480	512	PPV	10,4%
					NPV	98,0%
					+LR	1,74
					-LR	0,30

Fig. 4 12-Lead ECG assessment criteria for prediction of death among *Trypanosoma cruzi* seropositive in four different scenarios as reported in two cohort studies. *RBBB* right bundle branch block, *EV* ventricular extra systole; ¶ any of the following: P waves, Q/QS waves, pattern of ventricular hypertrophy, T wave or ST segment alterations, low QRS voltage, left axis deviation, atrioventricular blocks, short PR interval, ventricular conductions defects, repetitive atrial or junctional extra systole, ventricular extra systole, frequent, bigeminal, paired, multiform, or R-on-T phenomena, atrial rhythm or tachycardia, atrioventricular junction tachycardia, A-V dissociation, SA block or sinus arrest, atrial fibrillation or flutter. Borderline refers to any finding making an ECG reading considered not normal. *SENS* sensitivity, *SPEC* specificity, *PPV* positive predictive value, *NPV* negative predictive value, *+LR* positive likelihood ratio, *−LR* negative likelihood ratio

[RBBB] *and* ventricular extrasystole [VE] for Maguire et al., or an even more sensitive definition, any abnormalities for Mota et al.). Despite the relatively little statistical precision (203 people, 17 events for Maguire et al.; 512 people, 32 events for Mota et al.), sensitivity rates were 29.4% and 62.5% with their more specific definitions. When redefining to the most sensitive criteria (RBBB *or* EV for Maguire et al. and *any borderline* or abnormal ECG finding for Mota et al.), there are still 16–18% false-negative cases for mortality. In fact, sensitivities did not improve when using the broadest possible definition at a much higher cost in terms of false positives (specificity 73.6% for Maguire et al., going down to 51.5% for Mota et al.).

No doubt, ECG has value in itself to distinguish populations exposed to infection (i.e., when comparing seropositive with seronegative individuals) in areas of vector transmission [44] and, of course, seropositives at higher risk of death [42]. However, the issue is its limited sensitivity at individual level. Repeated ECGs to the same person over time may theoretically increase sensitivity further. But achieving certainty to rule out CCC based on serial ECGs will take longer than patients expect. In addition, it entails the risk of missing them from follow-up. A conventional ECG has therefore limited value for the purpose of informing individual decisions in this ruling-out scenario.

One way to enhance sensitivity while still having a one-time assessment is to increase the window of observation by using continuous ECG monitoring (cECGm) [37]. We are not aware of population- or community-based cohorts of seropositive individuals assessing cECGm, not necessarily over 24 h (as in a conventional Holter cECGm, but even shorter periods), that record patient-important outcomes over a follow-up. However, studies in apparently healthy individuals (seronegatives, from non-endemic areas for CCC) may help build some extrapolation around the presence of VE. Only around 3% of such individuals have more than 200 EV in 24 h, and less than 1% will record over 2000 [45, 46]. Thus, using 24-h monitoring to detect "clinically significant" numbers of EV in 24 h (say over 200 a day, or even 10 per hour in short-term recordings) may greatly enhance sensitivity while keeping specificity within reasonable limits. On that line, observing more than 10 EV an hour may be a reasonable threshold to discriminate a case of CCC in a seropositive. And other findings (that are virtually absent in healthy seronegative individuals) in a cECGm, such as ventricular tachycardia, sinus bradycardia, or atrioventricular blocks, will per se rule in CCC in this scenario.

cECGm, be this over 24 h or short term, may be less practicable than a single-strip, conventional 12-lead ECG (which is available even in rural areas of endemic countries). But a clinician could be more certain of the absence of CCC if a seropositive with a "negative" (i.e., no established rhythm or conduction abnormalities) conventional ECG *has also* no findings in a cECGm. Having much better diagnostic properties, cECGm would be thus advisable for those with a negative conventional ECG [47], and if in turn cECGm were negative, the procedure should be repeated every 3–5 years. Fortunately, for some ECG machines (or ECG monitors with recording capabilities), doing short-term cECGm is possible. We are also assuming that availability of 24-h Holter for a center implies access to a conventional ECG. Therefore, performing a serial assessment of ECG and then, if negative, a cECGm (or formal 24-h Holter monitoring) should be both useful and feasible.

Seeking to maximize our ruling-out capacity may also need to screen cardiac function abnormalities, other than those of rhythm and conduction. *T. cruzi* seropositive individuals in the "indeterminate" phase (i.e., with normal ECG and chest X-rays) may have abnormalities in the cardiac contractility or autonomic function [31]. Contractility may be assessed through standardized procedures in cardiology, whereas autonomic function has not reached that status. The variety of cardiac imaging options that are currently available may be both expensive and impracticable in this scenario.

B-type natriuretic peptides (BNP) prove to be sensitive enough by themselves to confidently rule out cardiac disease in population-based studies of seronegative populations [48, 49]. Further, these peptides are elevated among those with significant amount of EV under 24-h Holter ECG monitoring [46]. BNP have shown to have excellent association with cardiac function across different stages [50, 51] and response to change to guide treatment for patients with heart failure [52–54]. But once more, we are not aware of cohort studies assessing BNP values among *T. cruzi* seropositive asymptomatic (young or middle-aged) adults who are followed long-term for CCC outcomes. Thus, the value of these markers for our particular context of application remains unknown. However, by extrapolation from (a) cross-sectional studies of BNP across groups with different CCC stages [55, 56] and (b) a cohort study of BNP in seropositive individuals over 60 years [57], we may be confident that individuals with normal values may be very unlikely to develop clinical events in the midterm.

We therefore recommend that, for the goal of ruling out accurately CCC among *T. cruzi* seropositive individuals, BNP assessment be added to the series of ECG/cECGm. Some may argue that is not practical. However, every new infected case needs a serum sample in order to confirm serology. BNP may very well be measured when the serology result is positive and reported along with the confirmed serology to start classifying CCC stage at the onset of diagnosis. Other blood-derived biomarkers (of myocardial inflammation or necrosis, such as metalloproteinases or troponin, among others) may have a potential for excluding CCC cases or stratifying their risk (discussed in the next section). However, in this context they would add significant cost and may overlap, rather than improve diagnostic value [58, 59].

BNP thus may replace the need for assessment of cardiac images to assess ventricular function [60], and even add value for screening CCC [36]. In the absence of good-quality, longitudinal studies ascertaining the prognosis of those seropositive with normal BNP values (or comparing efficiency of newer biomarkers with BNP), we recommend adding the ECG/cECGm due to BNP to make sure a negative value corresponds to a CCC true-negative case. The left side of Fig. 5 shows a suggested use of those diagnostic tools in this prevention scenario.

Secondary prevention scenario: Once identified, an individual with CCC may be in a range of stages, with varying risk of mortality or nonfatal yet devastating outcomes. In the same line of the previous scenario, patients themselves, their relatives, and clinicians would want to make sure there are no signs of severe CCC. As a working definition here, we suggest here that severe CCC will be a stage where mortality risk exceeds 10% at 10 years. That would be a similar frame to what the

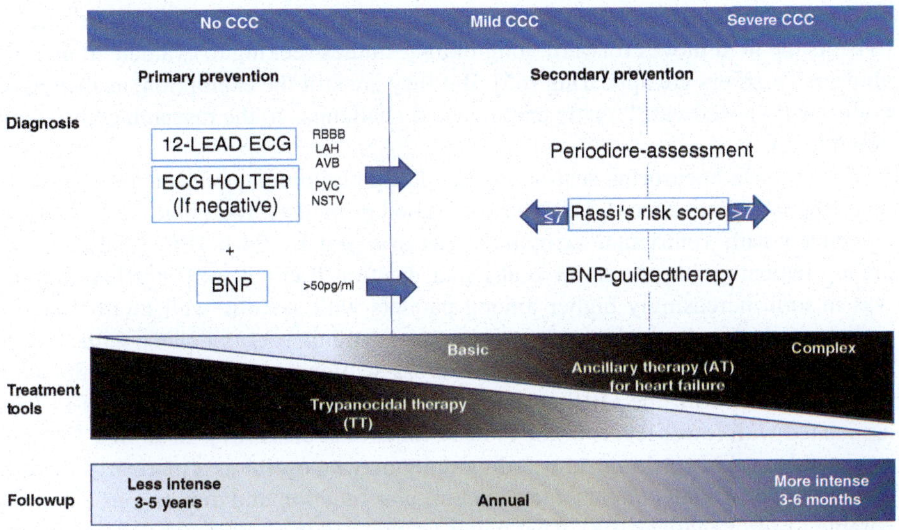

Fig. 5 Use of diagnostic tools in different phases of the natural history of chronic Chagas cardiomyopathy

Framingham risk score has traditionally used to divide intermediate from low risk [61]. We are setting two safeguards here, for the context of CCC. First, we are setting a lower threshold to define high risk (high risk for CCC equals to intermediate risk for atherosclerotic events), because CCC entails a higher proportion of lethal events. Second, we will propose periodic risk reassessments that will be much shorter than a 10-year estimate.

Generally speaking, individual markers of progress of CCC, from those based on electrocardiography (be this a conventional ECG, cECGm recordings, signal-averaged ECG, or heart rate variability) to diagnostic images (including echocardiography, cardiac magnetic resonance images) or biomarkers (including BNP), are not supported by strong evidence. What characterizes the field so far is the abundance of cross-sectional studies, with groups of seropositive participants (typically using Kuschnir's approach) and sometimes seronegatives as control group. These studies compare the values of the proposed markers across groups with increasing levels of cardiac involvement, as a surrogate for progression. In addition to cross-validate the traditional (ECG-based) staging, such design generates new insights and hypotheses on which candidate markers to test further. But the information coming out from these studies has limitations for clinical use [62]. Serving that purpose will require a process of validation. It starts from identifying the proposed marker as independently associated with the outcome, ideally in a prospective cohort study. A reliable process requires a representative sample, measuring the marker along with known correlates and confounding factors and recording a sizeable number of outcomes [63]. And, as it happened with more fortunate areas of medicine, any emerging prognostic information should come as a widely usable working definition (i.e., risk scores) for external validation in subsequent studies [64]. As the field moves forward, longitudinal studies seeking to evaluate more reliably such markers are appearing [65]. But they are still the exemption, as they have traditionally posed vast logistic and financial challenges to the research community around CCC.

For instance, measuring scar tissue with late gadolinium enhancement using cardiac magnetic resonance (CMR) may be a good prognostic tool. One study found it associated with ventricular arrhythmia (as assessed by 24-h Holter monitoring) among patients with preserved ventricular ejection fraction [66]. Cardiac fibrosis was in turn increasingly higher among patients with systolic wall motion abnormalities in that study. And in a triangulation, arrhythmia was associated with a lower left ventricular ejection fraction. Another cross-sectional study found the amount of scar associated with the QRS score of the electrocardiogram, which was in turn associated with lower levels of the NYHA functional class [67]. In similar fashion, strain and speckle-tracking echocardiography may be useful as a marker of cardiac abnormalities among patients whose ventricular function and dimensions are comparable to seronegatives [68]. That pattern correlates in another similar study with the degree of myocardial fibrosis, as detected by cardiac magnetic resonance [69]. While this information may be useful to refine staging of disease, it has not been converted into valid, useful prognostic tools. Using those markers to monitor progress or response to therapy would require even more work. However, as these findings fall into the category of sensitive diagnostic approaches to identify early CCC, they may become useful as surrogate outcome measures for future clinical trials for our prevention scenario.

The general exam of the body of evidence on individual markers leaves us, when applying the GRADE features of validity (risk of bias in the study design and conduct, effect size, inconsistency across results, imprecision of results, and selection bias in the studies included in the assessment) [70], judging the quality of evidence as generally low, or very low. Table 1 summarizes those features, describing the attributes weakening the evidence grading for those markers individually. Perhaps the better scenario is that of ECG as a tool in the primary prevention scenario (where the quality of evidence is moderate). But even so, the problems of applying an ECG-only stratification approach were discussed above.

As it happens in other fields, integrative risk scales may overcome deficiencies when making prognosis based on individual markers for complex diseases. Historically, echocardiography represented a turning point for disease staging and prognosis of Chagas disease [71]. In the past century, echocardiography emerged as a tool enabling clinicians to ascertain more accurately the heart dimensions and shape. In the context of CCC, that naturally challenged the use of cardiothoracic index from chest X-rays as criterion for staging. Such information, chiefly left ventricular diameters, ejection fraction, and the presence of apical aneurysms, have been tested and integrated to risk models. A consensus of experts summarized 12 cohort studies that have included echocardiography indices in their prognostic models, along with other demographic or clinical variables [72]. Only some of these highly valuable studies included a variety of stages of seropositive or CCC patients (less

Table 1 Strength of association for proposed individual markers of progression for CCC

Clinical scenario (target outcome)	Diagnostic tool	Prognostic Marker (usual cut-off points)	Risk of bias	Inconsistency	Imprecision	Size of association	Quality of evidence (GRADE)	References
Primary Prevention (progression)	12 LEAD ECG	RBBB				L	Moderate	Cardoso R, 2016; Marcolino 2015; Ribeiro 2014; Marques DS 2016,
		RBBB/LAH				L		
		LAH				M-L		
	Natriureticpeptides	>100 pg/ml (BNP)				M	Low	Echeverria 2017, Vilas 2008; Moreira 2009; Garcia Alvarez 2010; Talvani 2004
	CMR	Cardiac Fibrosis				N	Very Low	Uelledhal 2016, Torreao 2015;
	ECHOCARDIOGRAM STRAIN	Decrease of global longitudinal and radial Strain				S	Very Low	Barbosa 2013; Garcia Alvarez 2011; Barros 2015; Gomez 2016
Secondary Prevention (*Death*)	ECG Holter	VT				M	Low	Costa 2017; Benchimol 2007; Gadioli 2018; Silva 2015; Tassi 2014; Benchimol Barbosa 2013
		VPC				M		
		SDNN < 100				M		
	ECHOCARDIOGRAM	LVEF <50%				M	Low	Costa 2017; Barmosa 2014; Rassi 2014; Cedraz 2014; Garcia Alvarez 2010; Benchimol 2007
		LA Vol. (>72 mL/m2)				S		
		Apical anurism				M		
	SAECG	Various [a]				M	Very low	Benchimol 2013

RBBB **right bundle branch block,** ***LAH*** **left anterior hemiblock,** ***BNP*** **B-type natriuretic peptide,** ***ECG*** **electrocardiogram,** ***SDNN*** **standard deviation of RR,** ***VT*** **ventricular tachycardia,** ***VPC*** **ventricular premature contractions,** ***LVEF*** **left ventricular ejection fraction,** ***SAECG*** **signal-averaged electrocardiography,** ***CMR*** **cardiac magnetic resonance. Size of association:** ***S*** **small,** ***M*** **medium,** ***L*** **large,** ***N*** **not informed.**

[a]IVET (Intraventricular electrical transient) as part of the markers in the RIO risk score.

applicable results for those including only advanced cases). Unfortunately, most of those studies individually did not record sufficient number of clinical events (i.e., incidence of heart failure or symptomatic arrhythmias, hospitalization, or death for cardiac causes). The importance of that design feature is that it leads to very imprecise risk estimates (with wide confidence intervals) and that having less than ten events for predictor makes the models unstable, making replication difficult [73].

For those studies proposing risk scores, there is high clinical heterogeneity in the definition of both the exposure variables and the outcome events. In one prognostic study including mostly asymptomatic participants (recording 18 deaths and 34 heart failure events after 8 years), Viotti and colleagues developed a five-item risk score based on a multivariate risk model [74]. The score combines information from demographics (age over 50), ECG (presence of intraventricular conduction abnormalities), echocardiography (diameter of the left ventricle >40 mm or systolic dysfunction), Holter monitoring (sustained ventricular tachycardia), and prior exposure to trypanocidal treatment (which gives negative scoring) along with clinical deterioration over follow-up. The authors, however, defined as outcome the change in the clinical group (based on Kuschnir's scale level for each participant at inception) to derive the model. Assuming a nominally different outcome for each group may be controversial, particularly with the goal of deriving a single risk model. Further, we are not aware of validation studies for that scale so far. Because of the clinical heterogeneity, while the field develops more collaborative efforts toward combining patient-level data, we are left to examine the scores addressing all-cause or cardiac mortality.

In 2006, Rassi et al. proposed and validated a mortality risk score for CCC, based on 130 events in the original (derivation) cohort and 35 in its validation setting [75]. This tool combines clinical information (sex, New York Heart Association class) with cardiac images (chest X-rays and echocardiography), ECG, and 24-h Holter ECG monitoring. In the original description, it assigns up to 20 points and proposes three risk categories. Low-risk individuals (with less than 7 points) had a mortality rate of 10% at 10 years. For intermediate and high risk, they were 44% and 84%, respectively. Its classification properties are at least good (area under the curve 0.84), and its external validation analysis showed similar results (AUC 0.81). Other risk score alternatives including slightly different tools give different weights to the same markers, but used Rassi's risk score as refs. [76–78]. The discrimination ability of the newer scales did not surpass that of Rassi, as shown in Table 2. This makes Rassi's not only a good tool but also the only externally and consistently validated.

Despite these potential advantages, it is, surprisingly, underutilized. As far as we know, they are not studies reporting either doctor's acceptance or patterns of use of this risk scale in clinical practice. Neither have we seen the scale being used, for example, to describe other study populations in scientific papers, or as study outcome to assess its responsiveness to change. In this context, we encourage the use of Rassi's risk score to ascertain prognosis, and perhaps with less enthusiasm to monitor changes to treatment, until more studies are available.

We therefore recommend that every patient diagnosed with CCC be stratified with Rassi's risk score (Fig. 5). That implies, to attach strictly to what has been proven (until simpler or stronger tools emerge), to use the whole diagnostic battery

Table 2 Risk stratification scores in Chagas disease

Scale		Scale items (points)	Outcome	AUC	
				Derivation Cohort, year AUC (IC 95%) (events/total)	Validation Cohort, year AUC (IC 95%) (events/total)
RASSI	History Chest X-rays Echocardiogram ECG Holter 12-Lead ECG	NYHA III o IV = 5 Cardiomegaly = 5 Anomaly of the movement of the segmental or global wall = 3 NTV = 3 Low QRS voltage = 2 Male sex = 2	Mortality 10 years 0–6 (low risk) (10%) 7–11 (moderate risk) (44%) 12–20 (high risk) (84%)	Rassi, 2006 [75] AUC: 0.84 (0.79–0.89) 130/424	Rassi, et al. [75] AUC: 0.81 (0.72–0.90) 35/153 Ribeiro, et al. [79] AUC: 0.84 (0.72–0.96) 13/158 Benchimol-Barbosa, 2013 [76] AUC: 0.81 (0.69–0.99) 36/100 Silva, et al. [80] AUC: 0.77 (IC not reported) 8/45
RIBEIRO	Echocardiogram ECG Holter 12-Lead ECG/ SVEAG	LVEF < 50% = 1 NTV = 1 QRS > 133 ms/150 ms = 1	Mortality 5 years 1 (low risk) (1%) 2 (moderate risk) (20%) 3 (high risk) (50%)	Ribeiro, et al. [79] AUC: 0.92 (0.86 0.97) 13/184	
RIO	12-Lead ECG Echocardiogram ECG Holter SVEAG	Q-wave = 2 NTV = 2 SDNN < 100 ms = 1 IVET+ = 1	Mortality 10 years 1 (low risk) (16%) >1 (high risk) (50%)	Benchimol, 2013 [76] AUC: 0.92 (0.85–0.99) 36/100	
DE SOUZA	12-Lead ECG History 12-Lead ECG Echocardiogram	QT dispersión > 60 ms = 3 Prior syncope = 2 PVC = 1 LVEF < 35% = 1	Sudden mortality 5 years 0–2 (low risk) (1.5%) 3–4 (moderate risk) (25%) >5 (high risk) (51%)	De Souza, 2015 [78] AUC: 0.88 (0.82–0.94) 43/373	

ECG electrocardiogram, *SAECG* Signal-averaged electrocardiography, *NYHA* New York Heart Association class functional, *RBBB* right bundle branch block, *LAH* left anterior hemiblock, *SDNN* standard deviation of RR, *NVT* non-sustained ventricular tachycardia, *LVEF* left ventricular ejection fraction, *IVET+* intraventricular electrical transient.

(chest X-rays, ECG and 24-h Holter monitoring, and conventional echocardiography) at least at onset of diagnosis. We have no empirical data informing whether Rassi's risk score may be responsive to change so it can be used to monitor changes after treatment. However, since it integrates several criteria to assign 20 points, it has higher probability to change than single-marker, dichotomous or the 4-level Kuschnir's scale in this scenario.

Our suggestion, in keeping with the ruling-out perspective, is first to make sure first-known CCC patients found at low risk (<7 points) be reassessed annually. Perhaps those found with consistently (say twice on a row) below the threshold can delay further assessments to every 2 or 3 years. The minority at intermediate or high risk (a third of the population in Rassi's development and validation cohorts accounted for 90% of fatalities) may need reassessments every 6 months or even more often when severe.

1.3 The Challenge of Instating a Treatment

We have discussed earlier that, conceptually, the focus in the primary prevention scenario should be the risk factor, clearing the parasitic load in this case. But the field still faces numerous challenges to meet and validate the clinical value of that (apparently simple) concept. To start, parasitic load is not, unlike other infectious diseases, something that is even routinely measured yet. Moreover, the lack of this tool (quantitative polymerase chain reaction at the point of care) becomes a barrier for decision-making, among those asymptomatic seropositives. Such decisions should include when to consider an individual eligible for initiating, restarting, or even changing a trypanocidal therapy. That will greatly stimulate research on the relationship between parasitic load, parasite and host variability, and clinical progression [50, 81].

By today it is known that trypanocidal therapy (better established with benznidazole than nifurtimox) is associated after 60 days of treatment with higher clearance of parasitic load or antibodies [82]. This effect has been documented for randomized, placebo-controlled trials, for up to 5 years in adults [83] and 6 years in children [84]. It has also been demonstrated that treatment with antifungal agents produces parasitic clearance, but only while patients are taking their medication [83, 85, 86].

Unfortunately, whether getting off this parasitic load among seropositive is meaningful in terms of changing the natural history of CCC (i.e., to prevent the incidence of CCC or its complications) remains to be established. The evidence supporting trypanocidal therapy to modify patient-related, rather than parasite-related, outcomes is still insufficient. Despite great advances, especially in the last decade, observational studies yielded inconsistent, statistically imprecise results [82]. These studies have shown, in terms of pooled effect size for patient-related outcomes, an inconclusive signal toward reduction of mortality or progress of CCC. For inclusion in this analysis, these observational studies needed minimal quality requirements: a comparison with an untreated group and the record of clinical progression data (including mortality) after at least 4 years of follow-up. There were four studies providing data on clinical progression (106 events overall) and six studies on mortality (99 events in total).

Letting aside the inherent high risk of bias, these studies showed a potentially important reduction (26% for disease progression and 45% for mortality). However, the results were highly inconsistent across studies (heterogeneity I^2 statistics 66% for disease progression and 48% for mortality) and statistically imprecise (95% confidence intervals 0.32–1.73 for disease progression and 0.26–1.14 for mortality). Most of that heterogeneity comes from the study by Viotti [87], whose results seem inconsistent with the others, with a substantial effect size in favor of trypanocidal therapy.

This study, reporting a non-randomized treatment allocation but a blinded outcome assessment, expressed their results in terms of change in Kuschnir's clinical group. By making this design choice, authors were giving similar weight to developing a new ECG abnormality (change from 0 to 1 stage) and to developing cardiac enlargement (from 1 to 2 levels), or to changes in one or more disease stages over follow-up. In this study, despite matching their control group, there were differences in the baseline disease stage of participants reported as progressing. Among 12 people who evolved to more advanced stages in the treated group, six were stage 0 at baseline (three others were class I and three more were class II baseline). In contrast, in the untreated group, of 40 changing stage, 13 were classified as 0 at baseline (while 14 were class I and 13 were class II at baseline).

Another study recently published that met the requirements for inclusion of the previous analysis reported favorable results of treatment of similar magnitude to Viotti's [88]. Although this study did establish a similar type of patient-important events as outcome (there were 24, a relatively modest number), the results are challenged by an important selection bias. Only 15% of the cohort did not receive treatment, reflecting obvious preferences for prescribing TT in the study setting. Beyond the nonrandom allocation of treatment (and the retrospective nature of the analysis), the distribution of treatments shows by itself doctors' preferences. When deciding the prescription, they were far from the principle of equipoise [89, 90]. In fact, the treated group was on average 12 years younger than those left untreated, reflecting a discernible probability to develop clinical events and supporting the notion that doctors offer newer treatments to those likely to do better and comply with therapy [91].

The history of studies evaluating treatments has already shown important examples where observational studies yield different (sometimes opposite) results from experimental studies conducted later [92–94]. Even having randomized allocation of interventions, the number of participants, particularly the number of events recorded, is essential to obtain results that are both valid and statistically robust. The literature of a variety of fields, not even from neglected diseases, shows the abundance of clinical research results that are statistically fragile [95–98]. And there are painful lessons in cardiovascular medicine regarding misleading results coming from studies with small number of events [99–101].

The "Benznidazole Evaluation for Interrupting Trypanosomiasis" (BENEFIT) trial came as an attempt to overcome the limitations of previous observational studies. Despite its intrinsic merit (as the first large trial ever conducted in patients with

CCC, enrolling 2854 patients and recording 394 clinical events over the follow-up period), BENEFIT failed to demonstrate a statistically significant reduction in complications or death (HR 0.93, IC 0.81–1.07). The trial was designed to have 3000 participants, the number estimated to identify a 26% relative risk reduction in meaningful clinical events among those receiving benznidazole. Perhaps this effect size equated mechanically what is usual in the field of cardiology (typically a relative risk reduction of 20–25%). This magnitude of the effect could have been overoptimistic for a population with advanced CCC, probably at high risk of events, but with low responsiveness to TT. Also, leading investigators may have settled to the reality that a trial exceeding 3000 patients may not have been feasible.

But in addition to the effect size expectation and statistical boundaries in BENEFIT, some relevant design issues may have obscured a better performance of the treatment. These include the known intolerance to medication (13% did not complete the treatment), a foreseeable possibility of parasite resistance to medication in some areas for a geographically extensive study (benznidazole did not change parasitic load in the Colombia-El Salvador population, about 20% of the population), and the threat of reinfection (28% of participants lived in rural areas and 3% judged their livelihood as infested). The above three factors together left the trial with the capacity to demonstrate about 66% of the theoretical effect of the intervention (if we multiplied 0.87 due to intolerance, 0.80 for parasite resistance, and 0.95 for reinfection, assuming that it happened in 5% along the trial).

BENEFIT had a composite outcome of mortality and a variety of serious clinical complications (resuscitated cardiac arrest, insertion of a pacemaker or an implantable cardioverter-defibrillator, sustained ventricular tachycardia, cardiac transplantation, new heart failure, stroke or transient ischemic attack, or a systemic or pulmonary thromboembolic event). The direction of the results for all of these outcomes pointed consistently in favor of the intervention (i.e., relative effect size less than 1), although generally modest (0.85–0.95, or a relative risk reduction of 5–15%), with no statistical power for its detection. Only one of the secondary outcomes, hospitalization for cardiovascular causes, reached statistical significance.

The above discussion suggests that trypanocidal therapy may have a small (for the expectation) yet important clinical effect that BENEFIT failed to detect reliably. Table 3 shows an overview of the reported effect sizes by this randomized trial and the observational studies meeting the aforementioned criteria for inclusion. As discussed, observational studies testing benznidazole are likely overestimating the effects. But given the shortcomings of these studies, the variability of the results and the possible false-negative result of BENEFIT, the field will probably need a similar undertaking to establish the clinical value of parasite clearance for seropositive individuals. The field needs a new study testing better, safer trypanocidal therapy, which is pragmatic, geographically varied, using an efficient design—or even better, a cooperative, continued series of studies addressing simultaneously key questions on prognosis and therapy.

In practice, we should assume that we have now a demonstrated case for parasitic clearance (in general, admitting variations). Also that, based on the best evidence in hand, trypanocidal therapy points toward preventing progression or death in sero-

Table 3 Patient-related outcomes reported in selected studies testing Trypanocidal agents for chronic asymptomatic *Trypanosoma cruzi* infection

Trypanocidal agent	Outcome	Study author, year	Events/exposed (%) Active	Events/exposed (%) Control	Effect size estimates (IC 95%)
1. BZN alone	Progression of cardiomyopathy	Morillo, 2015 [83] (RTC)	148/1431 (10.3%)	157/1423 (11.0%)	0.93 (0.75–1.15)
		Viotti, 2006[a] [87]	6/294 (2.0%)	24/304 (7.8%)	0.24 (0.10–0.60)
		Fragata-Filho, 2016 [88]	10/263 (3.8%)	2/47 (4.2%)	0.89 (0.20–3.94)
	Mortality	Morillo, 2015 (RCT) [83]	246/1431 (17.2%)	257 /1423 (18.1%)	0.95 (0.81–1.11)
		Viotti, 2006[a] [87]	3/294 (1.0%)	12/304 (3.9%)	0.25 (0.07–0.9)
		Fragata-Filho, 2016 [88]	7/263 (2.6%)	5/47 (10.6%)	0.25 (0.08–0.75)
		Catalioti, 2001[a] [102]	2/74 (2.7%)	8/465 (1.7%)	1.59 (0.33–7.62)
2. NFX alone or with BZN	Progression of cardiomyopathy	Silveira, 2000[a] [103]	12/58 (20.6%)	6/39 (15.3%)	1.43 (0.49–4.21)
		Fabro, 2000[a] [104]	4/70 (5.7%)	10/130 (7.6%)	0.73 (0.22–2.41)
		Lauria-Pires, 2000[a] [105]	23/45 (51.1%)	21/46 (45.6%)	1.24 (0.55–2.84)
	Mortality	Silveira, 2000[a] [103]	5/58 (8.6%)	3/39 (7.6%)	1.13 (0.25–5.04)
		Fabro, 2000[a] [104]	2/70 (2.8%)	3/130 (2.3%)	1.25 (0.20–7.63)
		Gallerano, 2000[a] [106]	2/226 (0.8%)	37/668 (5.5%)	0.15 (0.04–0.64)
		Lauria-Pires, 2000[a] [105]	4/45 (8.8%)	3/46 (6.5%)	1.40 (0.29–6.63)

BZN benznidazole, *NFX* nifurtimox, *RCT* randomized clinical trial.
[a]Data abstracted from ref. [81]

positive. But in scenarios like this, while the tolerance of medication improves, with an intervention with probably a small effect, there should be caution on the risk-benefit ratio across the spectrum of CCC. Even more important, clinicians should leave room for a judicious, shared decision-making process with well-informed patients [107, 108]. With that in mind, meanwhile such a new trial reaches a "proof of concept" for trypanocidal therapy, we suggest to offer benznidazole for most cases, in most countries, for the purpose of parasite clearance. This should go along with sharing the uncertainty around the impact on patient-important outcomes, and the shortcomings on tolerance and potential resistance, and ascertaining individual responses. And certainly, as consumers of information to help their patients, clinicians should advocate for more and usable research.

On the side of preventing or treating CCC complications, there is also a special case for trypanocidal therapy among seropositives: that of women in childbearing age being diagnosed as seropositive. Observational studies have consistently found a strong effect of trypanocidal therapy for preventing congenital infection [109–112]. Having a large effect size is one of the criteria to strengthen the evidence, and the outcome being prevented here is highly valuable. Despite the benign course for the majority of infected mothers, the relatively low incidence of congenital infection [14, 113], and even the small number of CCC cases expected decades later, this is the case for a strong recommendation of trypanocidal therapy to promote parasitic clearance in future mothers.

2 Room for Ancillary (Non-trypanocidal) Therapy

Learning that a treatment modifies the course of a disease helps us understand its causes. Given the results of studies testing trypanocidal therapy, the field must acknowledge that its role for seropositive adults is modest at best. If parasites trigger pathophysiological mechanisms at onset, it is also likely that along the natural history of CCC, other processes take increasing importance in producing disease. Therefore, researchers and clinicians need to be open to new and alternative potentially synergistic treatment approaches. That includes at least two additional fronts other than improving the current trypanocidal therapy: manipulating the immune response through a variety of mechanisms and incorporating fully the variety of medications in hand targeting common elements in heart failure patients.

There are new avenues for treatment on the way. In the front of TT, ongoing studies are testing alternative drug dose regimens attempting to improve tolerance and testing new trypanocidal agents, combinations, and even more exciting, new medications directed to alternative targets. Vaccine studies are under way, while other research groups are trying to connect the immunopathology to the use of new biological drugs that are revolutionizing the oncology and rheumatology fields.

But there are also several alternatives currently available, which are largely overlooked. Cases of mild CCC may derive benefit from neurohormonal antagonists, which are typically reserved for the worse cases. Clinicians should take advantage of those missing opportunities to maximize the benefit. As a general principle, if CCC has a worse behavior compared with dilated cardiomyopathies, therapy should be administered earlier and with stronger intensity. But the situation in practice probably speaks of the opposite. We strongly recommend that prescription of proven, complementary, generic treatment for heart failure (i.e., ACE-Is, ARBs, or beta-blockers, in the first place) be actively promoted among physicians [114]. At any rate, decision-making on treatments for CCC (be this trypanocidal or ancillary therapy) should be well informed, balanced, and shared, so it always reflects the triangle of good judgment of scientific evidence, along with patients' values and preferences.

References

1. Pinto AY, Valente SA, Valente VC, Ferreira Junior AG, Coura JR. Acute phase of Chagas disease in the Brazilian Amazon region: study of 233 cases from Para, Amapa and Maranhao observed between 1988 and 2005. Rev Soc Bras Med Trop. 2008;41(6):602–14.
2. Noya BA, Diaz-Bello Z, Colmenares C, Ruiz-Guevara R, Mauriello L, Munoz-Calderon A, et al. Update on oral Chagas disease outbreaks in Venezuela: epidemiological, clinical and diagnostic approaches. Mem Inst Oswaldo Cruz. 2015;110(3):377–86.
3. Bern C, Martin DL, Gilman RH. Acute and congenital Chagas disease. Adv Parasitol. 2011;75:19–47.
4. Higuchi ML, Benvenuti LA, Martins RM, Metzger M. Pathophysiology of the heart in Chagas' disease: current status and new developments. Cardiovasc Res. 2003;60(1):96–107.
5. Validade do conceito de forma indeterminada de doença de Chagas. Rev Soc Bras Med Trop. 1985;18:46.
6. Dias JCP. The indeterminate form of human chronic Chagas' disease: a clinical epidemological review. Rev Soc Bras Med Trop. 1989;22:147–56.
7. Pinto Dias JC. Natural history of Chagas disease. Arq Bras Cardiol. 1995;65(4):359–66.
8. Bonney KM, Engman DM. Chagas heart disease pathogenesis: one mechanism or many? Curr Mol Med. 2008;8(6):510–8.
9. Marin-Neto JA, Cunha-Neto E, Maciel BC, Simoes MV. Pathogenesis of chronic Chagas heart disease. Circulation. 2007;115(9):1109–23.
10. Ferreira AW, de Avila SD. Laboratory diagnosis of Chagas' heart disease. Sao Paulo Med J. 1995;113(2):767–71.
11. Umezawa ES, Silveira JF. Serological diagnosis of Chagas disease with purified and defined Trypanosoma cruzi antigens. Mem Inst Oswaldo Cruz. 1999;94(Suppl 1):285–8.
12. Ribeiro AL, Sabino EC, Marcolino MS, Salemi VM, Ianni BM, Fernandes F, et al. Electrocardiographic abnormalities in Trypanosoma cruzi seropositive and seronegative former blood donors. PLoS Negl Trop Dis. 2013;7(2):e2078.
13. Abras A, Munoz C, Ballart C, Berenguer P, Llovet T, Herrero M, et al. Towards a new strategy for diagnosis of congenital Trypanosoma cruzi infection. J Clin Microbiol. 2017;55(5):1396–407.
14. Buekens P, Cafferata ML, Alger J, Althabe F, Belizan JM, Bustamante N, et al. Congenital transmission of Trypanosoma cruzi in Argentina, Honduras, and Mexico: an observational prospective study. Am J Trop Med Hyg. 2018;98(2):478–85.
15. Marcus JE, Webber BJ, Cropper TL, Wilson MC, Yun HC. Diagnostic evaluation of military blood donors screening positive for Trypanosoma cruzi infection. MSMR. 2018;25(2):16–9.
16. Massachusetts General Hospital. Case records of the Massachusetts General Hospital. Weekly clinicopathological exercises. Case 32–1993. A native of El Salvador with tachycardia and syncope. N Engl J Med. 1993;329(7):488–96.
17. Bimbi BJ, Unger P, Vandenbossche JL, Silance PG, Van LY. Chagas disease: don't forget it in Latin American patients with heart block! Acta Cardiol. 2014;69(2):206–8.
18. Cucunuba ZM, Manne-Goehler JM, Diaz D, Nouvellet P, Bernal O, Marchiol A, et al. How universal is coverage and access to diagnosis and treatment for Chagas disease in Colombia? A health systems analysis. Soc Sci Med. 2017;175:187–98.
19. Manne JM, Snively CS, Ramsey JM, Salgado MO, Barnighausen T, Reich MR. Barriers to treatment access for Chagas disease in Mexico. PLoS Negl Trop Dis. 2013;7(10):e2488.
20. Sanchez DR, Traina MI, Hernandez S, Smer AM, Khamag H, Meymandi SK. Chagas disease awareness among Latin American immigrants living in Los Angeles, California. Am J Trop Med Hyg. 2014;91(5):915–9.
21. Meymandi SK, Hernandez S, Forsyth CJ. A community-based screening program for Chagas disease in the USA. Trends Parasitol. 2017;33(11):828–31.
22. Herrador Z, Rivas E, Gherasim A, Gomez-Barroso D, Garcia J, Benito A, et al. Using hospital discharge database to characterize Chagas disease evolution in Spain: there is a need

for a systematic approach towards disease detection and control. PLoS Negl Trop Dis. 2015;9(4):e0003710.

23. Yamey G, Hotez P. Neglected tropical diseases. BMJ. 2007;335(7614):269–70.
24. Guyatt G, Montori V, Devereaux PJ, Schunemann H, Bhandari M. Patients at the centre: in our practice, and in our use of language. Evid Based Med. 2004;9(1):6.
25. Cucunuba ZM, Okuwoga O, Basanez MG, Nouvellet P. Increased mortality attributed to Chagas disease: a systematic review and meta-analysis. Parasit Vectors. 2016;9:42.
26. Capuani L, Bierrenbach AL, Pereira AA, Mendrone A Jr, Ferreira JE, Custer B, et al. Mortality among blood donors seropositive and seronegative for Chagas disease (1996–2000) in Sao Paulo, Brazil: a death certificate linkage study. PLoS Negl Trop Dis. 2017;11(5):e0005542.
27. Freitas HF, Chizzola PR, Paes AT, Lima AC, Mansur AJ. Risk stratification in a Brazilian hospital-based cohort of 1220 outpatients with heart failure: role of Chagas' heart disease. Int J Cardiol. 2005;102(2):239–47.
28. Shen L, Ramires F, Martinez F, Bodanese LC, Echeverria LE, Gomez EA, et al. Contemporary characteristics and outcomes in Chagasic heart failure compared with other nonischemic and ischemic cardiomyopathy. Circ Heart Fail. 2017;10(11):e004361.
29. Sabino EC, Ribeiro AL, Salemi VM, Di Lorenzo OC, Antunes AP, Menezes MM, et al. Ten-year incidence of Chagas cardiomyopathy among asymptomatic Trypanosoma cruzi-seropositive former blood donors. Circulation. 2013;127(10):1105–15.
30. Kuschnir E, Sgammini H, Castro R, Evequoz C, Ledesma R, Brunetto J. Evaluation of cardiac function by radioisotopic angiography, in patients with chronic Chagas cardiopathy. Arq Bras Cardiol. 1985;45(4):249–56.
31. Molina RB, Matsubara BB, Hueb JC, Zanati SG, Meira DA, Cassolato JL, et al. Dysautonomia and ventricular dysfunction in the indeterminate form of Chagas disease. Int J Cardiol. 2006;113(2):188–93.
32. Ribeiro AL, Moraes RS, Ribeiro JP, Ferlin EL, Torres RM, Oliveira E, et al. Parasympathetic dysautonomia precedes left ventricular systolic dysfunction in Chagas disease. Am Heart J. 2001;141(2):260–5.
33. Villar JC, Leon H, Morillo CA. Cardiovascular autonomic function testing in asymptomatic T. cruzi carriers: a sensitive method to identify subclinical Chagas' disease. Int J Cardiol. 2004;93(2–3):189–95.
34. Simoes MV, Pintya AO, Bromberg-Marin G, Sarabanda AV, Antloga CM, Pazin-Filho A, et al. Relation of regional sympathetic denervation and myocardial perfusion disturbance to wall motion impairment in Chagas' cardiomyopathy. Am J Cardiol. 2000;86(9):975–81.
35. de Almeida-Filho OC, Maciel BC, Schmidt A, Pazin-Filho A, Marin-Neto JA. Minor segmental dyssynergy reflects extensive myocardial damage and global left ventricle dysfunction in chronic Chagas disease. J Am Soc Echocardiogr. 2002;15(6):610–6.
36. Garcia-Alvarez A, Sitges M, Pinazo MJ, Regueiro-Cueva A, Posada E, Poyatos S, et al. Chagas cardiomyopathy: the potential of diastolic dysfunction and brain natriuretic peptide in the early identification of cardiac damage. PLoS Negl Trop Dis. 2010;21:4(9).
37. Pedrosa RC, Campos MC. Exercise testing and 24 hours Holter monitoring in the detection of complex ventricular arrhythmias in different stages of chronic Chagas' heart disease. Rev Soc Bras Med Trop. 2004;37(5):376–83.
38. Villar JC. Challenges in identifying subclinical chronic chagasic cardiomyopathy in a Colombian urban population. PhD thesis dissertation. 2008. ISBN: 9780494360392 0494360399. OCLC Number: 434557667. https://www.researchgate.net/publication/254623471_Challenges_in_identifying_subclinical_chronic_chagasic_cardiomyopathy_in_a_Colombian_urban_population.
39. Espinosa R, Carrasco HA, Belandria F, Fuenmayor AM, Molina C, Gonzalez R, et al. Life expectancy analysis in patients with Chagas' disease: prognosis after one decade (1973–1983). Int J Cardiol. 1985;8(1):45–56.
40. Pereira JB, Willcox HP, Coura JR. Morbidade da doença de Chagas: III. Estudo longitudinal, de seis anos, em Virgem da Lapa, MG, Brasil. Mem Inst Oswaldo Cruz. 1985;80:63–71.

41. Acquatella H, Catalioti F, Gomez-Mancebo JR, Davalos V, Villalobos L. Long-term control of Chagas disease in Venezuela: effects on serologic findings, electrocardiographic abnormalities, and clinical outcome. Circulation. 1987;76(3):556–62.
42. Maguire JH, Hoff R, Sherlock I, Guimaraes AC, Sleigh AC, Ramos NB, et al. Cardiac morbidity and mortality due to Chagas' disease: prospective electrocardiographic study of a Brazilian community. Circulation. 1987;75(6):1140–5.
43. Mota EA, Guimaraes AC, Santana OO, Sherlock I, Hoff R, Weller TH. A nine year prospective study of Chagas' disease in a defined rural population in northeast Brazil. Am J Trop Med Hyg. 1990;42(5):429–40.
44. Maguire JH, Mott KE, Lehman JS, Hoff R, Muniz TM, Guimaraes AC, et al. Relationship of electrocardiographic abnormalities and seropositivity to Trypanosoma cruzi within a rural community in northeast Brazil. Am Heart J. 1983;105(2):287–94.
45. Hingorani P, Karnad DR, Rohekar P, Kerkar V, Lokhandwala YY, Kothari S. Arrhythmias seen in baseline 24-hour Holter ECG recordings in healthy normal volunteers during phase 1 clinical trials. J Clin Pharmacol. 2016;56(7):885–93.
46. von Rotz M, Aeschbacher S, Bossard M, Schoen T, Blum S, Schneider S, et al. Risk factors for premature ventricular contractions in young and healthy adults. Heart. 2017;103(9):702–7.
47. Gontijo ED, Rocha MO, de OU T. Clinico-epidemiologic profile of Chagas disease patients attending an ambulatory referral center and proposal of a model of care for the Chagas patient under the perspective of a comprehensive health care. Rev Soc Bras Med Trop. 1996;29(2):101–8.
48. Davenport C, Cheng EY, Kwok YT, Lai AH, Wakabayashi T, Hyde C, et al. Assessing the diagnostic test accuracy of natriuretic peptides and ECG in the diagnosis of left ventricular systolic dysfunction: a systematic review and meta-analysis. Br J Gen Pract. 2006;56(522):48–56.
49. Balion C, Santaguida PL, Hill S, Worster A, McQueen M, Oremus M, et al. Testing for BNP and NT-proBNP in the diagnosis and prognosis of heart failure. Evid Rep Technol Assess (Full Rep). 2006;142:1–147.
50. Sanchez LV, Bautista DC, Corredor AF, Herrera VM, Martinez LX, Villar JC, et al. Temporal variation of Trypanosoma cruzi discrete typing units in asymptomatic Chagas disease patients. Microbes Infect. 2013;15(10–11):745–8.
51. Doust JA, Pietrzak E, Dobson A, Glasziou P. How well does B-type natriuretic peptide predict death and cardiac events in patients with heart failure: systematic review. BMJ. 2005;330(7492):625.
52. Porapakkham P, Porapakkham P, Zimmet H, Billah B, Krum H. B-type natriuretic peptide-guided heart failure therapy: a meta-analysis. Arch Intern Med. 2010;170(6):507–14.
53. McLellan J, Heneghan CJ, Perera R, Clements AM, Glasziou PP, Kearley KE, et al. B-type natriuretic peptide-guided treatment for heart failure. Cochrane Database Syst Rev. 2016;12:CD008966.
54. Li P, Luo Y, Chen YM. B-Type natriuretic peptide-guided chronic heart failure therapy: a meta-analysis of 11 randomised controlled trials. Heart Lung Circ. 2013;22(10):852–60.
55. Oliveira BM, Botoni FA, Ribeiro AL, Pinto AS, Reis AM, Nunes MC, et al. Correlation between BNP levels and Doppler echocardiographic parameters of left ventricle filling pressure in patients with Chagasic cardiomyopathy. Echocardiography. 2009;26(5):521–7.
56. Talvani A, Rocha MO, Cogan J, Maewal P, de Lemos J, Ribeiro AL, et al. Brain natriuretic peptide and left ventricular dysfunction in chagasic cardiomyopathy. Mem Inst Oswaldo Cruz. 2004;99(6):645–9.
57. Lima-Costa MF, Cesar CC, Peixoto SV, Ribeiro AL. Plasma B-type natriuretic peptide as a predictor of mortality in community-dwelling older adults with Chagas disease: 10-year follow-up of the Bambui Cohort Study of Aging. Am J Epidemiol. 2010;172(2):190–6.
58. Echeverria LE, Rojas LZ, Calvo LS, Roa ZM, Rueda-Ochoa OL, Morillo CA, et al. Profiles of cardiovascular biomarkers according to severity stages of Chagas cardiomyopathy. Int J Cardiol. 2017;227:577–82.
59. Clark EH, Marks MA, Gilman RH, Fernandez AB, Crawford TC, Samuels AM, et al. Circulating serum markers and QRS scar score in Chagas cardiomyopathy. Am J Trop Med Hyg. 2015;92(1):39–44.

60. Ribeiro AL, Teixeira MM, Reis AM, Talvani A, Perez AA, Barros MV, et al. Brain natriuretic peptide based strategy to detect left ventricular dysfunction in Chagas disease: a comparison with the conventional approach. Int J Cardiol. 2006;109(1):34–40.
61. Anderson TJ, Gregoire J, Pearson GJ, Barry AR, Couture P, Dawes M, et al. 2016 Canadian Cardiovascular Society guidelines for the management of dyslipidemia for the prevention of cardiovascular disease in the adult. Can J Cardiol. 2016;32(11):1263–82.
62. Sackett DL, Haynes RB. The architecture of diagnostic research. BMJ. 2002;324(7336):539–41.
63. Laupacis A, Wells G, Richardson WS, Tugwell P. Users' guides to the medical literature. V. How to use an article about prognosis. Evidence-Based Medicine Working Group. JAMA. 1994;272(3):234–7.
64. Guida P, Mastro F, Scrascia G, Whitlock R, Paparella D. Performance of the European System for Cardiac Operative Risk Evaluation II: a meta-analysis of 22 studies involving 145,592 cardiac surgery procedures. J Thorac Cardiovasc Surg. 2014;148(6):3049–57.
65. Cardoso CS, Sabino EC, Oliveira CD, de Oliveira LC, Ferreira AM, Cunha-Neto E, et al. Longitudinal study of patients with chronic Chagas cardiomyopathy in Brazil (SaMi-Trop project): a cohort profile. BMJ Open. 2016;6(5):e011181.
66. Tassi EM, Continentino MA, Nascimento EM, Pereira BB, Pedrosa RC. Relationship between fibrosis and ventricular arrhythmias in Chagas heart disease without ventricular dysfunction. Arq Bras Cardiol. 2014;102(5):456–64.
67. Strauss DG, Cardoso S, Lima JA, Rochitte CE, Wu KC. ECG scar quantification correlates with cardiac magnetic resonance scar size and prognostic factors in Chagas' disease. Heart. 2011;97(5):357–61.
68. Barbosa MM, Costa Rocha MO, Vidigal DF, Bicalho Carneiro RC, Araujo RD, Palma MC, et al. Early detection of left ventricular contractility abnormalities by two-dimensional speckle tracking strain in Chagas' disease. Echocardiography. 2014;31(5):623–30.
69. Gomes VA, Alves GF, Hadlich M, Azevedo CF, Pereira IM, Santos CR, et al. Analysis of regional left ventricular strain in patients with Chagas disease and normal left ventricular systolic function. J Am Soc Echocardiogr. 2016;29(7):679–88.
70. Guyatt GH, Oxman AD, Vist GE, Kunz R, Falck-Ytter Y, Alonso-Coello P, et al. GRADE: an emerging consensus on rating quality of evidence and strength of recommendations. BMJ. 2008;336(7650):924–6.
71. Acquatella H, Schiller NB, Puigbo JJ, Giordano H, Suarez JA, Casal H, et al. M-Mode and two-dimensional echocardiography in chronic Chages' heart disease. A clinical and pathologic study. Circulation. 1980;62(4):787–99.
72. Nunes MCP, Badano LP, Marin-Neto JA, Edvardsen T, Fernandez-Golfin C, Bucciarelli-Ducci C, et al. Multimodality imaging evaluation of Chagas disease: an expert consensus of Brazilian Cardiovascular Imaging Department (DIC) and the European Association of Cardiovascular Imaging (EACVI). Eur Heart J Cardiovasc Imaging. 2018;19(4):459–60.
73. Peduzzi P, Concato J, Kemper E, Holford TR, Feinstein AR. A simulation study of the number of events per variable in logistic regression analysis. J Clin Epidemiol. 1996;49(12):1373–9.
74. Viotti R, Vigliano C, Lococo B, Petti M, Bertocchi G, Alvarez MG, et al. Clinical predictors of chronic chagasic myocarditis progression. Rev Esp Cardiol. 2005;58(9):1037–44.
75. Rassi A Jr, Rassi A, Little WC, Xavier SS, Rassi SG, Rassi AG, et al. Development and validation of a risk score for predicting death in Chagas' heart disease. N Engl J Med. 2006;355(8):799–808.
76. Benchimol-Barbosa PR, Tura BR, Barbosa EC, Kantharia BK. Utility of a novel risk score for prediction of ventricular tachycardia and cardiac death in chronic Chagas disease – the SEARCH-RIO study. Braz J Med Biol Res. 2013;46(11):974–84.
77. Ribeiro AL, Rocha MO, Terranova P, Cesarano M, Nunes MD, Lombardi F. T-wave amplitude variability and the risk of death in Chagas disease. J Cardiovasc Electrophysiol. 2011;22(7):799–805.
78. de Souza AC, Salles G, Hasslocher-Moreno AM, de Sousa AS, Alvarenga Americano do Brasil PE, Saraiva RM, et al. Development of a risk score to predict sudden death in patients with Chaga's heart disease. Int J Cardiol. 2015;187:700–4.

79. Ribeiro AL, Cavalvanti PS, Lombardi F, Nunes Mdo C, Barros MV, Rocha MO. Prognostic value of signal-averaged electrocardiogram in Chagas disease. J Cardiovasc Electrophysiol. 2008;19(5):502–9. https://doi.org/10.1111/j.1540-8167.2007.01088.x.
80. Silva RRD, Reis MS, Pereira BB, Nascimento EMD, Pedrosa RC. Additional value of anaerobic threshold in a general mortality prediction model in a urban patient cohort with Chagas cardiomyopathy. Rev Port Cardiol. 2017;36(12):927–34. https://doi.org/10.1016/j.repc.2017.06.012.
81. Sabino EC, Ribeiro AL, Lee TH, Oliveira CL, Carneiro-Proietti AB, Antunes AP, et al. Detection of Trypanosoma cruzi DNA in blood by PCR is associated with Chagas cardiomyopathy and disease severity. Eur J Heart Fail. 2015;17(4):416–23.
82. Villar JC, Perez JG, Cortes OL, Riarte A, Pepper M, Marin-Neto JA, et al. Trypanocidal drugs for chronic asymptomatic Trypanosoma cruzi infection. Cochrane Database Syst Rev. 2014;5:CD003463.
83. Morillo CA, Marin-Neto JA, Avezum A, Sosa-Estani S, Rassi A Jr, Rosas F, et al. Randomized trial of benznidazole for chronic Chagas' cardiomyopathy. N Engl J Med. 2015;373(14):1295–306.
84. Andrade AL, Martelli CM, Oliveira RM, Silva SA, Aires AI, Soussumi LM, et al. Short report: benznidazole efficacy among Trypanosoma cruzi-infected adolescents after a six-year follow-up. Am J Trop Med Hyg. 2004;71(5):594–7.
85. Molina I, Prat J, Salvador F, Trevino B, Sulleiro E, Serre N, et al. Randomized trial of posaconazole and benznidazole for chronic Chagas' disease. N Engl J Med. 2014;370(20):1899–908.
86. Torrico F, Gascon J, Ortiz L, Alonso-Vega C, Pinazo MJ, Schijman A, et al. Treatment of adult chronic indeterminate Chagas disease with benznidazole and three E1224 dosing regimens: a proof-of-concept, randomised, placebo-controlled trial. Lancet Infect Dis. 2018;18:419–30.
87. Viotti R, Vigliano C, Lococo B, Bertocchi G, Petti M, Alvarez MG, et al. Long-term cardiac outcomes of treating chronic Chagas disease with benznidazole versus no treatment: a nonrandomized trial. Ann Intern Med. 2006;144(10):724–34.
88. Fragata-Filho AA, Franca FF, Fragata CS, Lourenco AM, Faccini CC, Costa CA. Evaluation of parasiticide treatment with benznidazol in the electrocardiographic, clinical, and serological evolution of Chagas disease. PLoS Negl Trop Dis. 2016;10(3):e0004508.
89. Freedman B. Equipoise and the ethics of clinical research. N Engl J Med. 1987;317(3):141–5.
90. Miller FG, Joffe S. Equipoise and the dilemma of randomized clinical trials. N Engl J Med. 2011;364(5):476–80.
91. Sackett D. Why did I become a clinician-trialist? J R Soc Med. 2015;108(8):325–30.
92. Collins R, MacMahon S. Reliable assessment of the effects of treatment on mortality and major morbidity, I: clinical trials. Lancet. 2001;357(9253):373–80.
93. MacMahon S, Collins R. Reliable assessment of the effects of treatment on mortality and major morbidity, II: observational studies. Lancet. 2001;357(9254):455–62.
94. Nelson HD, Humphrey LL, Nygren P, Teutsch SM, Allan JD. Postmenopausal hormone replacement therapy: scientific review. JAMA. 2002;288(7):872–81.
95. Walsh M, Srinathan SK, McAuley DF, Mrkobrada M, Levine O, Ribic C, et al. The statistical significance of randomized controlled trial results is frequently fragile: a case for a Fragility Index. J Clin Epidemiol. 2014;67(6):622–8.
96. Narayan VM, Gandhi S, Chrouser K, Evaniew N, Dahm P. The fragility of statistically significant findings from randomized controlled trials in the urologic literature. BJU Int. 2018;122(1):160–6.
97. Khan M, Evaniew N, Gichuru M, Habib A, Ayeni OR, Bedi A, et al. The fragility of statistically significant findings from randomized trials in sports surgery: a systematic survey. Am J Sports Med. 2017;45(9):2164–70.
98. Evaniew N, Files C, Smith C, Bhandari M, Ghert M, Walsh M, et al. The fragility of statistically significant findings from randomized trials in spine surgery: a systematic survey. Spine J. 2015;15(10):2188–97.
99. Durham J, Mackey WC. Perioperative beta-blockade in noncardiac surgery: a cautionary tale of over-reliance on small randomized prospective trials. Clin Ther. 2016;38(10):2302–16.

100. Yusuf S, Teo K, Woods K. Intravenous magnesium in acute myocardial infarction. An effective, safe, simple, and inexpensive intervention. Circulation. 1993;87(6):2043–6.
101. ISIS-4 (Fourth International Study of Infarct Survival) Collaborative Group. ISIS-4: a randomised factorial trial assessing early oral captopril, oral mononitrate, and intravenous magnesium sulphate in 58,050 patients with suspected acute myocardial infarction. Lancet. 1995;345(8951):669–85.
102. Catalioti F, Acquatella H. Mortality comparison of 5 years follow up in subjects with chronic Chagas disease with and without benznidazol treatment [Comparación de mortalidad durante seguimiento por 5 años en sujetos con enfermedad de Chagas crónica con y sin tratamiento de benznidazol]. Rev Biol Trop. 1998;27(Suppl):29–31.
103. Silveira CAN. Avaliação a longo prazo do tratamento específico da doença de chagas [PhD thesis]. Distrito Federal, Brazil: Faculdade de Medicina, Universidade de Brasília, 2000:p. 123.
104. Fabbro de Suasnábar D, Arias E, Streiger M, Piacenza M, Ingaramo M, Del Barco M, et al. Evolutive behavior towards cardiomyopathy of treated (nifurtimox or benznidazole) and untreated chronic chagasic patients. Rev Inst Med Trop Sao Paulo. 2000;42(2):99–109.
105. Lauria-Pires L, Braga M, Vexenat A, Nitz N, Simões-Barbosa A, Tinoco D, et al. Progressive chronic chagas heart disease ten years after treatment with anti-Trypanosoma cruzi nitroderivatives. Am J Trop Med Hyg. 2000;63(3, 4):111–8.
106. Gallerano R, Sosa R. Intervention study on the natural evolution of Chagas disease: evaluation of specific antiparasitic treatment [Estudio de intervencion en la evolucion natural de la enfermedad de Chagas: evaluacion del tratamiento antiparasitario especifico]. Rev Fac Cien Med Univ Nac Cordoba. 2000;57(2):135–62.
107. Hess EP, Hollander JE, Schaffer JT, Kline JA, Torres CA, Diercks DB, et al. Shared decision making in patients with low risk chest pain: prospective randomized pragmatic trial. BMJ. 2016;355:i6165.
108. Branda ME, Leblanc A, Shah ND, Tiedje K, Ruud K, Van HH, et al. Shared decision making for patients with type 2 diabetes: a randomized trial in primary care. BMC Health Serv Res. 2013;13:301.
109. Alvarez MG, Vigliano C, Lococo B, Bertocchi G, Viotti R. Prevention of congenital Chagas disease by Benznidazole treatment in reproductive-age women. An observational study. Acta Trop. 2017;174:149–52.
110. Murcia L, Simon M, Carrilero B, Roig M, Segovia M. Treatment of infected women of childbearing age prevents congenital Trypanosoma cruzi infection by eliminating the parasitemia detected by PCR. J Infect Dis. 2017;215(9):1452–8.
111. Moscatelli G, Moroni S, Garcia-Bournissen F, Ballering G, Bisio M, Freilij H, et al. Prevention of congenital Chagas through treatment of girls and women of childbearing age. Mem Inst Oswaldo Cruz. 2015;110(4):507–9.
112. Fabbro DL, Danesi E, Olivera V, Codebo MO, Denner S, Heredia C, et al. Trypanocide treatment of women infected with Trypanosoma cruzi and its effect on preventing congenital Chagas. PLoS Negl Trop Dis. 2014;8(11):e3312.
113. Torrico F, Alonso-Vega C, Suarez E, Rodriguez P, Torrico MC, Dramaix M, et al. Maternal Trypanosoma cruzi infection, pregnancy outcome, morbidity, and mortality of congenitally infected and non-infected newborns in Bolivia. Am J Trop Med Hyg. 2004;70(2):201–9.
114. Yancy CW, Jessup M, Bozkurt B, Butler J, Casey DE Jr, Colvin MM, et al. 2017 ACC/AHA/HFSA focused update of the 2013 ACCF/AHA guideline for the management of heart failure: a report of the American College of Cardiology/American Heart Association Task Force on Clinical Practice Guidelines and the Heart Failure Society of America. Circulation. 2017;136(6):e137–61.

Orally Transmitted Chagas Disease: Biology, Epidemiology, and Clinical Aspects of a Foodborne Infection

Belkisyolé Alarcón de Noya and Oscar Noya González

Abstract Chagas disease has been one of the great neglected infectious pathologies, affecting not only developing countries but also the most vulnerable sectors of the population with greater intensity. In the spectrum of its mechanisms of infection, the most forgotten, unknown, and underestimated is the oral infection. Despite being the usual form of infection among mammals and the increasing frequency of oral *Trypanosoma cruzi*-transmitted outbreaks since 1967, many aspects about the physiopathology, the histotropism, clinical evolution, evasion mechanisms of *Trypanosoma cruzi*, the consequent host immune response, and the therapeutic efficacy are still unknown. However, nowadays we already know, and are discussed in this chapter, the risk factors, the clinical aspects, the incriminated vectors, the ideal diagnostic methods, and the therapeutic approach primarily applicable to the acute phase.

1 Introduction

American trypanosomiasis is one of the most important zoonoses of the New World since its causative agent *Trypanosoma cruzi* infects all mammals that are exposed to this protozoan. Its original mechanism of transmission between the wild and, more recently, domestic reservoirs has been fundamentally the oral route, since the fur

B. Alarcón de Noya (✉)
Facultad de Medicina, Sección de Inmunología, Instituto de Medicina Tropical, Universidad Central de Venezuela, Caracas, Venezuela

O. Noya González (✉)
Facultad de Medicina, Sección de Biohelmintiasis, Instituto de Medicina Tropical, Universidad Central de Venezuela, Caracas, Venezuela

Centro para Estudios sobre Malaria, Instituto de Altos Estudios, Dr. Arnoldo Gabaldón, Instituto Nacional de Higiene Rafael Rangel, MPPS, Aragua, Venezuela

J. M. Altcheh, H. Freilij (eds.), *Chagas Disease*, Birkhäuser Advances in Infectious Diseases, https://doi.org/10.1007/978-3-030-00054-7_11

and the thickness of the skin are a physical barrier almost insurmountable for its penetration. Hence, it is considered that the oral route is the usual way of infection in different species. Today we know the intrinsic mechanisms of the cellular biology of this parasite, by its more efficient penetration through the gastric mucosa [1]. Perhaps the exception to this route of penetration is represented by marsupials which seem to be the ancestral reservoirs, among which *Didelphis marsupialis*, which has the capacity to house vertebrate and vectors cycles, the latter in their odoriferous glands. Under situations of stress, they are able to vaporize their secretions containing metacyclic trypomastigotes that can infect remotely other animals through mucous membranes such as the conjunctiva [2].

The participation of humans in the epidemiology of this hemoparasite is related to the arrival of Asian migrants that populated America, and their infection has been demonstrated by molecular biology techniques in mummies of different Amerindian ethnic groups, at least for 7000 years [3]. Unlike wild mammals, humans and in particular Amerindian ethnic groups have a skin devoid of mechanical barriers such as fur, which allow the easy penetration of metacyclic trypanosomes through small fissures and even at the stinging point of their vectors, which eliminated trypanosomes through the excreta. For these reasons, humans are the only species in which the cutaneous route is considered a natural via of infection. *Trypanosoma cruzi*, the causal agent of this infection, multiplies intracellularly in the vertebrate host causing damage especially in the heart, the target organ, causing Chagas disease (ChD).

When Carlos Chagas published the first report of a new disease in 1908 [4], it was recognized that this parasitic tropical disease was acquired through the skin [5]. Typical entry-site symptoms and clinical signs (Romaña sign, cutaneous indurations known as "chagomas") were described by this outstanding Brazilian physician and researcher. Other entry ports have been recognized (blood transfusions, vertical transmission, organ transplantation, laboratory accidents), and since 1960 the oral route has been recognized as one of the causes of foodborne diseases [6]. This fact is one of the reasons why this mode of infection has been neglected and minimized by health authorities, as rare accidents of secondary importance. However, the increased number of oral ChD (OChD) reports especially originating from the Amazon region and in four other countries (Venezuela, Bolivia, Colombia, and French Guyana) [7] has elicited and increased interest in this particular form of infection. In Venezuela, 14 outbreaks have been reported—ten are recorded in Alarcón de Noya et al. [8] and the other four in Ruiz-Guevara et al. [9]—being the first outbreak occurred in 2007, the largest registered in Latin America [10]. In September 19, 2017, the Colombia Health Ministry (Ministerio de Salud y Protección Social) warned about the situation of a family of Venezuelans who had three children died with acute ChD, one of 9 months in Táriba (Táchira State), Venezuela, and the other two children of 5 and 9 years old died in the North of Santander, Colombia. The two grandparents had the same symptoms. All belonged to the same family and came from the same locality (https://www.minsalud.gov.co/Paginas/Intensificadas-medidas-de-salud-p%C3%BAblica-por-muertes-de-venezolanos-por-Chagas.aspx). Due to the simultaneity in related persons of the same family, and the identification of acute ChD, this episode could be identified as the 15th outbreak of oral transmission of *T. cruzi* in Venezuela.

2 Particularities of the Orally Transmitted Chagas Disease

Although transcutaneous transmission has been considered the most important transmission mechanism until now, in Venezuela in the last 11 years, only a total of ten patients have been reported [11], while in that interim, there have been 15 OChD outbreaks affecting 273 people. Furthermore, among these ten cases, four were members of the same family, and possibly they were infected by the oral route [8, 11]. This decline in prevalence coincides with that observed in the rest of Latin America, as a result of the integration of the Chagas disease control programs of the Latin American countries, promoted by PAHO (Central America, Andean, and Southern Cone initiatives) [12]. However, in contrast to the decreased transmission shown in these figures, an increase in acute cases of orally acquired infection has been observed in the last two decades [13, 14].

There are multiple reasons that could explain this change in the transmission patterns in the endemic region of Latin America. Among them, the following are worth highlighting:

2.1 *Bioecological Factors*

1. Human migrations: great demographic changes have occurred in this continent such as the reversal in the percentages of rural vs. urban population, which decreased from 70.5% in 1950 to 21% in 2013 [15, 16]. This allowed the migration of infected populations to the periphery of the cities, which today make up the belts of misery and that, consequently, correspond to recently invaded and deforested areas, now absent from the natural wild reservoirs that were the original source of food for the triatomines. These demographic changes in some way forced changes in the ethology of vectors, which were now forced to feed on domestic reservoirs (rodents, dogs, and cats) and humans. In short, a rural pathology now becomes a pathology of urban predominance in some countries, as is the case of Venezuela [17].
2. Triatomine vectors: domiciliation is one of the major bioecological changes that have occurred in this disease. It has been reported by a series of species that are not necessarily good vectors through the cutaneous route (*Triatoma dimidiata*, *Panstrongylus rufotuberculatus*, *Rhodnius stali*, *Eratyrus mucronatus*, and *Panstrongylus geniculatus*). It is the case of *Panstrongylus geniculatus*, which nevertheless has revealed to be a species of less importance for the transmission through the skin, since the speed factor in the defecation reflex that is essential for skin transmission, is irrelevant for the contamination of food and therefore

Fig. 1 Homemade or artisan fresh (unpasteurized) juices are the main source of infection with metacyclic trypomastigotes from completely blended triatomines or their excretions. *Panstrongylus geniculatus* is the species most frequently incriminated as the source of infection (drawing by Alberto Monteagudo)

for the oral route [18, 19]. Both nymphal stages and adults can be the source of infection, either through their depositions on food or by the fact of being blended during the processing, particularly of fruits, in the form of artisanal juices (guava, mango, and juice of sugarcane) [17, 20] (Fig. 1).

3. Reservoirs: a total of 180 species of 25 families of infected mammals have already been described as potential hosts for *T. cruzi* (marsupials, monkeys, bats, horses, dogs, cats, rats, raccoons, skunks, coyotes, etc.) [21, 22]. Among them, *Didelphis marsupialis* is the most important synanthropic reservoir of the continent due to its wide geographical distribution, high levels of natural infection with *T. cruzi*, and the presence of infective metacyclic trypomastigotes in the perianal glands similar to triatomines and also able to infect through oral, nasal, ocular, and skin routes of infection [2]. The rat, *Rattus rattus*, deserves special attention as three basic conditions have enabled it to become a very efficient urban reservoir for *T. cruzi*. Firstly, rats are very susceptible to *T. cruzi* infection maintaining the parasite for a long period of time; secondly, they have a high urban population density, tending to increase with the garbage in the cities, and thirdly, they are a preferred food source of *Panstrongylus geniculatus* [23, 24].

2.2 Cultural Patterns

This is a fundamental aspect to understand some of the OChD outbreaks, in which the habits of nutrition and customs of the population in the different social strata have conditioned the expansion of outbreaks of this serious pathology.

The consumption of homemade, street, and school-canteen fruit juices is frequent in Latin America, from which these outbreaks have been generated. Fruits such as guava, açai, mango, pineapple, and sugarcane have been incriminated as vehicles of the outbreaks. These juices either came from palm trees where triatomines are present, which were then processed without checking or without pasteurizing the juice, or the contamination occurred in the kitchens where adults and nymphs wander at night from any of the triatomine species that can contaminate with their defecations with metacyclic trypomastigotes or they are blended in the preparation process.

Unlike skin contamination in which triatomines excrete a few microliters of urine containing between 3000 and 4000 metacyclic trypomastigotes per microliter, in the case of the contamination with an entire triatomine, it has been estimated that a *Triatoma infestans* can contain 684,000 trypomastigotes in the intestine [25]. It explains the high parasitic load that a patient can receive, having a mechanism of penetration more efficient than the skin; in consequence the morbidity and mortality are much higher.

2.3 Other Environmental Factors: Vegetation, Housing, Brightness, and Climate

Perhaps the deforestation in the periphery of the cities is the main component for the bioecological changes, because it introduces very important changes on the ecology of the vectors and their habits and in particular the domiciliation of vectors of previously wild behavior. The parasites move from a wide variety of species of wild reservoirs to only four urban species (dogs, cats, rodents, and humans), limiting the number of parasitic variants in a sort of screening or funnel decreases the genetic variability of the variants that affect the domestic and peridomestic species as well as humans.

These other factors also have an important role in the epidemiological changes that are occurring in this disease. In relation to vegetation, it has already been mentioned how deforestation or the simple fact of living near it increases the risk as a factor of its transmission. This is what is called "border phenomenon," which has been one of the important factors in the acute outbreaks of the Caracas area, since all of the oral outbreaks occurred near the forest.

The houses most likely to be infested by triatomines are mostly inhabited by marginalized populations located in slums ("favelas," "barrios," "villas miseria," "ranchos," etc.) in which most of the walls are not plastered with numerous crevices in which vectors can enter, hide, and multiply [17].

The periods of drought (March–May) and the increase of the luminosity in the communities have been evidenced as risk factors additional to those already mentioned, as the invasions of the vectors increased. In the latter case, the streetlights and houses are powerful attractors of the triatomines, in particular of *Panstrongylus geniculatus* [18].

What characterizes oral transmission and what differentiates it from cutaneous transmission

- Usually seen in a form of outbreaks.
- Can be transmitted by any triatomine species and also by bedbugs.
- Metacyclic and sanguineous trypomastigotes are likely infectious.
- The entire triatomine with very high parasite load could be the infectious source.
- Infection source: complete triatomines or their feces and *D. marsupialis* anal excretions (food contamination).
- Occurs in rural as well as urban areas.
- There is no previous history of cutaneous manifestations.
- It affects all social status.
- Patients often deny having seen triatomines.
- It is the most common infection route in animal species other than humans.

All the previously described items should make clinicians to suspect that they are facing an OChD outbreak, which should be managed as an emergency of clinical severity and high mortality.

3 Clinical Aspects of the Acute Cases in Orally Transmitted Chagas Disease

The suspicion of oral transmission of ChD can almost only be made during the acute phase of the disease or when a chronically case is found to be related to people who suffered the acute form of the infection.

The acute condition of oral infection with *T. cruzi* is different from the clinical manifestations of other acute cases due to other mechanisms of infection. In the case of blood transfusion, organ transplantation, and congenital transmission, the immediate antecedent is essential for raising the suspicion of a possible acute infection by *T. cruzi*. In the transcutaneous vector transmission, there is usually a previous history of contact with triatomines, and even the patient usually brings the insect full of blood found in the bed or in the house.

The epidemiological antecedent most frequently described by patients is their association with a meal or a drink shared with friends and family, who also agree with a similar symptomatology. It can take days or weeks to establish the network of people exposed simultaneously and who should be evaluated as soon as possible to avoid clinical severity and mortality [23, 26, 27]. Among the antecedents, there is no recognition of contact with triatomines nor the entrance door of the parasite (Romaña and/or chagoma signs) as it can appear in acute cases due to transcutaneous vectorial infection [28]. The incubation period is 3–22 days in oral transmission [10, 29]. In Chacao, the maximum incubation period was determined in 11 days, as one of the infected teachers joined their work just 11 days before the onset of symptoms [20].

1. Clinical presentation of acute cases: In the review of several outbreaks of OChD (two in Colombia, a compilation from Brazil, and two from Venezuela) [20] and the most numerous and recent outbreaks reported in Colombia [29] which in total add up to 434 acute cases, the predominant symptomatology has been fever, headache, facial and lower limb edema (Fig. 2), abdominal pain, diarrhea, myalgias, arthralgias, and asthenia (Table 1).

   The clinical picture of the acute phase of oral transmission is that of a systemic infection with cardiac involvement as the heart is the target organ. Frequent cardiological manifestations result in chest pain, palpitations, and dyspnea, which are reasons for hospitalization. Cardiomegaly (Fig. 2) and heart failure can occur as a result of severe arrhythmia or cardiac tamponade due to pericardial effusion leading the patient to a fatal outcome [33]. The most frequent findings on the EKG and ECHO can be seen in Table 2.

   Regarding the laboratory findings observed in hematology and blood chemistry, there are no particular findings. During the first microepidemic episode, troponin was evaluated only in hospitalized patients, and it was elevated in 8/11 (73%) of patients; the globular sedimentation rate was altered in 57%, the C-reactive protein in 87.5%, the lactate dehydrogenase in 88.8%, and leukocytosis in seven patients [36]. During the second large outbreak, laboratory exams were made to 43 out of 88 infected persons. They showed anemia (78%), elevated troponin (50.6%), leukocytosis (>10,000) (69%), and elevated transaminases 26% AST and 51.6% ALT, and 2.3% of the patients showed elevated creatinine [33].
2. Severity and mortality. Oral infection with *T. cruzi* has a faster and more severe course than the cutaneous transmission, probably due to the size of the inoculum to which the patient's immune system faces and also better efficacy of parasites to penetrate the gastric mucosa and accessing to systemic circulation. The severity of symptoms and mortality depend on time between the onset of symptoms, the confirmed diagnosis, and treatment delivery. Mild and severe cases were more frequent as much as 87.2% in 227 persons orally infected from ten outbreaks registered in Venezuela from which only 12.7% of the confirmed cases

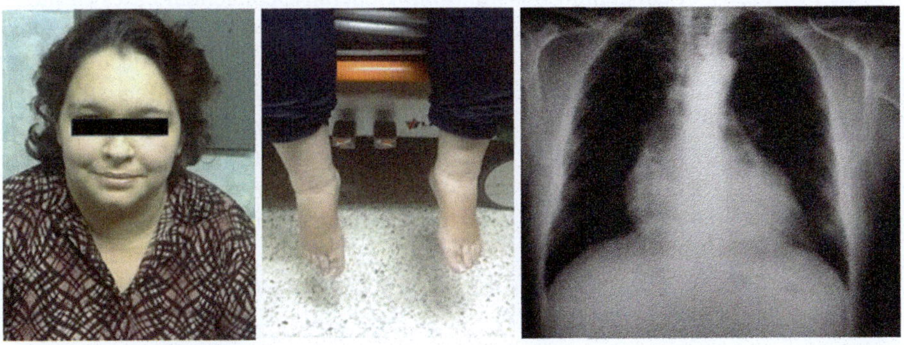

Fig. 2 Facial, lower limb edemas and pericardial effusion in patients with oral Chagas disease

Table 1 Predominant symptomatology in 434 acute oral cases of Chagas disease registered in Brazil, Colombia, and Venezuela

Symptomatology	Frequency (%)
Fever	80–100
Headache	25–93
Facial edema	28–70
Edema in lower limbs	4–57
Abdominal pain	22–50
Diarrhea	2–30
Myalgias, arthralgias	11–85
Asthenia	10–73

n = Brazil 181 [13]; Venezuela 192: 103 [10] and 89 [30]; Colombia 61: 11 [31], 10 [32], and 40 [29].

Table 2 Frequency (%) of EKG and ECHO alterations in acute oral *Trypanosoma cruzi*-infected patients from four outbreaks

	Brazil [34]	Colombia [32]	Venezuela [30, 35]
ST-T changes	100		37.8[a]
Right branch blockage	25	40	1.9[a]
Atrial or ventricular fibrillation	8.3	20	3[a]
Left ventricular hypertrophy			17.5[b]
Pericardial effusion	36.4		81.3[b]

[a]Marques et al. [35]
[b]Alarcón de Noya [30]

were asymptomatic [8]. In relation to mortality, it varies from 0% to 100% being the highest frequency in two outbreaks in Colombia where death occurred in regions of Magdalena and Santander with 5/13 (38.5%) and 3/3 (100%) in 2003 and 2008, respectively [33]. From ten cited outbreaks in Venezuela, ten persons died from 249 infected persons, with 4% mortality, being predominantly children, pregnant woman and her stillborn, and a woman that had given birth 2 months before and was still breastfeeding.

3. Vertical transmission. One of the most feared consequences in oral infection is the possibility of congenital transmission. Generally, congenital cases of mothers in the chronic phase of *T. cruzi* infection are asymptomatic, and if the mother is not screened for infection, the congenital case may go unnoticed into the chronic phase. Different is the case of acute infection when it affects the pregnant woman. Vertical transmission in acute patients with oral transmission is manifested in any of its forms as spontaneous abortion [34, 37], death of the fetus in the uterus [38], and the birth of an infected newborn [39]. The dissemination of the parasite reaches all the organs of the mother, including the placenta, infiltrating it from where it reaches the fetus to spread and affect all organs as evidenced by the presence of the parasite by direct microscopic examination of

the tissues or by PCR [38]. The treatment of the pregnant woman with Chagas disease is contraindicated [40]; however each acute case in pregnant women should be evaluated because not giving treatment puts the mother at risk [37].

4 Laboratory Diagnosis

Once there is epidemiological and clinical suspicion, confirmation of the probable case of acute infection by oral transmission of *T. cruzi* must be done. In isolated cases that are usually the index cases, one of the frequent diagnostic errors is the confusion with malaria, although it has a clinic and a different pattern of fever. Nonetheless, when performing the thick and thin blood smears, it is possible to discriminate these two hemoparasites and apply the appropriate treatment [26]. The parasitological diagnosis should be made in all patients with epidemiological and clinical suspicion, but cases present unexpectedly, and sometimes the disease can affect large groups making it difficult to comply with the entire diagnostic protocol.

1. Immunological diagnosis: ELISA tests should be applied to the entire study population, searching simultaneously for IgG and IgM antibodies. Following the PAHO/WHO guidelines, two serological diagnostic tests of different bases [41] should be tested. Tests such as indirect hemagglutination, direct agglutination, and immunofluorescence have been used for this purpose. An example of this approach is the flowchart used in our two large oral outbreaks, in which diagnosis of the cases was initiated by the ELISA test and thereafter indirect hemagglutination or Western blot were used as confirmatory methods [30, 42].
2. Parasitological diagnosis: To confirm that we are facing an acute outbreak of oral transmission ChD, it is necessary to identify *T. cruzi* in at least one case of all clinically and epidemiologically related cases. Thick and thin blood smears are used as traditional techniques, although fresh examination of a drop of blood under a coverslip is very useful because it allows for immediate visualization of mobile trypomastigotes. Nevertheless, the best and rapid diagnostic test for the acute phase is microhematocrit [43], due to its parasite concentrating effect, which can be observed at the level of the buffy coat after centrifugation of the capillary tube (Fig. 3). In reference laboratories, it is important to obtain the parasitic isolate to allow for subsequent molecular studies and to know the circulating genotype [44, 45], so that inoculation in mice and blood culture could be performed in all possible patients.
3. Molecular diagnosis: Upon suspicion of acute cases of ChD, all blood samples should be taken from the first contact, even without previous serological diagnosis, since in some occasions there is no new opportunity to take a second sample from the patient, or by the time the serology result arrives, the patient is already receiving treatment. The first blood sample should be of enough volume to allow pretreatment testing with polymerase chain reaction (PCR) and animal culture in order to obtain *T. cruzi* DNA for molecular studies. The follow-up studies of outbreaks carried out by the Tropical Medicine Institute in Caracas tested for the

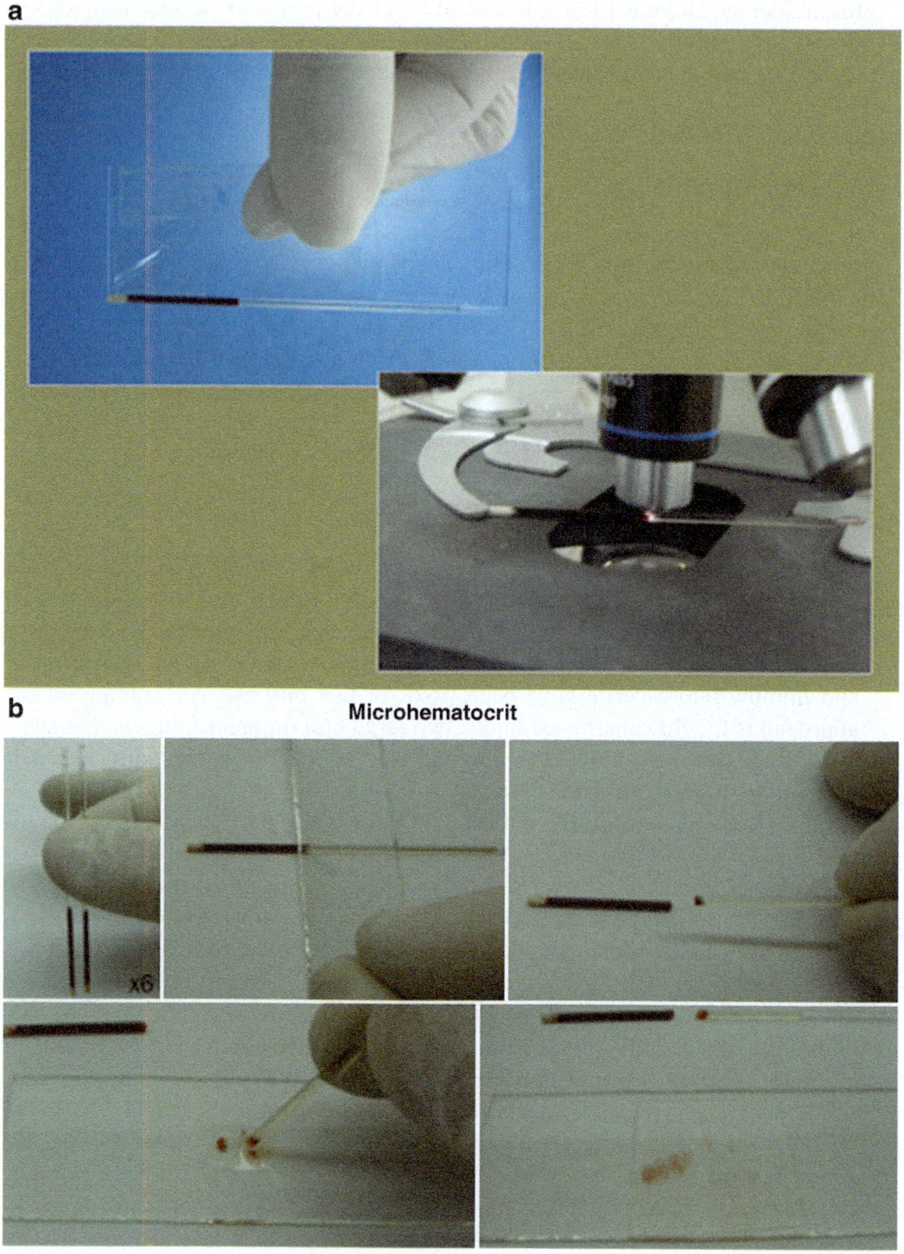

Fig. 3 (**a**) Microhematocrit technique, for the rapid demonstration of blood trypomastigotes in acute infected patients. Mobile parasites are concentrated on the top of the buffy coat. (**b**) Blood sample obtained with the microhematocrit capillary could be examined under the coverslip to observe motile trypomastigotes and, after fixation, must be stained with Giemsa

presence of parasitic DNA [30, 42]. Oral ChD acute cases have shown the highest parasite load by PCR when compared to vectorial acute and chronic cases [46, 47]. PCR can be considered more as an indicator of treatment failure rather than a marker of cure [48]. Molecular studies also allow to determine the genotype. Studies by Diaz et al. analyzing parasite DTUs from six Colombian OChD outbreaks found that these molecular biomarkers were not strictly correlated with clinical forms of the disease [49].

4. Diagnostic strategies depend on the size of outbreak. The first urban ChD outbreak in Venezuela allowed to compare the performance of different diagnostic methods used under medical emergency conditions and their contribution to accurately confirm the acute infected persons [42]. A complete parasitological and immunological evaluation was performed in the initial small group of patients, confirming the etiology of the outbreak in a non-endemic area for ChD. Thereafter, only conventional serology methods, including ELISA (IgM/IgG) and IHA, were applied in all individuals at risk. The selection of these methods was based on the speed of implementation, sensitivity, and specificity. The use of fresh blood smears, Giemsa stains, and PCR assays was restricted to a very limited number of cases because these techniques are considered cumbersome and time-consuming in an urgent situation. The parasitological outcomes were different between the severely ill group and the subset of mildly symptomatic individuals. The parasitemia was lower in the second group, probably due the previous use of different antibiotics that may have inhibited the parasite's growth in culture. In severely symptomatic patients, fresh blood smears were more sensitive for detecting parasites than Giemsa-stained smears, suggesting that the former technique should be routinely applied in people with long-lasting fever of unknown origin. When *T. cruzi* trypomastigotes are detected in peripheral blood by any technique during an outbreak, it is necessary to investigate all symptomatic and asymptomatic cases suspected of sharing the same epidemiological risk in order to prevent increased morbidity and mortality. The importance of simultaneous screening for specific IgM and IgG antibodies by ELISA can confirm the acute condition of the infection. The detection of IgM was very specific in that study, as there were no false positives detected when comparing with the results obtained by ELISA IgG test. In spite of the application of all these serologic techniques, the diagnosis of ChD was inconclusive in 16 individuals who were positive in only one of the immunological tests, not complying with the international criteria of identification of a Chagas case. Due to the epidemiological link, these individuals were treated with nifurtimox. Subsequently, seven of them tested positive for PCR.

These experiences reinforced the widespread use of conventional serology methods detecting IgM and the use of all available diagnostic techniques to confirm infection. For the second largest outbreak in Venezuela, blood culture, serology, and PCR were carried out during the first contact to the probable infected persons. In this study more parasite isolates and PCR were recorded [30]. In conclusion, all available parasitological, immunological, and molecular procedures are essential for identifying the causal agent of the T. cruzi oral transmission.

5 Treatment and Follow-Up

1. Routine treatment: In ChD, the same drugs are used for both the acute and chronic phases: nifurtimox (NFX) (Lampit®, Bayer), a nitrofuran, and benznidazole (BNZ) (Rochagan®, Roche), a nitroimidazole [50]. Due to better tolerance, BNZ is considered the first drug of choice; however, individual tolerance varies widely, and NFX represents an alternative. The choice depends mostly on the international availability [51]. For the acute phase, treatment is recommended for all patients, regardless of the mode of transmission and age of the patient [40]. BNZ is indicated at 6 mg/kg/day during 60 days in three daily doses and NFX, at 8 mg/kg/day during 90 days divided into two daily doses. When necessary, antiallergic, antineuritic, or gastric protectors are prescribed. During the follow-up of the treatment, laboratory tests that include complete hematology, urea and creatinine, aminotransferases, and bilirubin are recommended [52].
2. Adverse events of treatment during the acute phase in patients infected by the oral route. The undesirable effects of the two unique available compounds to treat ChD are usually caused by oxidative and reductive damage on host tissues that limit considerably their use [50]. A cross-sectional analytical study was carried out in Venezuela on 122 patients who received 176 courses of treatment with any of the two anti-*T. cruzi* drugs (113 with NFX and 63 with BNZ). Treating symptomatic persons and taking into account the drug side effects, it was important to know how much of the adverse effects were caused by the disease and how much by the medication. When the occurrence of side effects was compared among individuals with confirmed acute ChD and suspected cases, no statistically significant difference ($p = 0.363$) was found in relation to the drug used, age, or date of treatment. From the 176 treatment, 79% had one or more adverse effects which predominated in adults (97.8%) as compared to children (75.5%), and the risk of having side effects was significantly higher for NFX in comparison with BNZ. Four adults and a child treated with NFX had severe side effects (pulmonary infarction, facial paralysis, neutropenia, blurred vision, and bone marrow hypoplasia) warranting hospitalization and drug suspension. Abdominal pain, hyporexia, weight loss, headache, nausea, and lymphocytosis were frequent reports for NFX, whereas skin rash, neurosensory effects, hyporexia, fatigue, pyrosis, abdominal pain, and eosinophilia were observed with BNZ. Frequency and severity of side effects during treatment of acute oral infection by *T. cruzi* demand direct supervision and close follow-up, even in those asymptomatic patients, to prevent life-threatening situations [52].
3. Posttreatment follow-up. The efficacy of BNZ and NFX varies with the stage of the disease, the patient's age, geographic area, and the *T. cruzi* isolate involved. A drawback of the studies assessing their efficacy in any phase, age, or geographical area is the lack of a marker to define cure. Current recommendations rely on the switch of serology from positive to negative, but this may take many years, limiting its use in clinical practice. Detection of *T. cruzi* DNA in peripheral blood only serves as a tool to identify treatment failure because a negative result does not mean absence of infection [48]. Nonetheless, based on serology, these drugs are more efficient during acute phase in which negative seroconver-

sion varies between 50% and 80% in comparison with 8% and 20% efficacy in chronic phase patients [48, 53]. Also efficacy is different in children and adults. Treatment with BZ for 60 days in children aged 6–12 years with early or indeterminate phase of ChD in Brazil and Argentina had a cure efficacy of 56% and 62%, monitored by seroconversion after 36 and 48 months after treatment, respectively [54, 55]. Treatment with NFX for 90 days of 100 patients with acute oral ChD showed a therapeutic failure that ranged from 69% to 78%, after 3 years of follow-up in Venezuela [56]. A similar situation was for 60 days of BNZ to 80 persons infected by the oral route from Chichiriviche outbreak during the same observational time (Alarcón de Noya et al., unpublished data). In spite of the treatment in our studies being administered under supervision, during acute phase, to children, the effectiveness of these two drugs seems to be lower than the treatment administrated during the acute phase of the cutaneous mechanism of infection as that described by Cançado in 2002, who observed cure in 76% of patients with acute phase after 13- to 21-year follow-up [57].

6 Prophylaxis and Control Measures in Foodborne Chagas Disease

The control measures that are recommended to prevent ChD orally acquired are those implemented by the general population in their own home and those implemented by the Ministry of Health of each country, ranging from legal regulations for the production, packaging, preservation, and distribution of food, particularly those that are to be consumed without cooking for public institutions (school canteens, nursing homes, factory canteens, or public and private companies) [58].

Individual and family prophylaxis measures are based on avoiding the triatomine-food contact, through physical barriers, preventing food from being exposed to contact with infected vectors and marsupials. The first one is based on preventing the entrance of the triatomines through the use of mosquito nets in windows and doors, ceilings, and sealed walls, as well as the spraying of walls with residual insecticides such as fenitrothion. The second one is by improving the housing with fritted walls and the use of paintings with insecticides that prevent the domiciliation of vectors, as has been observed with *Panstrongylus geniculatus*, originally a wild species. Likewise, protect the food either by covering it or placing it in hermetic containers especially at night, avoiding the exposure to the evacuations of the triatomines or the urine and excretions of *Didelphis marsupialis*. Being the schools where the biggest outbreaks have occurred, precautions must be taken on the hygiene of beverages and solid foods that are served in canteens, as in nursing homes as well as in military barracks, restaurants, etc. The supervisory role not only by the directors of the schools and companies but also the users or their representatives (parent associations, unions, etc.) is essential [58].

Other special prophylactic measures are based on avoiding the consumption of palm fruits such as açai or manaca for the production of juices in the Amazon region, since in the clusters of this fruit, the triatomines frequently inhabit. The triatomines are crushed and their content released into the drink during preparation, sometimes being consumed far from the production site. This way has been called distantiae transmission [59], and contamination of artisanal açai juices is the most frequent mechanism of transmission in the Brazilian Amazon [60], where the majority of oral outbreaks have occurred in America.

The viability of trypomastigotes in beverages for human consumption varies according to the type of fruit, which has been studied by Añez et al. [61], who observed that parasites survive several hours in milk, orange, pineapple, papaya, sugarcane, and guava juices, after preparation and subjected to cooling in a refrigerator. Pasteurization of juices would be one of the prophylactic measures, particularly in those juices of industrial production and even artisanal.

It has been shown that blood trypomastigotes also have infective capacity by the oral route [62]; the consumption of meats or blood derivatives of raw wild animals, especially hunting, should be avoided.

Finally, the role of education from the school level to the general public on food hygiene is very particular about the risks of street foods which hygienic norms and preparation procedures are unknown.

References

1. Yoshida N. Molecular mechanisms of *Trypanosoma cruzi* infection by oral route. Mem Inst Oswaldo Cruz. 2009;104:101–7.
2. Deane MP, Lenzi HL, Jansen A. *Trypanosoma cruzi*: vertebrate and invertebrate cycles in the same mammal host, the opossum Didelphis marsupialis. Mem Inst Oswaldo Cruz. 1964;79:113–5.
3. Guhl F, Jaramillo C, Yockteng R, Cárdenas-Arroyo F. *Trypanosoma cruzi* DNA in human mummies. Lancet. 1997;349:1370.
4. Chagas C. Nova tripanosomiase humana, Estudo sobre a morfologia e o ciclo evolutivo do *Schizotrypanum cruzi*. Gen.Sp. Agente etiologico de uma nova entidade morbida do homem. Mem Inst Oswaldo Cruz. 1909;1:159–218.
5. Rassi A Jr, Rassi A, Marin Neto JA. Chagas disease. Lancet. 2010;375:1388–402.
6. Robertson LJ, Devleesschauwer B, Alarcón de Noya B, Noya González O, Torgerson PR. *Trypanosoma cruzi*: time for international recognition as a foodborne parasite. PLoS Negl Trop Dis. 2016;10(6):1–6.
7. Ruiz-Guevara R, Noya O, Alarcón de Noya B. Documented outbreaks of foodborne Chagas disease. In: Alarcón de Noya B, Noya O, Robertson LJ, editors. *Trypanosoma cruzi* as a foodborne pathogen, Springer briefs in food, health and nutrition. New York, NY: Springer; 2015. p. 53–71.
8. Alarcón de Noya B, Díaz-Bello Z, Colmenares C, Ruiz-Guevara R, Mauriello L, Muñoz-Calderón A, Noya O. Update on oral Chagas disease outbreaks in Venezuela: epidemiological, clinical and diagnostic approaches. Mem Inst Oswaldo Cruz. 2015;110(3):377–86.
9. Ruiz-Guevara R, Muñoz-Calderón A, Alarcón de Noya B, Redondo C, Pulido L, Beitia Y, Mauriello L, Rojas D, Muñoz J, Navarro E, Reyes-Lugo M. Brote familiar de Enfermedad de Chagas por transmisión oral en Yaguapita, Estado Miranda, Venezuela. Bol. Mal. Salud Amb. En prensa. 2018.
10. Alarcón de Noya B, Díaz-Bello Z, Colmenares C, Ruiz-Guevara R, Mauriello L, Zavala-Jaspe R, Suárez JA, Abate T, Naranjo L, Paiva M, Rivas L, Castro J, Márques J, Mendoza I, Acquatella H, Torres J, Noya O. Large urban outbreak of orally-acquired acute Chagas disease, at a school in Caracas, Venezuela. J Infect Dis. 2010;201:1308–15.

11. Añez N, Crisante G, Parada H. Nuevos casos de enfermedad de Chagas en el Occidente de Venezuela. Salus. 2007;11:87–90.
12. OMS/OPS/TDR. Reporte sobre la Enfermedad de Chagas. Grupo de Trabajo Científico. Eds. Felipe Guhl and Janis Lazdins. 2007. pp 1–96.
13. Pinto AYN, Valente SA, Valente VC, Ferreira AG, Coura JR. Fase aguda da doença de Chagas na Amazônia brasileira. Estudo de 233 casos do Pará, Amapá e Maranhão observados entre 1988 e 2005. Rev Soc Bras Med Trop. 2008;41:602–14.
14. Shikanai-Yasuda MA, Carvalho NB. Oral transmission of Chagas disease. Clin Infect Dis. 2012;54:845–52.
15. Briceño-León R. La Enfermedad de Chagas en las Américas: una perspectiva de ecosalud. Cad Saude Publica. 2009;25:S71–82.
16. World Bank. 2013. http//ata.worldbank.org
17. Noya O, Alarcón de Noya B, Robertson L. Epidemiological factors related in foodborne transmission of Chagas disease. In: Alarcón de Noya B, Noya O, Robertson LJ, editors. *Trypanosoma cruzi* as a foodborne pathogen, Springer Briefs in Food, Health and Nutrition. New York, NY: Springer; 2015. p. 41–51.
18. Reyes-Lugo M. *Panstrongylus geniculatus* Latreille 1811 (Hemiptera: Reduviidae: Triatominae), vector de la enfermedad de Chagas en el ambiente domiciliario del centro-norte de Venezuela. Rev Biomed. 2009;20:180–205.
19. Coura JR, Albajar-Viñas P, Junqueira ACV. Ecoepidemiology, short history and control of Chagas disease in the endemic countries and the new challenge for nonendemic countries. Mem Inst Oswaldo Cruz. 2014;109:856–62.
20. Alarcón de Noya B, Noya O. Clinical aspects in foodborne Chagas disease. In: Alarcón de Noya B, Noya O, Robertson LJ, editors. *Trypanosoma cruzi* as a foodborne pathogen, Springer Briefs in Food, Health and Nutrition. New York, NY: Springer; 2015. p. 33–40.
21. Guhl F. Enfermedad de Chagas: realidad y perspectivas. Rev Biomed. 2009;20:228–3.
22. Bern C, Kjos S, Yabsley MJ, Montgomery SP. *Trypanosoma cruzi* and Chagas disease in the United States. Clin Microbiol Rev. 2011;24:655–81.
23. Díaz-Bello Z, Zavala-Jaspe R, Reyes-Lugo M, Colmenares C, Noya-Alarcón O, Noya O, Herrera L, Alarcón de Noya B. Urban *Trypanosoma cruzi* oral transmission: from a zoonotic founder focus to the largest microepidemic of Chagas disease. SOJ Microbiol Infect Dis. 2016;4(1):1–9.
24. Alarcón de Noya B, Noya-González O. An ecological overview on the factors that drives to *Trypanosoma cruzi* oral transmission. Acta Trop. 2015;151:94–102.
25. Shaub GA. *Trypanosoma cruzi*: quantitative studies of development of two strains in small intestine and rectum of the vector *Triatoma infestans*. Exp Parasitol. 1989;68:260–27.
26. Martín A, Alarcón de Noya B, Montero R, Rojas C, Garrido E, Ruiz-Guevara R, Díaz-Bello Z. Epidemia de Chagas agudo adquirido por vía oral en una escuela de Caracas: descripción del caso índice. Arch Venez Ped Puericult. 2009;72(3):97–100.
27. Alarcón de Noya B, Díaz-Bello Z, Colmenares C, Ruiz-Guevara R, Noya O. Enfermedad de Chagas de transmisión oral: vinculación del caso índice con una microepidemia urbana en Venezuela. Bol Mal Salud Amb. 2010;60:135–8.
28. Morocoima A, Tineo Brito EJ, Ferrer E, Herrera L, Nuñez M. Enfermedad de Chagas en el estado Anzoátegui, Venezuela: registro de un caso agudo y caracterización parasitológica y molecular del aislado. Bol Mal Salud Amb. 2008;48(2):121–6.
29. Zuleta-Dueñas LP, López-Quiroga ÁJ, Torres-Torres F, Castañeda-Porras O. Posible transmisión oral de la enfermedad de Chagas en trabajadores del sector de los hidrocarburos en Casanare, Colombia 2014. Biomédica. 2017;37:218–32.
30. Alarcón de Noya B, Colmenares C, Díaz-Bello Z, Ruiz-Guevara R, Medina K, Muñoz-Calderón A, Mauriello L, Cabrera E, Montiel L, Losada S, Martínez J, Espinosa R, Abate T. Orally-transmitted Chagas disease: epidemiological, clinical, serological and molecular outcomes of a school microepidemic in Chichiriviche de la Costa, Venezuela. Parasite Epid Control. 2016;1:188–98.
31. Ríos JF, Arboleda M, Montoya AN, Alarcón EP, Parra-Henao GJ. Probable outbreak of oral transmission of Chagas disease in Turbo, Antioquia. Biomédica. 2011;31:185–95.

32. Hernández LM, Ramírez A, Cucunubá Z, Zambrano P. Brote de Chagas agudo en Lebrija, Santander, 2008. Rev Observ Salud Púb Santander. 2009;4:28–33.
33. Noya O, Ruiz-Guevara R, Díaz-Bello Z, Alarcón de Noya B. Epidemiología y clínica de la transmisión oral de *Trypanosoma cruzi*. Rev Esp Epidem: XI workshop on chagas disease, Barcelona Spain. 2015. pp. 23–34.
34. Bastos CJC, Aras R, Mota G, Reis F, Dias JP, Silva de Jesús R, Silva Freire MG, de Araújo E, Prazeres J, MF RG. Clinical outcomes of thirteen patients with acute Chagas Disease acquired through oral transmission from two urban outbreaks in northeastern Brazil. PLoS Negl Trop Dis. 2010;4(6):e711.
35. Marques J, Mendoza I, Noya B, Acquatella H, Palacios I, Marques-Mejias M. ECG manifestations of the biggest outbreak of Chagas disease due to oral infection in Latin-America. Arq Bras Cardiol. 2013;101:249–54.
36. Alarcón de Noya B, Veas J, Ruiz-Guevara R, Martín A, Rojas C, Machado I, Telo C, Henao L, Díaz-Bello Z, Noya O. Evaluación clínica y de laboratorio de pacientes hospitalizados durante el primer brote urbano de Enfermedad de Chagas de transmisión oral en Venezuela. Rev Patol Trop. 2013;42(2):177–86.
37. Suárez J, B de Suárez C, Alarcón de Noya B, Espinosa R, Chiurillo MA, Villaroel A, De Martín F, Paiva M, Díaz-Bello Z, Valderrama E, Estrada D, Vivas E. Enfermedad de Chagas sistémico en fase aguda por transmisión oral: diagnóstico integral de un caso autopsiado. Gac Méd Caracas. 2010;118(3):212–22.
38. Alarcón de Noya B, Pérez-Chacón G, Díaz-Bello Z, Dickson S, Muñoz-Calderón A, Hernández C, Pérez Y, Mauriello L, Moronta E. Description of an oral Chagas disease outbreak in Venezuela, including a vertically transmitted case. Mem Inst Oswaldo Cruz. 2017;112:569–71.
39. Pinto AY, Valente VC, Valente SA, Figueiras AC. Congenital Chagas disease due to acute maternal *Trypanosoma cruzi* infection transmitted by the oral route. Rev Pan-Amaz Saude. 2011;2(1):89–94.
40. Andrade JP, Marin-Neto JA, Paola AA, Vilas-Boas F, Oliveira GM, Bacal F, Bocchi EA, Almeida DR, Fragata Filho AA, Moreira da MC, Xavier SS, Oliveira Junior WA, Dias JC, Sociedade Brasileira de Cardiologia. I Diretriz Latino Americana para o Diagnóstico e Tratamento da Cardiopatia Chagásica. Arq Bras Cardiol. 2011;97(2 Suppl 3):1–48.
41. WHO. Control of Chagas disease. Second report of the WHO expert committee. WHO technical report series no 905. Geneva: World Health Organization; 2002.
42. Alarcón de Noya B, Díaz-Bello Z, Colmenares C, Zavala-Jaspe R, Abate T, Contreras R, Losada S, Artigas D, Mauriello L, Ruiz-Guevara R, Noya O. The performance of laboratory tests in the management of a large outbreak of orally transmitted Chagas disease. Mem Inst Oswaldo Cruz. 2012;107:893–8.
43. Freilij H, Muller L, González-Cappa M. Direct micromethod for diagnosis of acute and congenital Chagas' disease. J Clin Microbiol. 1983;18:327–30.
44. Díaz-Bello Z, Thomas MC, López MC, Zavala-Jaspe R, Noya O, Alarcón de Noya B, Abate T. *Trypanosoma cruzi* genotyping supports a common source of infection in a school-related oral outbreak of acute Chagas disease in Venezuela. Epidemiol Infect. 2014;142(1):156–62.
45. Rueda K, Trujillo JE, Carranza JC, Vallejo GA. Transmisión oral de *Trypanosoma cruzi*: una nueva situación epidemiológica de la enfermedad de Chagas en Colombia y otros países suramericanos. Biomedica. 2014;34:631–41.
46. Duffy T, Cura CI, Ramirez JC, Abate T, Cayo NM, Parrado R, Díaz-Bello Z, Velazquez E, Muñoz-Calderon A, Juiz NA, Basile J, Garcia L, Riarte a NJR, Ocampo SB, Yadon ZE, Torrico F, Alarcón de Noya B, Ribeiro I, Schijman AG. Analytical performance of a multiplex real-time PCR assay using TaqMan probes for quiantification of *Trypanosoma cruzi* satellite DNA in blood samples. PLoS Negl Trop Dis. 2013;7(1):e2000.
47. Ramírez JC, Cura CI, da Cruz Moreira O, Lages-Silva E, Juiz N, Velázquez E, Ramírez JD, Alberti A, Pavia P, Flores-Chávez MD, Muñoz-Calderón A, Pérez-Morales D, Santalla J, Marcos da Matta Guedes P, Peneau J, Marcet P, Padilla C, Cruz-Robles D, Valencia E, Crisante GE, Greif G, Zulantay I, Costales JA, Alvarez-Martínez M, Martínez NE, Villarroel

R, Villarroel S, Sánchez Z, Bisio M, Parrado R, Maria da Cunha Galvão L, Jácome da Câmara AC, Espinoza B, Alarcón de Noya B, Puerta C, Riarte A, Diosque P, Sosa-Estani S, Guhl F, Ribeiro I, Aznar C, Britto C, Yadón ZE, Schijman AG. Analytical validation of quantitative real-time PCR methods for quantification of Trypanosoma cruzi DNA in Blood samples from Chagas disease patients. J Mol Diagn. 2015;17:605–15.
48. Sales PA, Molina I, Fonseca Murta SM, Sánchez-Montalvá A, Salvador F, Correa-Oliveira R, Martins Carneiro C. Experimental and clinical treatment of Chagas disease: a review. Am J Trop Med Hyg. 2017;97(5):1289–303.
49. Díaz ML, Leal S, Mantilla JC, Molina-Berríos A, López-Muñoz R, Solari A, Escobar P, González-Rugeles CI. Acute Chagas outbreaks: molecular and biological features of *Trypanosoma cruzi* isolates, and clinical aspects of acute cases in Santander, Colombia. Parasit Vectors. 2015;8:608–22.
50. Urbina JA, Docampo R. Specific chemotherapy of Chagas disease: controversies and advances. Trends Parasitol. 2003;19(11):495–501.
51. Coura JR, Castro SL. A critical review of Chagas disease chemotherapy. Mem Inst Oswaldo Cruz. 2002;97(1):3–24.
52. Alarcón de Noya B, Ruiz-Guevara R, Noya O, Castro J, Ossenkopp J, Díaz-Bello Z, Colmenares C, Suárez JA, Noya-Alarcón O, Naranjo L, Gutiérrez H, Quinci G, Torres J. Long-term comparative pharmacovigilance of orally transmitted Chagas disease: first report. Expert Rev Anti-Infect Ther. 2017;15(3):319–25.
53. Guedes PM, Silva GK, Gutierrez FR, João S, Silva JS. Current status of Chagas disease chemotherapy. Expert Rev Anti-Infect Ther. 2011;9(5):609–20.
54. de Andrade AL, Zicker F, de Oliveira RM, Almeida Silva S, Luquetti A, Travassos LR, Almeida IC, de Andrade SS, de Andrade JG, Martelli CM. Randomized trial of efficacy of benznidazole in treatment of early *Trypanosoma cruzi* infection. Lancet. 1996;348:1407–13.
55. Sosa Estani S, Segura EL, Ruiz AM, Velazquez E, Porcel BM, Yampotis C. Efficacy of chemotherapy with benznidazole in children in the indeterminate phase of Chagas disease. Am J Trop Med Hyg. 1998;59:526–9.
56. Alarcón de Noya B, Díaz-Bello Z, Colmenares C, Muñoz A, Ruiz-Guevara R, Balzano L, Campelo R, Abate T, Maniscalchi MT, Ramírez JL, Noya O. Eficacia terapéutica en el primer brote de transmisión oral de la enfermedad de Chagas en Venezuela. Biomedica. 2011;31(3):64–5.
57. Cançado JR. Long term evaluation of etiological treatment of Chagas disease with benznidazole. Rev Inst Med Trop Sao Paulo. 2002;44:29–37.
58. Robertson L, Noya O. Prophylactic measures and implementation of control measures in food-borne Chagas disease. In: Alarcón de Noya B, Noya O, Robertson LJ, editors. *Trypanosoma cruzi* as a foodborne pathogen, Springer briefs in food, health and nutrition. New York, NY: Springer; 2015. p. 81–8.
59. Xavier SC, Roque ALR, Bilac D, de Araújo VAL, Neto SF, da Silva LFC, Jansen AM. *Distantiae* transmission of *Trypanosoma cruzi*: a new epidemiological feature of acute Chagas disease in Brazil. PLoS Negl Trop Dis. 2014;8(5):e2878.
60. Valente SAS, Vera da Costa Valente VC, Neto HF. Considerations on the epidemiology and transmission of Chagas disease in the Brazilian Amazon. Mem Inst Oswaldo Cruz. 1999;94(I):395–8.
61. Añez N, Crisante G, Romero M. Supervivencia e infectividad de formas metacíclicas de *Trypanosoma cruzi* en alimentos experimentalmente contaminados. Bol Malariol Salud Amb. 2009;49:91–6.
62. Dias GB, Gruendling AP, Marquez S, Gomes ML, Toledo MJ. Evolution of infection in mice inoculated by the oral route with different developmental forms of *Trypanosoma cruzi* I and II. Exp Parasitol. 2013;135:511–7.

Gastrointestinal Chagas Disease

Ênio Chaves de Oliveira, Alexandre Barcelos Morais da Silveira, and Alejandro O. Luquetti

Abstract The chronic phase of Chagas disease is presented in three clinical forms: indeterminate (no clinical manifestations), cardiac disease, and megaviscera syndromes. The latter comprise up to 18% of the infected individuals and most often present a compromise of the esophagus and colon. The physiological function of these organs depends on a perfect coordination/synchronization of muscular constriction and relaxation waves, in order to push a rather hard material (alimentary bolus and feces) through their cavity, and a coordinated transposition of two sphincters. When this function is hampered, a progressive increase in the diameter of both organs is produced, termed megaesophagus and megacolon. A clue for the cause of this dysfunction is the selective destruction of parasympathetic plexus neurons by the etiological agent of the disease, *Trypanosoma cruzi*. Besides motor alterations, secretory and absorptive functions may be affected. Why these features are observed in only some of the infected is not clear. A markedly discrete geographical distribution of digestive Chagas disease cases below the equatorial line suggests it may be due to the type of circulating *T. cruzi* lineages (TcII and TcV). Different incidences according to gender and age are also seen. A number of neurotransmitters and neuropeptides have been linked to Chagas disease-associated megaviscera syndromes. Dysphagia and obstipation are clinical hallmarks of this disease, but serological

Ê. C. de Oliveira
Faculdade de Medicina, Núcleo de Estudos da Doença de Chagas) (NEDoC), Hospital das Clinicas and Surgery Department, Universidade Federal de Goiás, Goiânia, Brazil

A. B. M. da Silveira
Department Human Anatomy, Instituto de Ciências Biomédicas, Universidade Federal de Uberlândia, Uberlândia, Brazil

A. O. Luquetti (✉)
Núcleo de Estudos da Doença de Chagas (NEDoC), Hospital das Clinicas and Instituto de Patologia Tropical e Saúde Pública, Universidade Federal de Goiás, Goiânia, Brazil
e-mail: luquetti@ufg.br

J. M. Altcheh, H. Freilij (eds.), *Chagas Disease*, Birkhäuser Advances in Infectious Diseases, https://doi.org/10.1007/978-3-030-00054-7_12

diagnosis is necessary to exclude other possible causes, like idiopathic megaviscera. Association with cardiac involvement is observed in 20–30% of the cases, and 40% of the cases exhibit both megaesophagus and megacolon. Progression from mild to severe disease is seen in some of the cases, in which surgery is ultimately required. Fecaloma and volvulus are common complications of megacolon. In recent years, new methods have significantly improved post-surgery prognosis. Alternative treatments with botulinum toxin and by mechanical dilatation are indicated for specific cases. Other hollow viscera may be involved, albeit at lower frequencies, such as the stomach, duodenum, gallbladder, ureter, bladder, and others. However, these are usually also associated with megacolon and/or megaesophagus.

1 Introduction

T. cruzi produces overt disease in only half of the infected individuals. The other half will remain chronically infected in the indeterminate or asymptomatic form of the disease for life. The reasons for this are not clearly known. The evolution to symptomatic forms of the disease is slow, with an estimated 2% of the patients progressing into cardiac and/or digestive compromise every year. Cardiac manifestations are the most frequent, and they have been detected since the discovery of the disease in 1909. Conversely, the association between Chagas disease and megaviscera was only acknowledged 47 years later by Koberle [1] who unequivocally showed denervation of the myenteric plexus in infected individuals with megaesophagus and megacolon. The reasons for this delay include low frequency of cases, geographical differences, lack of parasites in the scrutinized lesions at the time of examination, similarity with idiopathic megaviscera, and difficulties to reproduce these syndromes in experimental animal models. Nevertheless, megaesophagus and megacolon had been described in Chagas disease endemic areas (Central Brazil) by historians in 1823 [2] and 1857 [3], long before the discovery of the disease, and in frequencies that recalled that of *T. cruzi* infection. Also, mummies with megacolon were found [4].

1.1 Pathophysiology

The parasite destroys, mainly during the acute phase, the nervous intramural plexuses of hollow viscera, with an unpredictable and uneven distribution. Coordinated motor activity of the digestive tract is directed by myenteric plexuses, and it is essential for the alimentary bolus and feces to traverse it. Furthermore, both the esophagus and the colon have a sphincter at their ends, which need to coordinately open in order for the material to pass through. This explains the compromise of mainly the esophagus and the distal segment of the colon. With a lower frequency, dilation of other segments of the digestive tract, i.e., the megagastria, megaduodenum, megajejunum, dilated gallbladder, and urinary tract dilations, may be found, nearly always

Table 1 Frequency of megaesophagus and megacolon in Central Brazil

Megaviscera involved	*n*	%
Megaesophagus	1183	47.6
Megacolon	288	11.6
Megaesophagus + megacolon	1013	40.8
Total	2484	100.0

In all cases diagnosis was performed by barium swallow and barium enema. All patients had four serological positive tests. Data from Núcleo de Estudos da doença de Chagas (NEDoC), Federal University of Goias, Goiania, Brazil

Table 2 Association of megaesophagus and megacolon in Central Brazil

	Colon		Esophagus			
Primarily considered megaviscera	Normal	Mega	Normal	Mega	Total	Association, %
Megacolon	–	–	288	1013	1301	77.9
Megaesophagus	1183	1013	–	–	2196	46.1
Group I/II	696	521	–	–	1217	42.8
Group III/IV	487	492	–	–	979	50.3

Megaesophagus cases were divided in non-severe (compensated, groups I and II) and severe (groups III and IV). In all cases diagnosis was performed by barium swallow and barium enema. All patients had four serological positive tests. Data from Núcleo de Estudos da doença de Chagas (NEDoC), Federal University of Goias, Goiania

associated with megaesophagus and/or megacolon. In fact, in endemic areas, megacolon-megaesophagus association is frequent (~40%, Table 1).

When we consider the degree of association of both main megasyndromes, almost all megacolon cases are associated with megaesophagus, but megaesophagus cases present lower frequency of association with megacolon. The latter seems to be related to the severity of megaesophagus: 43% of the patients with compensated megaesophagus have megacolon, while half of those with severe megaesophagus also exhibit megacolon (Table 2), indicating a widespread compromise of the enteric nervous system.

Koberle [5] attributed the development of megaviscera after variable periods of time to the natural aging and progressive loss of neurons in the enteric nervous system. According to the serial count of neurons in the wall of hollow viscera, the denervation required to develop megaesophagus is at least of 90% and for megacolon 55%. These figures may be attained in elderly noninfected individuals (presbyesophagus) but are seen in young *T. cruzi*-infected subjects due to an accelerated destruction of such cells.

1.2 Geographical Differences

Geographical differences in incidence and pathology are recognized and are probably related to the differential distribution of *T. cruzi* variants. At least six different discrete typing units (DTU) of this parasite have been described (TcI to TcVI), out of which three are the most clinically relevant [6]. TcI is found mainly in endemic regions above the equatorial line, including part of Brazil, Colombia, Venezuela, Central America, and Mexico. Even though sporadic cases of megaesophagus and megacolon were described, examples of megaviscera are rare in this area. Human infection with *T. cruzi* TcII is mainly found in Central Brazil, where most of the megaviscera cases are seen, with a higher proportion of megaesophagus (Table 1 [7]). TcV is found in humans all across the Southern Cone, including south of Brazil. Cases of megacolon and megaesophagus have been reported, but in lower proportions than in Central Brazil. Interestingly, in Chile, the number of patients with Chagas disease-associated megacolon is higher than that of megaesophagus patients [8].

T. cruzi-related megaviscera should be distinguished from other similar clinical entities, such as high-altitude megacolon (Andean megacolon), occurring in Peru. This is in fact a type of dolichocolon (increase in the length of the colon, without diametral enlargement), causing volvulus, and for which maize-based diet has been appointed as the main cause.

1.3 Epidemiology

For symptomatic individuals, megaesophagus usually appears before cardiopathy, while megacolon has a later onset than cardiac symptoms. Both may evolve differently, from slight alterations that remain so for decades to early severe organ dilation [9]. The reasons behind these different patterns are not clear. Apart from age-associated variation, gender differences are remarkable: megaesophagus is clearly more frequent in males, especially in its more severe forms [10]. This gender difference is less clear in chronic Chagas cardiopathy, even though male patients tend to present more severe evolution. Megacolon patients are predominantly female. Clinical manifestations also differ: it is unusual for a patient with megaesophagus to have no complaints. Dysphagia is the clinical hallmark, present in nearly all patients, and is the main reason leading them to medical consultation and treatment. On the contrary, up to 1/3 of the megacolon cases, diagnosed by barium enema, present no obstipation [11]. This has consequences on treatment: patients only undergo surgery if severe obstipation is present, irrespective of the radiological findings. Other reasons that oblige them for medical attention are volvulus and fecaloma, frequent in megacolon patients. Conversely, some *T. cruzi*-infected subjects with severe obstipation and positive serology have no dilation, but need to be treated. Clearly, the physiopathology of megacolon deserves more investigation. The occurrence of megaviscera during the acute phase, or shortly after it, has been described, but it is infrequent. Acute phase Chagas disease is seldom diagnosed in endemic areas nowadays, as vector-borne transmission has been effectively controlled by insecticides in most endemic countries.

1.4 Other Causes of Megaviscera

Cases with megaviscera may have other etiologies, i.e., idiopathic megaesophagus, caustic megaesophagus, cancer, and others. Congenital megacolon (Hirschsprung disease) is seen mainly in children, and megacolon may also be acquired by traumatisms and alimentary habits, as in severe psychiatric disturbances. For all these reasons, epidemiological history should be thoroughly enquired, endoscopy should be performed, and serological tests should be asked for [9]. Idiopathic cases of megaviscera are seen also in endemic countries, in similar frequencies as in non-endemic regions, which are rare (1 in 100,000). Interestingly, association of chagasic megaesophagus with cancer has been recorded (up to 2%), but the association of chagasic megacolon has not, suggesting a protective effect of the parasite [12, 13].

1.5 Serological Diagnosis and Megaviscera

Search for anti-*T. cruzi* antibodies is mandatory for all cases of megaviscera. Diagnosis should be performed by at least two serological tests of different principles, as recommended by the World Health Organization [14]. Any of the conventional methods (indirect immunofluorescence, indirect hemagglutination, or ELISA) are suitable, and a second test may be performed with rapid tests or non-conventional methods (see Chap. 2.4). Antibody titers may be evaluated, and they are generally high. Parasitological tests are not appropriate, as parasitemia is usually low during the chronic phase of Chagas disease. The presence of megaviscera strongly suggests Chagas disease, not only in endemic regions but around the globe, due to large migratory movements implying a dispersion of infected Latin American natives residing in non-endemic countries. The predictive value of these syndromes is very high in endemic areas, in the order of 95%, as seen in Table 3 [15]. Obviously, these values are lower in non-endemic areas, and subjects with a positive association may have lived in endemic countries for some time in the past.

Table 3 Positive serology in megaesophagus and megacolon

	Serology				
	Positive		Negative		
Megaviscera	*n*	%	*n*	%	Total
Megaesophagus	2902	94.6	167	5.4	3069
Megacolon	1747	94.3	105	5.7	1852
Both megasyndromes	1013	98.6	14	1.4	1027
Total	5662		286		5948

Data from Núcleo de Estudos da doença de Chagas (NEDoC), Federal University of Goias, Goiania, Brazil

1.6 Some Features on Immune System and Enteric Nervous System in Chagasic Megaviscera

In 1916, the first indication of the existence of the digestive form of Chagas disease arose by the observation of patients with dysphagia who required the aid of water to complete food intake. Even the ingestion of fluids could be difficult, requiring them to be taken in small doses. This phenomenon, without pathogenic explanation at the time, was denominated "choking sickness" [16].

Megacolon is characterized as intestinal dilation associated with an inflammatory infiltrate. This infiltrate consists mainly of CD3⁺ T lymphocytes, CD20⁺ B lymphocytes, and also natural killer cells, macrophages, and mast cells [17, 18]. Pathogen-specific B and T lymphocytes take part in adaptive immunity. An effective T cell response requires adequate stimulation by other host cells (Fig. 1).

Constipation is a typical symptom of megacolon, and both the clinical and anatomical diagnoses are usually late, after dilatation appears. Macroscopically, lack of motor coordination, sphincter achalasia, and distension caused by the accumulation of fecal contents result in chagasic megacolon [19]. According to Tafuri [20], a progressive lesion in the plexus is likely to occur, which aggravates according to the development of the megacolon. Considering the stasis of fecal content, one of the main factors leading to megacolon, we may state that the accumulation of fecal content in the lumen provokes compression of the mucosa and consequent dilation of the organ. The changes resulting from compression cause the mucosa to undergo ischemia, favoring the diffusion of the inflammatory process through the nervous plexus and muscle layers. Muscle cells are also affected as a consequence of the

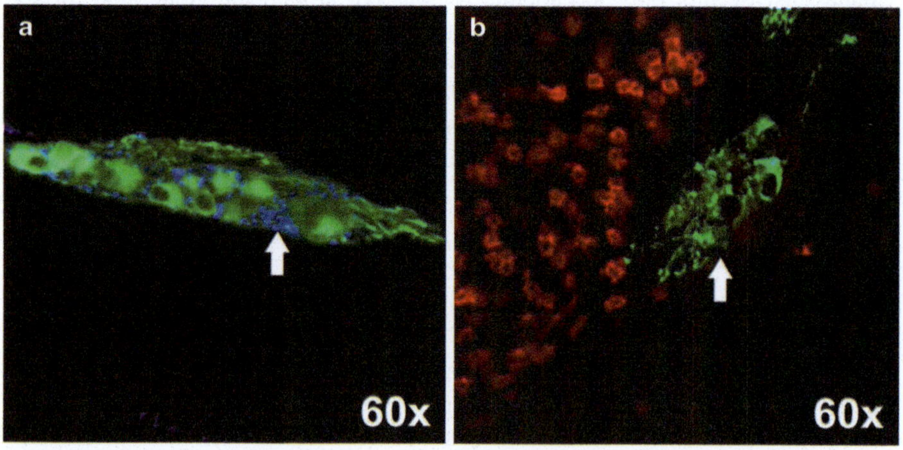

Fig. 1 Relation among neurons (green), serotonin (blue), and CD8⁺ lymphocytes (red). (**a**) Noninfected individual presented serotonin near to neuronal ganglia associated with low concentrations of CD8⁺ lymphocytes. (**b**) Chagasic patient with megacolon displaying high incidence of CD8⁺ lymphocytes adjacent with neuronal ganglia combined with serotonin absence. Micrography courtesy of Dr. A.B. Morais da Silveira

greater contraction effort, due to the greater resistance of the medium, with the development of hypertrophy over time. Since the submucosal plexus is closely related to muscle cells, it is easy to understand how the inflammatory process can aggravate the process of neuronal destruction.

Studies from the last decades indicate that, in addition to inflammatory processes, the lesions in Chagas disease depend on the presence of *T. cruzi* DNA and the parasite itself, albeit in small numbers. A close correlation between the presence of parasite antigens and the intensity of the inflammatory infiltrate has been demonstrated [21]. It is now accepted that the inflammatory process is primarily responsible for the destruction of the enteric nervous system (ENS) components [17, 18].

1.7 Enteric Neurons, Neuropeptides, and Other Markers

The ENS organization is similar in humans and other mammals. The physiological control exerted by the enteric neurons on motility, secretion, and other digestive processes and the mechanism of action of drugs that affect neurotransmission are similar between different species [22]. It has approximately the same number of nerve cells than the spinal cord, around 200–600 million neurons, which demonstrates its great importance [23]. Such neurons can be identified by function, morphology, and neurochemical correlation. Functionally, they can be divided into excitatory motor neurons, inhibitory motor neurons, interneurons, and intrinsic primary afferent neurons. More than 30 potential neurotransmitters affecting neuronal, muscle, and epithelial cell activity are present in the ENS. Moreover, a single neuron may harness several neurotransmitters, in addition to other neuro-specific proteins. The combination of chemical attributes related to neural function and their locations in the nerve circuit provide a chemical code by which neurons may be identified. In general, more than one substance contributes to the transmission process [24]. The myenteric plexus is a network of small neuronal ganglia, which are interconnected by nerve bundles between the internal and external muscular layers of the gastrointestinal tract. This plexus forms a continuous network around the section and throughout the extension of the digestive system (Fig. 2). The lymph nodes found in this plexus vary in size and shape, depending on the portion of the intestine and the animal species under analysis [25].

Intrinsic primary afferent neurons (IPAN) are numerous, approximately 500 by square millimeter in length of the small intestine and are best identified by immunohistochemical staining of the intracellular protein calretinin. They are transducers of physiological stimuli, including mucosal villus movement, intestinal muscle contraction, and chemical changes in intestinal contents. IPANs are the first neurons in intrinsic reflexes that influence the patterns of gut secretion and motility. Therefore, they are directly sensitive to mechanical and chemical stimuli from the intestinal mucosa, and the sum of synaptic events caused by the transmission of IPAN results in the activation of numerous interneurons and motor neurons [26, 27].

Excitatory motor neurons innervate the longitudinal and circular smooth muscle and the muscular mucosa of all digestive tract. The primary transmitter of these

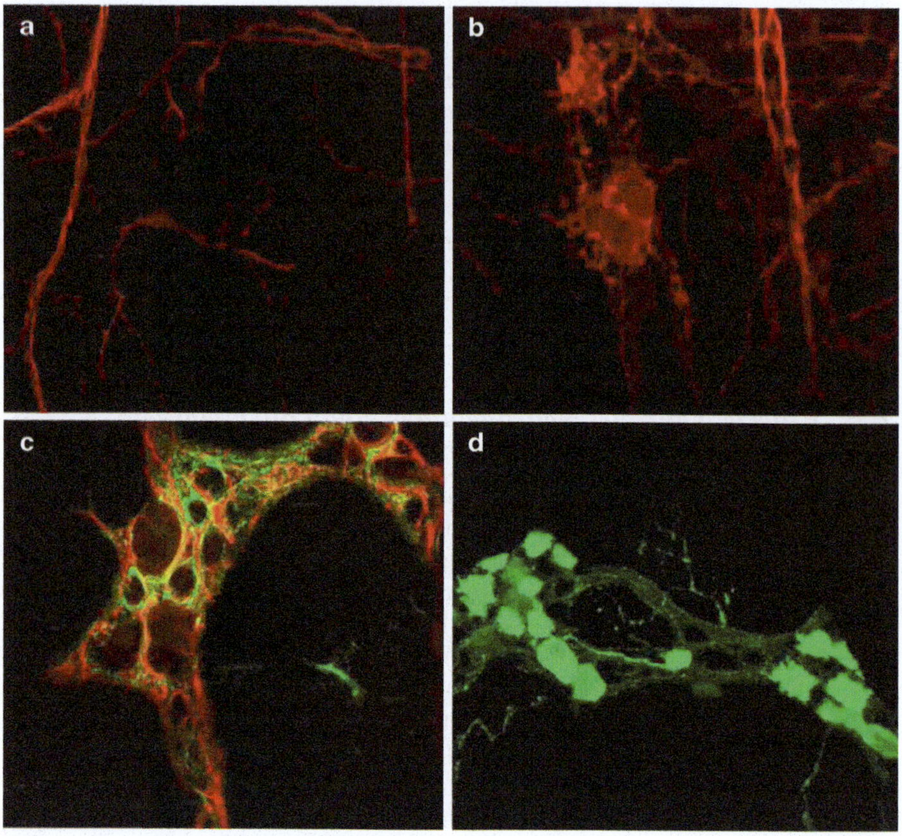

Fig. 2 Whole mount fluorescent immunohistochemistry of myenteric plexus in a colon sample from a Chagas disease patient. (**a**) Presence of nerve fibers emerging from the myenteric plexus toward the muscular layers; (**b**) interconnection between neuronal bodies in the myenteric plexus and its relationship with nerve fibers; (**c**) relationship between nerve fibers and neural bodies (red) and enteric glial cells (green); (**d**) nervous ganglion in the myenteric plexus where nerve fibers are observed in close relation with neurons. Micrography courtesy of Dr. A.B. Morais da Silveira

neurons is acetylcholine (ACh) which acts on muscle cells through muscarinic receptors. The major marker of excitatory motor neurons is the precursor enzyme of ACh, choline acetyltransferase (ChAT). Tachykinins, represented by substance P (SP), neurokinin A, neuropeptide K, and neuropeptide gamma (γ), contribute to excitatory transmission, but play a secondary role compared to ACh [22, 24].

In contrast, inhibitory motor neurons release a combination of transmitters that contribute to relaxation of the gastrointestinal tract. The primary neurotransmitter of these neurons is nitric oxide (NO), which receives a secondary contribution from other substances such as the vasoactive intestinal peptide (VIP) and adenosine triphosphate (ATP). It is possible that there is more than one primary transmitter for this neural subclass. However, the relative roles of these transmitters differ between regions and species. The main marker of inhibitory motor neurons is the NO-producing enzyme nitric oxide synthase (NOS) [22, 24].

In addition to neurotransmitters, neuropeptides also have considerable activity on the immune system, influencing ENS activities through the secretion of various types of compounds. Substance P is a protein that, in addition to its function as neuromodulator, has a pro-inflammatory action on immune cells. It stimulates lymphocyte proliferation, lymphocyte trafficking through lymph nodes, and IL-2 production. In addition, substance P acts as a natural killer (NK) cell activator and has chemotactic action for mast cells, macrophages, and neutrophils [28].

VIP has anti-inflammatory effect, by inhibiting the response of NK cells and T lymphocytes, as well as the production of IL-2 and IL-4 by these cells and antigen presentation. On the other hand, it also stimulates macrophage chemotaxis and IL-5 production by lymphocytes [29]. The pro-inflammatory effects of NO are not evident under acute physiological conditions. Among its anti-inflammatory effects, we can mention the inhibition of neutrophil adhesion, cyclooxygenase activity, cytokine formation, and bone reabsorption [30].

Currently, denervation is accepted as one of the causes of development of chagasic megacolon [5, 26]. Neuronal destruction in acute Chagas disease is due to the great concentration of the parasite in the tissue, but in the chronic phase, specific segments, such as the stomach, small intestine, and colon [31], are also involved in this inflammatory process [32, 33]. The interrelationship between the nervous, endocrine, and immune systems is very important for the understanding of the intestinal compromise and may be definitive for the determination of clinical manifestations and for the development of inflammatory processes in the intestine [34]. The earliest descriptions relating changes in the myenteric plexus and Chagas disease date back to 1930. However, they still lacked anatomopathological evidence. This was provided by Koeberle, starting in 1953, by his quantitative studies on neurons of the esophagus.

Some studies have confirmed neuronal destruction in the myenteric plexus of chagasic patients and demonstrated a relation with the inflammatory process in portions of the gastrointestinal tract [18, 32, 33]. Later it was evidenced that this neuronal destruction could be selective, that is, some neuron types would be preferentially destroyed [26]. The question then arises as to what would be the regenerative process in the different neuron types of the ENS, whether a type is actually destroyed or its immunoreactive area is reduced because it did not present a satisfactory regenerative process.

The regeneration process was found to be greatly increased in VIP- and NOS-positive neurons. This indicates that the organism activates regeneration processes as an attempt to compensate for the loss of inhibitory neurons caused by the parasite.

These basic results are in agreement with the clinical observation of relaxation capacity loss in the muscular layers of the colon in Chagas disease patients who develop megacolon. This produces alterations in the motility of the colon, leading to fecal accumulation and, consequently, to organ dilation. Incoordination of the rectosigmoid segment, hyperactivity to cholinergic stimuli, and achalasia of the internal anal sphincter are the most common symptoms in chagasic megacolon [31].

1.8 Etiological Treatment

Etiological treatment (see Chap. 2.5) may be prescribed in chagasic patients with megaviscera, but in the case of severe megaesophagus, surgical treatment should be performed in advance, in order to allow the drug to be ingested and absorbed.

2 Megaesophagus

2.1 Epidemiology

Idiopathic megaesophagus is a rather unusual entity, with a prevalence of 1/100,000 in global population. In Latin America, the prevalence of megaesophagus is higher due to chagasic etiology. The pathogeny, physiopathology, symptoms, evolution, and treatment of this entity are similar to the idiopathic one, the main difference being the presence of antibodies against the parasite, as well as the association with cardiopathy and/or megacolon in some cases. The term "mega" may be misleading since the anectasic forms have no dilatation of the esophagus.

Geographical distribution of chagasic megaesophagus has been recognized and is limited to the south of the Amazon River, with a higher prevalence in Central Brazil. This has been linked to the *T. cruzi* DTU (TcII) preferentially isolated from humans in this large area [7]. In the Southern Cone, the prevalence of megacolon is higher.

Age range is wide, and, in our case studies' history, we have recorded patients from 2 years old to over 100 years old. It usually appears earlier than the other clinical forms, most often between 20 and 40 years of age. Our case studies comprise more than 3200 cases, beginning in 1975. In the last decades (1990–2010), the case distribution is shifting to older age, and patients usually consult after some years after symptoms have appeared [35] (Table 4). Gender has been found to correlate

Table 4 Distribution of 2925 cases of megaesophagus and relation with age (by the time of first consultation) and gender

	Female		Male		Total	
Age group (years)	*n*	%	*N*	%	*N*	%
<10	6	**66.7**	3	33.3	9	0.3
11–20	19	34.5	36	**65.5**	55	1.9
21–30	82	46.3	95	**53.7**	177	6.1
31–40	159	43.6	206	**56.4**	365	12.5
41–50	295	45.3	356	**54.7**	651	22.3
51–60	361	47.4	401	**52.6**	762	26.1
61–70	299	47.8	327	**52.2**	626	21.4
> 70	126	45.0	154	**55.0**	280	9.6
Total	1347	46.1	1578	53.9	2925	100.0

Data from Núcleo de Estudos da doença de Chagas (NEDoC), Federal University of Goias, Goiania, Brazil

The values in bold represents P<0.05

Table 5 Distribution by radiological groups of 2475 non-treated cases with megaesophagus and relation with gender

	Female		Male		Total	
Radiological group	*n*	%	*N*	%	*N*	%
I	391	60.4	256	39.6	647	26.1
II	414	46.6	477	**53.5**	891	36.0
III	216	36.9	370	**63.1**	586	23.7
IV	113	32.2	238	**67.8**	351	14.2
Total					2475	100.0

Non-treated means not submitted to dilatation by pneumatic balloon or to surgery previously that may change the classification of group. Data from Núcleo de Estudos da doença de Chagas (NEDoC), Federal University of Goias, Goiania, Brazil
The values in bold represents P<0.05

with the severity of the involvement, being the female more frequent among the less severe (anectasic) cases, similar frequencies for both sexes in group II, and male predominance in the decompensated severe groups (III and IV) (Table 5). Male predominance is observed in all age groups (Table 4).

2.2 *Clinical Findings*

Dysphagia is a common complaint among these patients; it is generally progressive and initially implies solid and cold food. As it worsens, patients begin to suffer regurgitation during meals and later regurgitation while laying down, in those cases with advanced megaesophagus. Patients also present heartburn, hiccups, cough, ptyalism, and constipation. Enlargement of salivary glands may be found in physical exam.

Barium swallow is the main complementary evaluation, as it allows to see the degree of dilation and evaluate the stomach. In advanced cases, it may be necessary to substitute the affected esophagus by a gastric tube.

Endoscopy is necessary in all cases before surgical treatment. When this is the first exam performed on the patient, an experienced endoscopist may suspect megaesophagus diagnosis in patients within groups with early phase pathology, as group I or II. Esophageal cancer and other esophageal diseases may be excluded by endoscopy.

Esophagus manometry is not a routine exam. It gains relevance in cases of symptomatic patients without dilation or in patients who have undergone surgery and report recurrent symptoms. The usual findings are lower esophageal sphincter achalasia and uncoordinated contractions of the esophageal body.

2.3 *Classification: Groups I–IV*

Rezende et al. [36] proposed a radiological classification of chagasic megaesophagus in four groups (Fig. 3):

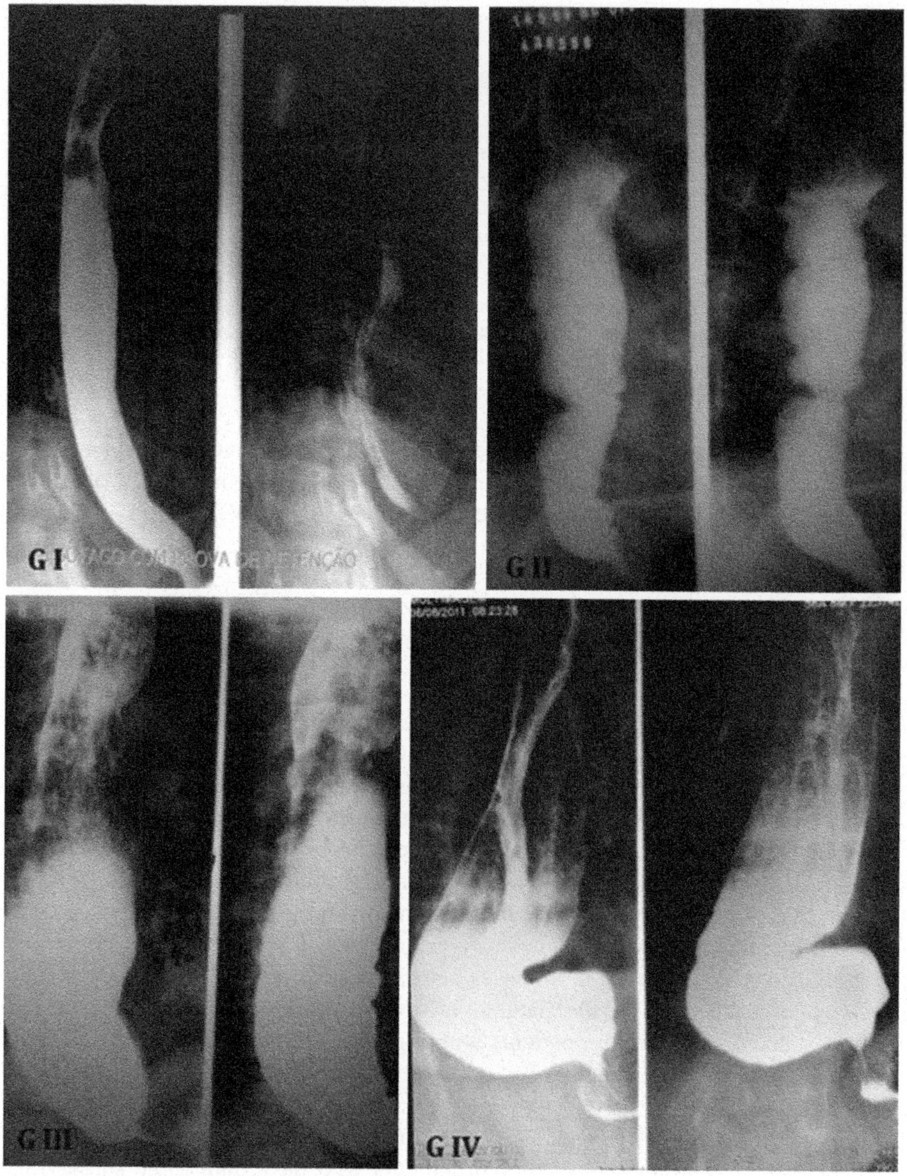

Fig. 3 Megesophagus classification [36]. X-ray images courtesy of Dr. Ê. Chaves de Oliveira

- Group I—anectasic and with barium transit delay in the esophagus. One minute after barium swallow, a barium column in the lower esophagus and an air column above are visible.
- Group II—small dilation and frequent tertiary uncoordinated contractions.
- Group III—major dilation, may present tertiary contractions.
- Group IV—most severe form, dolichomegaesophagus with dilation and sigmoid-like aspect in radiography.

2.4 Differential Diagnosis

The differential diagnosis should mainly consider esophagus neoplasia. Idiopathic megaesophagus and congenital megaesophagus are less frequent, but should also be taken into account.

2.5 Treatment

Most cases of initial megaesophagus (group I) do not require treatment. Patients with mild dysphagia may be treated with prokinetic drugs. Prokinetics such as cisapride, metoclopramide, and domperidone may be administered 15 min before meals for relief of dysphagia and regurgitation.

Surgical treatment is properly indicated in patients with megaesophagus groups II, III, and IV. We usually divide groups II and III as non-advanced and group IV as advanced.

Several different surgical techniques were developed to treat megaesophagus, but the main procedure performed in group II and III patients is cardiomyotomy with anti-reflux valve. The most practiced method in many centers in Brazil is the Heller-Pinotti operation. This surgical technique has low outcome complication rates, and most patients improve their symptoms [37]. Laparoscopic surgery may be carried out according to surgeon experience, with outcome as good as that of open surgery [38]. The laparoscopic approach has become the gold standard practice, and it is considered the procedure of choice for the treatment of achalasia [39].

Treatment of group IV patients is controversial. Some authors advocate the Serra-Doria operation owing to its low mortality and morbidity [40], while others suggest that the best approach is esophagectomy, whereby the sick esophagus must be removed and replaced by the stomach or colon [41, 42]. Esophagectomy has a higher morbidity, even in experienced hands. Robotic surgery has been introduced to treat chagasic megaesophagus [43] presenting the same advantages of this technology for other diseases [44].

Balloon dilation was commonly used some decades ago to treat megaesophagus groups I and II patients. Nonetheless, follow-up studies revealed that the relief of symptoms was temporary, and most of them needed additional surgical treatment. Besides, the dilated cardia evolved with fibrosis after two or more sessions of dilation making surgery more difficult. Therefore, dilation was deprecated as first-line treatment for megaesophagus, and currently it is reserved only to patients with severe comorbidities, such as chronic obstructive pulmonary disease, cardiopathy, or pregnancy.

Botulinum toxin has been used as treatment of non-severe megaesophagus, with good transitory results [45].

Recently, a new endoscopic technique was reported for the treatment of achalasia: per oral endoscopic myotomy (POEM) [46]. It is a minimally invasive endoscopic therapy that has not been used yet to treat chagasic megaesophagus.

3 Megacolon

Colon involvement in Chagas disease may be as frequent as megaesophagus, but diagnostic approaches make it less reported in statistic records. Megacolon is the dilatation of the colon caliber over 6.3 cm, as measured by barium enema radiographic exam [47].

3.1 *Epidemiology*

Chagasic megacolon is more frequent after the fifth decade of age, with slight predominance in the female (Table 6). It is frequent in Central Brazil and rare in Northern Brazil. A similar clinical entity has been recognized, Andean megacolon, described as elongated colon (dolichocolon), with similar radiological aspect to chagasic megacolon. However, Andean megacolon patients have negative serology for Chagas disease. These patients usually live in high altitudes in the Andes [48].

3.2 *Pathophysiology*

The pathophysiology of colon dilation is similar to that of megaesophagus, as ENS neuron destruction is required (Fig. 4).

After infection, *T. cruzi* shows a tropism to ENS neurons and destroys them. Myenteric and submucous plexuses' damage occurs with different intensities. Some patients display broader neuron destruction in one plexus than the other, and this seems to influence the clinical symptoms. Destruction of myenteric neurons leads to colon enlargement, while destruction of submucous neuron cells is responsible for dysmotility, and, consequently, the patient suffers from chronic constipation [49].

Table 6 Distribution of 1748 cases of chagasic megacolon and relation with age and gender

	Female		Male		Total	
Age group (years old)	*N*	%	*N*	%	*n*	%
<10	1		0	0.0	1	0.1
11–20	5	29.4	12	70.6	17	1.0
21–30	42	**56.8**	32	43.2	74	4.2
31–40	112	**57.7**	82	42.3	194	11.1
41–50	190	**54.0**	162	46.0	352	20.1
51–60	259	**52.4**	235	47.6	494	28.3
61–70	193	48.6	204	51.4	397	22.7
> 70	108	49.3	111	50.7	219	12.5
Total	910	52.1	838	47.9	1748	100.0

All patients were diagnosed by barium enema and had at least four positive serological tests. Data from Núcleo de Estudos da doença de Chagas (NEDoC), Federal University of Goias, Goiania, Brazil

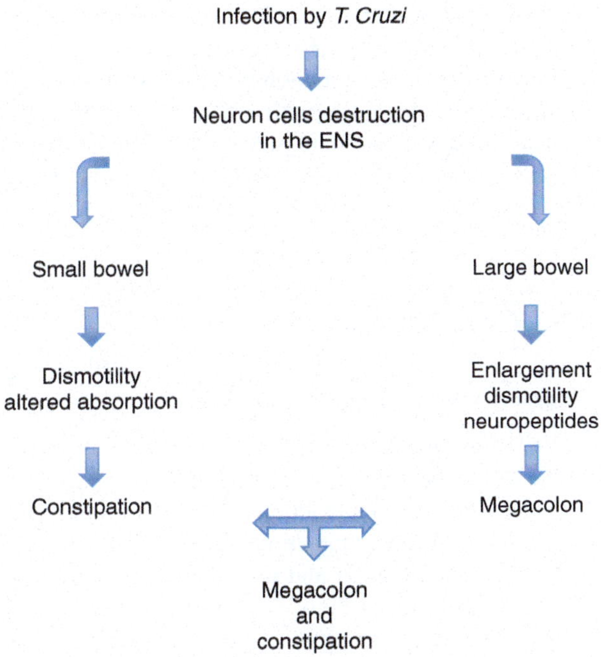

Fig. 4 Pathophysiology of chagasic megacolon. *ENS* enteric nervous system

Besides ENS damage in the colon, the small intestine ENS seems to be damaged in acquired megacolon and may also play a role in clinical symptoms [50].

The reason behind the dilation occurring only in the sigmoid portion of the colon is not clear. It is thought to be related to the proximity of the anal sphincters. Internal anal sphincter achalasia has been reported by some authors. Nevertheless, these findings were not consistent with studies by our group and others [51, 52]. Besides there are many patients with megacolon without constipation, showing that constipation and dilation are different factors that interact only sometimes [49, 53]. The rectum may be dilated or not. The role of dilated rectum in constipation is not clear but seems to be a non-relevant factor [54].

3.3 Clinical Findings

The main complaint from patients with megacolon is constipation, which is usually long-lasting and severe. Patients frequently report constipation ranging from 10 to 60 days, spanning several years. Abdominal cramps, abdominal distention, flatulence, scybalous-type feces, and straining are frequent as well.

On physical examination, patients with megacolon usually present slight abdominal distention, sometimes tenderness on palpation. After a long period without evacuation, the fecal mass may be felt as a moldable mass when the abdominal wall is pressed which, upon release of this pressure, produces a sensation similar to the detachment of a nurse tape from the skin, owing to the colon wall displacement from the fecal mass (i.e., Gersuny sign).

Rectum digital examination is mandatory for all patients with obstipation, mainly for differential diagnosis with other diseases. Diagnosis is based in epidemiology, clinical findings, and complementary exams (barium enema and serology for Chagas disease, Fig. 5). Differential diagnosis should discard any other cause of slow transit and constipation, as neoplasias or diverticular disease, for example.

3.4 Complications

Megacolon has three main complications:

- Volvulus—the torsion of the colon over its own axis, usually occurring just above the rectum-sigmoid transition (Fig. 6).
- Fecaloma or fecal impaction—after 10 or more days without evacuation, patients may present fecal impaction. The mass of solid and hard feces may accumulate in the sigmoid colon and may be felt by digital examination (lower fecaloma), but sometimes it cannot be thereby reached (high fecaloma). Most times, the feces are removed by slow instillation with saline solution into the rectum (Fig. 7).
- Colonic perforation—hard feces may erode the colonic wall and perforate it. Also, ischemic points due to prolonged volvulus may evolve toward necrosis and puncture (Fig. 7).

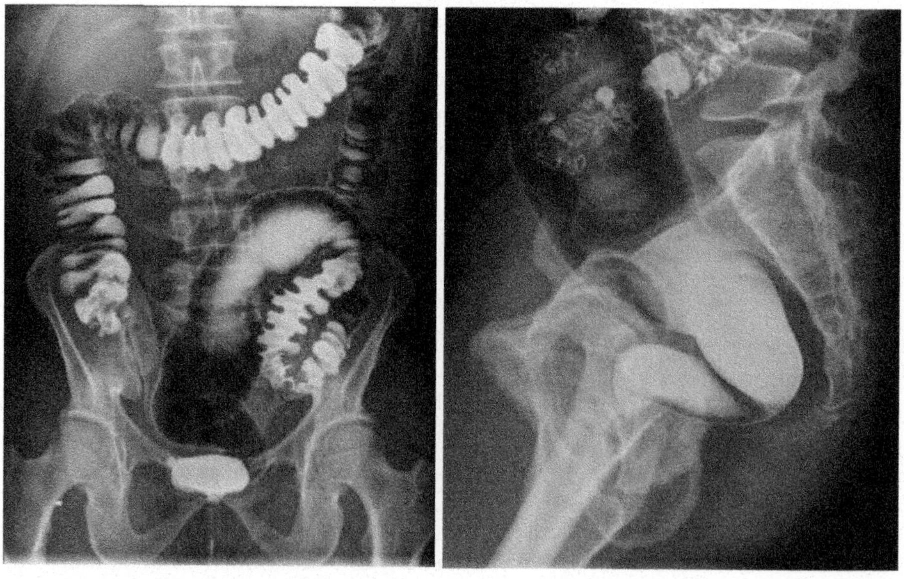

Fig. 5 Barium enema showing a dilation of the sigmoid colon and rectum. X-ray images courtesy of Dr. Ê. Chaves de Oliveira

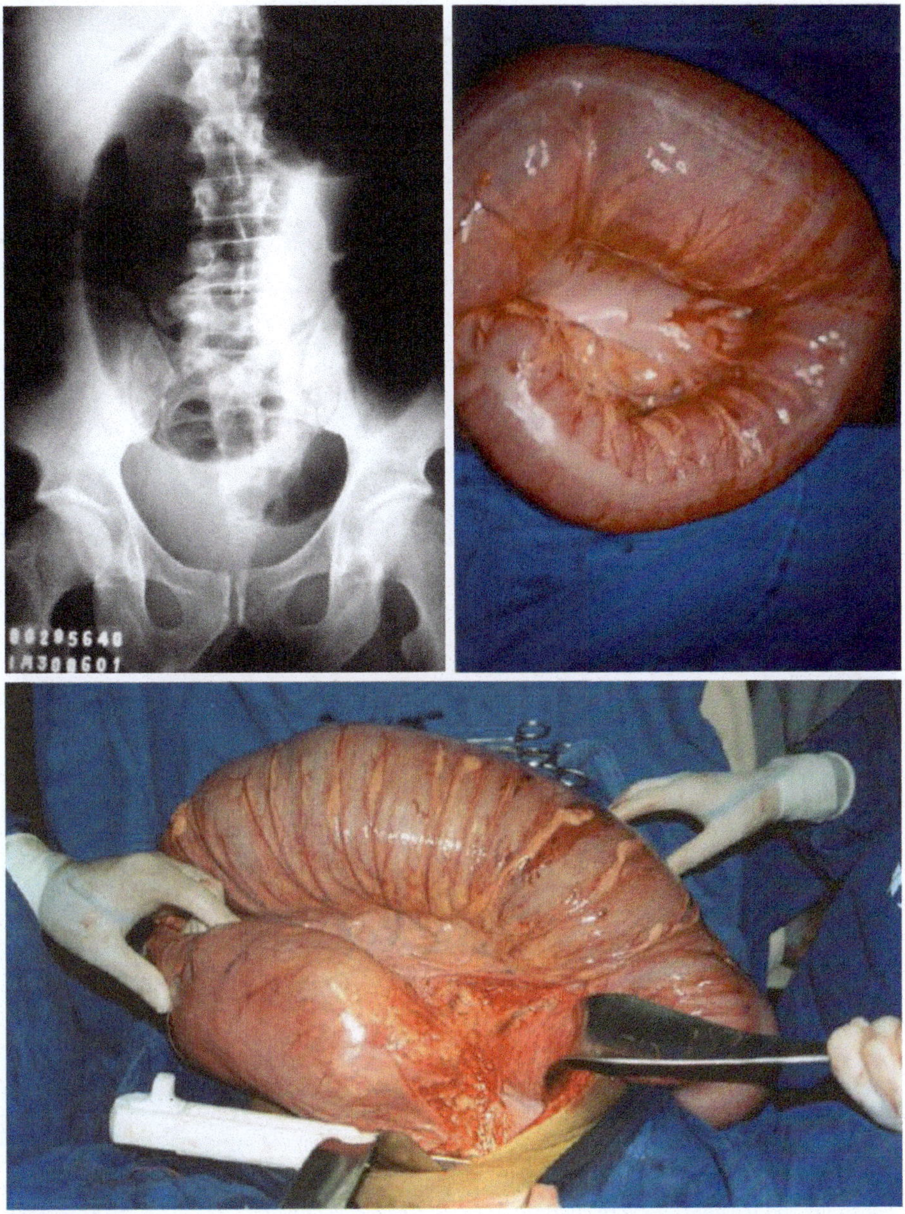

Fig. 6 Sigmoid volvulus: abdominal plain X-ray and surgical treatment. Pictures courtesy of Dr. Ê. Chaves de Oliveira

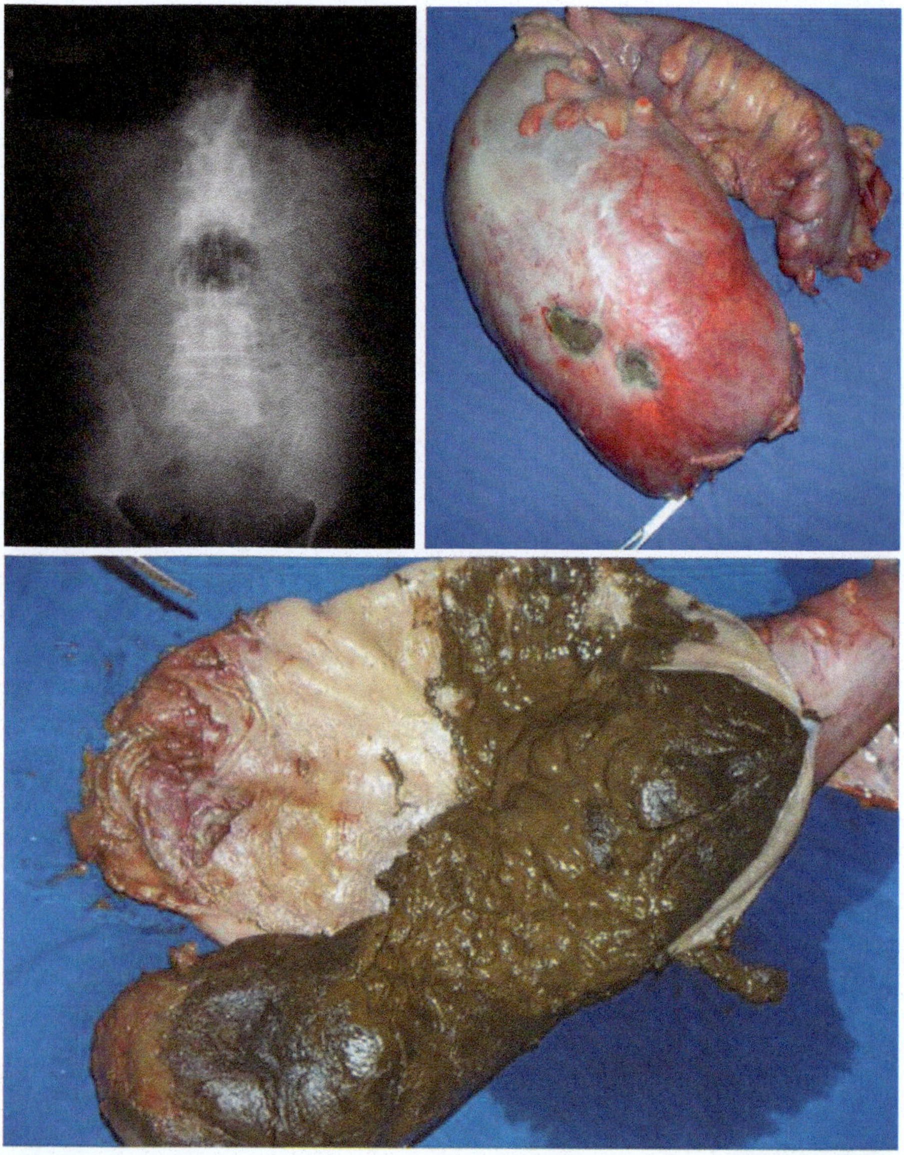

Fig. 7 Plain abdominal X-ray showing large fecaloma; colon perforation due to fecaloma; partial colectomy and fecaloma. Picture courtesy of Dr. Ê. Chaves de Oliveira

3.5 Treatment

Patients with megacolon without constipation (nearly one in three) do not need medication. Advice about possible complications of dilated bowel, such as unexpected fecal impaction or volvulus, should be given. Preventive surgery may be indicated according to age, comorbidities, or labor characteristics.

Constipated patients with megacolon should undergo left colectomy, with removal of the dilated colonic portion. Historically, many surgical techniques have been applied to these cases, but two have been employed the most:

- Duhamel operation—due to similarities between acquired and idiopathic megacolon, authors used the same technique described by Duhamel [55, 56]. This technique was modified by Haddad [57] and has been carried out on megacolon patients with satisfactory results [58].
- Low anterior resection is primarily indicated for patients without megarectum. It has been made simpler by the introduction of mechanical suture. This technique may be performed by laparotomy and laparoscopy.

The choice of technique should consider the presence of megarectum. Patients with megarectum are easier to operate by the Duhamel procedure, because there is a great difference in the calibers of the rectum and the normal proximal colon. Duhamel operation is a construction of a side-to-side anastomosis, so the caliber is not an obstacle [59].

Early outcomes are similar for both methods, although fecal incontinence is more frequent in the Duhamel procedure. Functionally, both techniques are equivalent [60].

4 Other Digestive Involvements

Despite a much lower incidence than megaesophagus and megacolon, other viscera may be enlarged in Chagas disease, nearly always associated with the former.

Megagastria may be present in up to 20% of patients with megaesophagus and is due to intrinsic denervation of the stomach. In cases of gastric emptying difficulty, pyloric muscle hypertrophy may be seen [15].

Duodenum is frequently dilated at the bulb, with eventual compromise of the whole arcade. Jejunum or ileum are rarely involved. An abnormal increase in the absorption of glucose has been described. Gallbladder and the Oddi sphincter are seldom involved as well. Parotid hypertrophy is seen in some patients with megaesophagus.

5 Conclusion

Megaesophagus and megacolon are the main digestive manifestations in individuals infected with *T. cruzi*. Other organs may be affected, but generally associated with the former. In regions endemic for Chagas disease, or in patients who used to live there, chagasic etiology should be investigated by serology. Of note, it is confirmed in more than 95% of the cases. Less than 20% of the infected people will develop megaviscera syndromes, with a marked geographical distribution, probably related to the parasite subspecific variants circulating in humans (DTU TcII or TcV). Contrasted radiological diagnosis is necessary, and the degree of involvement may be important for the decision of an appropriate treatment. Surgery is advised in patients with advanced compromise. Fecaloma and volvulus are frequent in subjects with megacolon. The association of megaviscera with cardiac compromise is seen in around 30% of the cases and needs to be investigated, among other reasons, to account for surgical risk. Surgical advances in the last decades have significantly improved success and prognosis of these approaches.

References

1. Koberle F. Patologische befunde an den muskularen hoholorganen bei der experimentellen Chagaskrankheit. Zentralkbl Allg Path Path Anat. 1956;95:321–9.
2. Spix JB, Martius CF. Reise in Brasilien. Munchen. Traduction in portuguese. 1823. Rio de Janeiro: Ed. Itatiaia, EDUSP; 1981. p. 97–240.
3. Kidder DP, Fletcher JC. Brazil and the brazilians. Philadelphia: Ed. Childs & Peterson; 1857. p. 416–8.
4. Ferreira LF, Reinhard KJ, Araújo A. Fundamentos da paleoparasitologia. Rio de Janeiro: Edit. Fiocruz; 2011. p. 437–53.
5. Koberle F. The causation and importance of nervous lesions in American trypanosomiasis. Bull WHO. 1970;42:739–43.
6. Zingales B, Miles MA, Campbell DA, Tibayrenc M, Macedo AM, Teixeira MM, Schijman AG, Llewellyn MS, Lages-Silva E, Machado CR, Andrade SG, Sturm NR. The revised *Trypanosoma cruzi* subspecific nomenclature: rationale, epidemiological relevance and research applications. Infect Genet Evol. 2012;12:240–53.
7. Luquetti AO, Miles MA, Rassi A, de Rezende JM, De Souza AA, Povoa MM, Rodrigues I. *Trypanosoma cruzi*: zymodemes associated with acute and chronic Chagas' disease in central Brazil. Trans R Soc Trop Med Hyg. 1986;80:462–70.
8. Atias A, Neghme A, Aguirre Mackay L, Jarpas S. Megaesophagus, megacolon and Chagas disease in Chile. Gastroenterology. 1963;44:433.
9. de Rezende JM, Luquetti AO. Chagasic Megavisceras. In: Pan American Health Organization, editor. Chagas' disease and the nervous system. Scientific publication no. 547. Washington, DC: Pan American Health Organization; 1994. p. 149–71.
10. Luquetti AO. Megaesôfago e anticorpos anti-*Trypanosoma cruzi*. Rev Goiana Med. 1987;33:1–16.
11. Rassi A, Rezende JM, Moreira H, Ximenes CA, Luquetti AO, Lousa L, Ferrioli Filho F. Associação de cardiopatia, megaesôfago e megacolo na fase crônica da doença de Chagas. Rev Soc Bras Med Trop. 1986;19. (Suppl. 2:29.
12. Oliveira EC, Leite MS, Miranda JA, Andrade AL, Garcia SB, Luquetti AO, Moreira H. Chronic *Trypanosoma cruzi* infection associated with low incidence of 1,2-dimethylhydrazine-induced colon cancer in rats. Carcinogenesis. 2001;22:737–40.
13. Garcia SB, Aranha AL, Garcia FR, Basile FV, Pinto AP, de Oliveira EC, Zucoloto S. A retrospective study of histopathological findings in 894 cases of megacolon: what is the relationship between megacolon and colonic cancer? Rev Inst Med Trop Sao Paulo. 2003;45:91–3.
14. World Health Organization. Control of Chagas disease WHO technical report series N° 905. Second report of the WHO Expert Committee. Geneva: World Health Organization; 2002.
15. Rassi A, Rezende JM, Luquetti AO, Rassi A Jr. Clinical phases and forms of Chagas disease. In: Telleria J, Tibayrenc M, editors. American Trypanosomiasis. Chagas disease. One hundred years of research. 2nd ed. Amsterdam: Elsevier; 2017.
16. Chagas C. Processos patogênicos da tripanosomíase americana. Mem Inst Oswaldo Cruz. 1916;8:5–37.
17. Corbett CE, UJr R, Prianti MG, Habr-Gama A, Okumura M, Gama-Rodrigues J. Cell-mediated immune response in megacolon from patients with chronic Chagas' disease. Dis Colon Rectum. 2001;44:993–8.

18. da Silveira AB, Adad SJ, Correa-Oliveira R, Furness JB, D'Avila Reis D. Morphometric study of eosinophils, mast cells, macrophages and fibrosis in the colon of chronic chagasic patients with and without megacolon. Parasitology. 2007;134:789–96.
19. Campos JV, Tafuri WL. Chagas enteropathy. Gut. 1973;14:910–9.
20. Tafuri WL, Maria TA, Lopes ER. Myenteric plexus lesions in the esophagus, jejunum and colon of chronic chagasic patients. Electron microscopy study. Rev Inst Med Trop Sao Paulo. 1971;13:76–91.
21. Almeida HO, Teixeira VP, Gobbi H, Rocha A, Brandao MC. Inflammation associated with cardiac muscle cells parasitized by *Trypanosoma cruzi*, in chronic Chagas' disease patients. Arq Bras Cardiol. 1984;42:183–6.
22. Furness JB, Young HM, Pompolo S, Bornstein JC, Kunze WA, McConalogue K. Plurichemical transmission and chemical coding of neurons in the digestive tract. Gastroenterology. 1995;108:554–63.
23. Furness JB, Costa M. Types of nerves in the enteric nervous system. Neuroscience. 1980;5:1–20.
24. Furness JB. The enteric nervous system: normal functions and enteric neuropathies. Neurogastroenterol Motil. 2008;20(Suppl 1):32–8.
25. Gabella G. Ultrastructure of the nerve plexuses of the mammalian intestine: the enteric glial cells. Neuroscience. 1981;6:425–36.
26. da Silveira AB, D'Avila Reis D, de Oliveira EC, Neto SG, Luquetti AO, Poole D, Correa-Oliveira R, Furness JB. Neurochemical coding of the enteric nervous system in chagasic patients with megacolon. Dig Dis Sci. 2007;52:2877–83.
27. Furness JB, Jones C, Nurgali K, Clerc N. Intrinsic primary afferent neurons and nerve circuits within the intestine. Prog Neurobiol. 2004;72:143–64.
28. Cruvinel W de M, DJr M, Araujo JA, Catelan TT, de Souza AW, da Silva NP, Andrade LE. Immune system – part I. Fundamentals of innate immunity with emphasis on molecular and cellular mechanisms of inflammatory response. Rev Bras Reumatol. 2010;50:434–61.
29. Kodali S, Ding W, Huang J, Seiffert K, Wagner JA, Granstein RD. Vasoactive intestinal peptide modulates Langerhans cell immune function. J Immunol. 2004;173:6082–8.
30. Schmidt K, Klatt P, Mayer B. Uptake of nitric oxide synthase inhibitors by macrophage RAW 264.7 cells. Biochem J. 1994;301:313–6.
31. Rezende JM. Rev Med Chil. 1979). [Chagas disease of the digestive tract (author's transl);107:71–2.
32. da Silveira AB, Arantes RM, Vago AR, Lemos EM, Adad SJ, Correa-Oliveira R, D'Avila Reis D. Comparative study of the presence of *Trypanosoma cruzi* kDNA, inflammation and denervation in chagasic patients with and without megaesophagus. Parasitology. 2005;131:627–34.
33. da Silveira AB, Lemos EM, Adad SJ, Correa-Oliveira R, Furness JB, D'Avila Reis D. Megacolon in Chagas disease: a study of inflammatory cells, enteric nerves, and glial cells. Hum Pathol. 2007;38:1256–64.
34. Sato H, Leo N, Katakai Y, Takano J, Akari H, Nakamura S, Une Y. Prevalence and molecular phylogenetic characterization of *Trypanosoma* (Megatrypanum) minasense in the peripheral blood of small neotropical primates after a quarantine period. J Parasitol. 2008;94:1128–38.
35. Souza DHS, Vaz MGM, Fonseca CR, Luquetti A, Rezende Filho J, Oliveira EC. Current epidemiological profile of chagasic megaesophagus in Central Brazil. Rev Soc Bras Med Trop. 2013;46:316–21.
36. Rezende JM, Lauar KL, Oliveira AR. Aspectos clinicos e radiologicos da aperistalsis do esôfago. Rev Bras Gastroenterol. 1960;12:247–62.
37. Aquino JL, Said MM, Pereira DA, Leandro-Merhi VA, Nascimento PC, Reis VV. Early and late assessment of esophagocardioplasty in the surgical treatment of advanced recurrent megaesophagus. Arq Gastroenterol. 2016;53:235–9.
38. Asti E, Sironi A, Lovece A, Bonavina G, Fanelli M, Bonitta G, Bonavina L. Health-related quality of life after laparoscopic Heller myotomy and Dor fundoplication for achalasia. Surgery. 2017;161:977–83.
39. Allaix ME, Patti MG. Toward a tailored treatment of Achalasia: an evidence-based approach. J Laparoendosc Adv Surg Tech. 2016;26:256–63.

40. Ponciano H, Cecconello I, Alves L, Ferreira BD, Gama-Rodrigues J. Cardiaplasty and Roux-en-Y partial gastrectomy (Serra-Dória procedure) for reoperation of achalasia. Arq Gastroenterol. 2004;41:155–61.
41. Herbella FA, Aquino JL, Stefani-Nakano S, Artifon EL, Sakai P, Crema E, Andreollo NA, Lopes LR, de Castro Pochini C, Corsi PR, Gagliardi D, Del Grande JC. Treatment of achalasia: lessons learned with Chagas' disease. Dis Esophagus. 2008;21:461–7.
42. Pochini Cde C, Gagliardi D, Saad Júnior R, de Almeida RF, Corsi PR. Esophagectomy with gastroplasty in advanced megaesophagus: late results of omeprazole use. Rev Col Bras Cir. 2015;42:299–304.
43. Zilberstein B, Franciss MY, Genovesi A, Volpe P, Domene CE, Barchi LC. Pioneer robotic Serra-Doria Operation for recurrent Achalasia After Heller's cardiomyotomy: a "new quondam" procedure. J Laparoendosc Adv Surg Tech A. 2017;27:524–8.
44. Rebecchi F, Allaix ME, Morino M. Robotic technological aids in esophageal surgery. J Vis Surg. 2017;8:3–7.
45. Brant C, Moraes-Filho JP, Siqueira E, Nasi A, Libera E, Morais M, Rohr M, Macedo EP, Alonso G, Ferrari AP. Intrasphincteric botulinum toxin injection in the treatment of chagasic achalasia. Dis Esophagus. 2003;16:33–8.
46. Uppal DS, Wang AY. Update on the endoscopic treatments for achalasia. World J Gastroenterol. 2016;22:8670–83.
47. Gladman MA, Dvorkin LS, Scott MS, Lunniss PJ, Williams NS. A novel technique to identify patients with megarectum. Dis Colon Rectum. 2007;50:621–9.
48. Anand AC, Sashindran VK, Mohan L. Gastrointestinal problems at high altitude. Trop Gastroenterol. 2006;27:147–53.
49. Oliveira EC, Menezes JG, Cardoso VK, Luquetti AO, Neto SG, Garcia SB. The Relationship between megacolon and constipation in Chagas' disease. Neurogastroenterol Motil. 2009;21(Suppl 1):5.
50. Bafutto M, Luquetti AO, Gabriel Neto S, Penhavel FAS, Oliveira EC. Constipation is related to small bowel disturbance rather than colonic enlargement in acquired chagasic megacolon. Gastroenterol Res. 2017;10:213–7.
51. Oliveira EC, Gabriel Neto S, Bafutto M, Luquetti AO. False-negative rectoanal inhibitory reflex in acquired megacolon. Annual scientific meeting. American Society of Colon and Rectum Surgeons.
52. Cavenaghi S, Felicio OCS, Ronchi LS, Cunrath GS, Melo MMC, Netinho JG. Prevalence of rectoanal inhibitory reflex in chagasic megacólon. Arq Gastroenterol. 2008;45:128–31.
53. Castro C, Hernandez EB, Rezende JM, Prata A. Radiological study on megacolon cases in an endemic area for Chagas disease. Rev Soc Bras Med Trop. 2010;43:562–6.
54. Shafik A, Mostafa RM, Shafik I, El-Sibai O, Shafik AA. Functional activity of the rectum: a conduit organ or a storage organ or both. World J Gastroenterol. 2006;12:4549–52.
55. Duhamel B. New operation for congenital megacolon: retrorectal and transanal lowering of the colon, and its possible application to the treatment of various other malformations. Presse Med. 1956;64:2249–50.
56. Oliveira AB. Tratamento cirúrgico do megacólon pela operação de Duhamel. Rev Paul Med. 1963;63:283–304.
57. Haddad J, Raia A, Correa Neto A. Abaixamento retro-retal do colon com colostomia perineal no tratamento do megacólon adquirido. Rev Assoc Med Bras. 1965;11:83–8.
58. Reis Neto JA. Resultados tardios da operação de Duhamel no tratamento do megacólon adquirido. Rev Assoc Med Bras. 1970;18:57–62.
59. Salerno G, Sinnatamby C, Branagan G, Daniels IR, Heald RJ, Moran BJ. Defining the rectum: surgically, radiologically and anatomically. Color Dis. 2006;8. (Suppl. 3:5–9.
60. Gabriel Neto S, Oliveira EC, Ramos GC, Gabriel AG, Habr-Gama A, Zilberstein B, Luquetti AO. Treatment of megacolon: low anterior resection vs. Duhamel procedure evaluated by colonic transit time. Dis Colon Rectum. 2009;50:798.

Chagas Disease in Immunosuppressed Patients

Adelina R. Riarte, Marisa L. Fernandez, Claudia Salgueira, and Javier Altclas

Abstract This chapter is a brief update on immunosuppression and Chagas disease (CD) and refers to the pharmacological immunosuppression in solid organ transplantation and hematopoietic stem cell transplantation and to the biological immunosuppression induced by the human immunodeficiency virus (HIV).

In transplantation a review is made of historical and current immunosuppressive drugs and of the keys in the diagnosis for recipients and donors with Chagas disease and their management during the follow-up after transplantation. Likewise, the main events published in the different heart, kidney, liver, and bone marrow transplants in Chagas disease in endemic and non-endemic countries are described. The events corresponding to non-endemic countries have been published, due to the intense migratory flow from the endemic countries to the non-endemic (developed) countries in Europe, Canada, Australia, and the United States. This chapter also covers published reports of the main transplants, key situations in the diagnosis, as well as conclusions.

A. R. Riarte (✉)
Instituto Nacional de Parasitología Dr. M Fatala Chaben, Administración de Laboratoriose Institutos de Salud (ANLIS) Dr. CG Malbran [Parasitology Institute–Ministry of Health], Ministerio de Salud, Buenos Aires, Argentina

M. L. Fernandez
Instituto Nacional de Parasitología Dr. M Fatala Chaben, Administración de Laboratoriose Institutos de Salud (ANLIS) Dr. CG Malbran [Parasitology Institute–Ministry of Health], Ministerio de Salud, Buenos Aires, Argentina

Centro Municipal de Patología Regional Argentina-Medicina Tropical [Tropical Medicine and Regional Pathology Centre], Hospital de Infecciosas FJ Muñiz [Infectious Diseases Hospital], Buenos Aires, Argentina

C. Salgueira · J. Altclas
Infectious Diseases Unit, Sanatorio de la Trinidad Mitre, Buenos Aires, Argentina

Infectious Diseases Unit, Sanatorio Anchorena, Buenos Aires, Argentina
e-mail: jaltclas@intramed.net

J. M. Altcheh, H. Freilij (eds.), *Chagas Disease*, Birkhäuser Advances in Infectious Diseases, https://doi.org/10.1007/978-3-030-00054-7_13

In HIV and Chagas, a historical review is made since the appearance of AIDS in the world and its characterization as a pandemic. The great and rapid advances in the research of AIDS are a paradigm on how to make progress in order to control a disease. The appearance of new treatments meant a change in the clinical profile and a change in the life expectancy of AIDS patients. With respect to Chagas disease and AIDS, the reactivation of CD in AIDS patients constitutes an AIDS-defining event. This chapter presents different clinical CD-AIDS scenarios, the diagnosis and treatment, as well as a vision of the future.

1 Immunosuppression and Chagas Disease

The Chagas disease (CD) produced by *Trypanosoma cruzi* is mainly transmitted by vectors in endemic areas, by blood transfusion, congenitally from an infected mother to the fetus and, more recently, orally. In addition, a new form of transmission came to occupy a relevant epidemiological place via different organs and tissues infected by *Trypanosoma cruzi* to non-Chagas disease recipients.

Two clearly different phases characterize CD: the acute phase, usually symptomatic in children, and the chronic phase in which most people remain in an indeterminate phase under some classification [1]. Historically, it was estimated that 30% of these people evolve to the symptomatic phase with heart or digestive (megaviscera) involvement or both.

Some 20–25 years ago, the interaction of pharmacological or biological immunosuppression, such as organs and hematopoietic stem cell transplantations (HSCT) and HIV infection, has modified the natural evolution of Chagas disease, challenging the health system for the generation of new algorithms for diagnosis and treatment. Likewise, the intense migratory flows from endemic countries from Latin America to European countries, Australia, Japan, Canada, and the United States have transformed Chagas disease in a worldwide health problem. Recent evaluations of this migratory impact in those non-endemic countries in different periods consider that 5.2% of Latin American migrants in Spain, 2.9% in other European countries, 2% in the United States, 3.8% in Australia, and 3.5% in Canada are potentially infected by *T. cruzi* [2].

1.1 The Scenario of Immunosuppression in Transplantation and Chagas Disease

The reactivation of Chagas disease in the context of immunosuppression was described about 25 years ago, with organ and HSCT being the field where medical practice has focused on organ acceptance and monitoring during follow-up. The risk of reactivation in these populations seems to be related to immunosuppression severity.

In heart transplant, two immunosuppressive drugs, such as mycophenolate (MMF) and azathioprine (AZA), were compared to assess the impact on reactivation risk and benefits. Regarding other immunosuppressive drugs, the main experience is with steroids [3].

1.1.1 Immunosuppressive Drugs Used in Transplantation

1. 1960s: 6-mercaptopurine, AZA. AZA in combination with steroids.
2. 1980s: cyclosporine.
3. 1994: MMF, tacrolimus.
4. Currently tacrolimus has gradually replaced cyclosporine, and MMF has replaced AZA and sirolimus.
5. At present monoclonal antibodies are also used for GVHD and rejection: basiliximab, daclizumab, and muromonab.

The introduction of new immunosuppressive therapies during the last decade had a significant impact on treatment strategies, and this has been associated with clinical benefits in hematological diseases, solid tumors, etc. However, published information regarding the reactivation of Chagas in patients with neoplasms remains scarce [4, 5]. The emergence of new immunosuppressive drugs could encourage the reporting of new reactivation cases.

In organ transplant patients, the risk of reactivation varies according to the transplanted organ and the degree of immunosuppression. The reactivation rate was recently compared in heart transplants under two immunosuppressive treatments. The reactivation rate was lower in those who received AZA versus MMF, the latter having a sixfold increased risk of Chagas reactivation during the first 2 years after transplantation [6].

1. TNF inhibitors are adalimumab, infliximab, certolizumab, and etanercept.
2. Other biological drugs as abatacept, anakinra, alefacept, and efalizumab.

1.1.2 Biological Therapies Used in Rheumatology

The biological therapy has improved the management and prognosis of autoimmune diseases, but also its use produced adverse effects such as anaphylaxis and infections among other processes [7]. Anti-TNFs have been associated with an increased risk of opportunistic infections. Only three cases of Chagas disease reactivation in patients with psoriasis receiving biological therapies have been reported: one with infliximab, one with adalimumab, and the third with etanercept [7–9].

There is little information available on the need to perform serological screening of Chagas disease before starting biological therapy. This screening in endemic and non-endemic countries and close monitoring will allow to have greater certainty about the evolution of the interaction between the biological therapy, the diseases in which it applies, and its impact on the course of Chagas disease. The effect of immunomodulatory drugs on the dynamics of *T. cruzi* parasitemia in humans is unknown, with only experimental evidence in animal models with contradictory results.

1.1.3 Targeted Therapies Used in Oncohematology

1. Monoclonal antibodies: rituximab-ofatumumab-obinutuzumab (CD20), alemtuzumab (CD52), brentuximab (CD30) and blinatumomab (CD19).
2. Purine analogs: clofarabine-fludarabine-cladribine.
3. Inhibitors of proteosomes: bortezomib and carfilzomib.
4. Hypomethylating agents: azacitidine-decitabine.
5. Alkylating agent: bendamustine.

Reactivation of Chagas disease was recently reported in a patient with follicular lymphoma treated with cyclophosphamide, doxorubicin, vincristine, prednisone, and rituximab for six cycles [10]. Reactivation with conventional chemotherapy has also been described in a patient with non-Hodgkin's lymphoma living in an endemic area, who received cyclophosphamide, daunorubicin, vincristine, prednisone, and bleomycin for 6 months [11]. Reactivation also has been described in a patient with non-Hodgkin lymphoma as early as 10 days after treatment with dexamethasone (16 mg/day) [12].

Reactivation should be considered in patients with chronic Chagas disease receiving intense immunosuppression.

1.1.4 Targeted Therapies Used in Oncology

Targeted therapies act on specific molecular targets, blocking the growth and spread of cancer by interfering with a specific target involved in the growth, progression, and spread of cancer. Most drugs are monoclonal antibodies or small molecules:

1. Monoclonal antibodies: trastuzumab, bevacizumab, rituximab, and cetuximab.
2. Tyrosine kinase inhibitors: dasatinib, ibrutinib, imatinib, and nilotinib.
3. Inducers of apoptosis: bortezomib.
4. Inhibitors of angiogenesis: bevacizumab, sunitinib, and thalidomide.
5. mTOR inhibitors: rapamycin, temsirolimus, and everolimus.
6. Checkpoint inhibitors: ipilimumab, nivolumab, and pembrolizumab.

The continuous emergence of new immunosuppressants requires being aware of the serological status before the start of drug therapy. Reactivation should be considered in patients with chronic Chagas disease, and neoplasms due to these diseases require intensive or long-term treatment with immunosuppressive drugs [5].

In sum, it is possible at present that immunosuppressant drugs have an incidence in the reactivation of Chagas disease in cases of solid organ transplantation. Different immunosuppressants and their severity through their targets involved have different impacts on the course of Chagas disease. In HSCT, the risk of reactivation varies depending on whether it is an autologous or allogeneic transplant. In cases of other drugs used for different pathologies, few cases of reactivation have been described, and it is expected that more cases will be reported with the experience of their use.

Monitoring and early treatment help reduce potentially severe and/or fatal cases.

As in organ transplantation, a relationship between immunosuppressant drugs and the risk of Chagas disease reactivation in allogeneic HSCT has been suggested, but to the low number of cases, further studies could contribute to this knowledge [13–15].

1.2 Key Criteria in the Diagnosis of Chagas Disease and Transplantation

Every patient, including transplant recipients or living or cadaveric donors who have lived in endemic countries, Latin American migrants to non-endemic countries for *T. cruzi* around the world, as well as those with epidemiological background of endemicity, must have a serological diagnosis of *T. cruzi* infection. This diagnosis facilitates subsequent monitoring of the infection. The serological diagnosis [16] is based on different tests, of which at least two must be reactive to describe the patient as infected by *T. cruzi*. The pairs of tests, which are usually quantitative, must include techniques for the recognition of different antibody types. The most usual serological tests include enzyme-linked immunosorbent assay (ELISA), indirect immnunoflourescence assay (IFA), indirect hemagglutination (IHA), etc. either the one that is commercially available or the one developed "in-house."

Two clinical outcomes can occur in this association of Chagas disease and transplantation in which immunosuppression plays a decisive role—the reactivation of chronic Chagas disease in the recipient or the transmission of *T. cruzi* infection from the graft of a living or deceased infected donor to a non-Chagas recipient. Both clinical events should be considered an acute process and treated as such. The reactivation of Chagas disease is characterized by an exacerbation of parasitemia that can be detected by direct parasitological methods, usually the Strout test on peripheral blood and on the cerebrospinal fluid when the CNS is involved or *T. cruzi* parasites impact on different organs. The most common ones are the skin, in the form of unifocal or multifocal nodular dermatitis, usually in the chest, abdomen, or lower limbs. In this case and in other cases of organ involvement, if peripheral blood tests for *T. cruzi* parasites are negative, a biopsy of the lesion will help identify the inflammatory damage associated with typical intracellular *T. cruzi* amastigotes.

In cases of transmission by the *T. cruzi*-infected graft from a donor, both serological and parasitological methods have an essential role in the diagnosis of *T. cruzi* infection. Early reactive serological conversion and/or in peripheral blood *T. cruzi* detection using the Strout test [17] will provide a positive diagnosis. Like Chagas disease reactivation in recipients, nodular dermatitis may also occur in cases of *T. cruzi* transmission. There may be involvement of other organs or tissues in both acute processes: reactivation or transmission. With prompt diagnosis and therapy administration, according to a monitoring schedule, morbidity and mortality decline.

Current recommendations state that chronic Chagas disease recipients or non-Chagas recipients with potential transmission from the infected graft should be monitored at regular intervals after transplant for potential parasite increase in peripheral blood or cerebrospinal fluid or *T. cruzi* amastigotes detection in tissue biopsies [18–20].

1.3 *Use of DNA Polymerase Chain Reaction (PCR) for Diagnosis*

Different studies were recently conducted to determine the value of PCR as an early and sensitive method for the detection of clinical events in the immunosuppression and Chagas disease scenario. As early as in 2000, a first study suggested the value of PCR as a predictor of reactivation in a heart transplant patient, showing DNA amplification by mini-circle fragment of the kinetoplastid DNA (k-DNA) and satellite sequence of the nuclear *T. cruzi* DNA (Sat-DNA). These positive PCRs occurred 41 days before positive parasitemia by Strout test and skin lesions [21]. Later on, in 2007 a similar protocol was applied in ten patients with Chagas cardiomyopathy undergoing heart transplantation and monitored using conventional methods and PCR with the same two abovementioned assays. Five of these patients were positive by PCR 38–85 days before clinical reactivation, in four of them by skin lesions, and the others by myocarditis [22]. A limitation of this study was that PCR was positive in a heart transplant patient who was not reactive for chronic Chagas disease during follow-up. In 2011, an international report evaluated the different performances of the PCR procedures currently used for the detection of *T. cruzi* in peripheral blood in different countries of the Americas and Europe. This study was an initial step in the standardization and validation of PCR in the diagnosis of *T. cruzi*. Despite the great variability, four PCR assays, two of which were real-time PCR (qPCR), showed a good performance [23]. In line with the above research, another study was conducted in 2012 mainly aimed at determining the diagnostic value of qPCR in the interaction of Chagas disease recipients and transplantation and also in the transmission of *T. cruzi* infection from the graft to non-Chagas recipients. The study used three qPCRs with different targets: the first one targeting the Sat-DNA, the second one targeting the k-DNA (both demonstrated high performance in previous report), and a third qPCR assay targeting the small subunit ribosomal RNA (18S rRNA) gene of *T. cruzi*. This study was able to detect reactivation of chronic Chagas disease and a *T. cruzi* infection transmitted by the infected graft, following a CDC protocol. At that time, the protocol considered that a blood sample should be positive for *T. cruzi* if the three real-time PCR assays previously described were positive and if any sample that was only positive for the k-DNA and/or TCZ assays TaqMan and negative for the 18S rRNA TaqMan assay was "inconclusive." This study also showed a preliminary approach on higher sensitivity of PCR in buffy coat vs peripheral blood alone [24]. In 2013 a report showed the value of k-DNA-PCR in early diagnosis and its comparison with conventional parasitological and serological techniques in non-Chagas recipients transplanted with *T. cruzi*-infected donors. In five out eight recipients, *T. cruzi* infection was detected at 68.4 days (range 36–98 days) vs Strout test or serological methods [25]. In addition, for the first time, serial measurements of qPCRs were evaluated in this type of patients, showing a median of 343.8 eq parasites/mL (IQ range 164.7–826.6) in the follow-up, being the different times of diagnosis 93 days (IQ range 56–103 days) vs 121 days (range IQ 76–235 days) for qPCR and serology, respectively.

More recently, in 2015, a new international study was carried out with 26 PCR laboratories in 14 countries to evaluate two qPCR strategies—SAT-DNA and k-DNA—using TaqMan probes, for the potential detection and quantification of parasite loads in blood samples from Chagas disease patients. The authors suggest that, although k-DNA had a higher clinical sensitivity to detect low parasite loads, it also has the possibility of false-positive results; thus SAT-DNA qPCR, and not k-DNA qPCR, should be the qPCR-based method of choice in different clinical scenarios; in our case, in immunosuppression and Chagas disease. Nevertheless, the clinical sensitivity of both qPCRs ranged between 80% and 85% [26]. This would not be enough to consider it an exclusive alternative in the diagnosis of *T. cruzi* infection. However, in certain clinical scenarios, the optimization of the qPCR (i.e., the quantification of parasite load) by means of diagnostic algorithms such as vertical transmission, post-treatment monitoring, and immunosuppression and Chagas disease, would be a valuable contribution of this tool to Chagas disease.

1.4 Different Scenarios in Chagas Disease and Transplant

1.4.1 Heart Transplantation

The first pioneering and most experienced studies on heart transplantation and Chagas disease were conducted in Brazil [27]. This work has allowed access to transplants for patients with end-stage heart Chagas disease in Brazil. This cardiomyopathy is between the second and the third causes of heart transplantation in that country. Prior to these studies, Chagas disease was an absolute contraindication for cardiac transplantation.

Heart transplant in Chagas disease went through different phases that allowed clarifying a progressive improvement in the application of this therapy. Certain essential criteria and the results were strengthened with that development. Survival of patients with Chagas disease was significantly better in comparison with idiopathic and ischemic cardiomyopathy (p=0.027) during short- and long-term follow-up. Chagas disease patients were always younger than in other etiologies and with less comorbidities. Critically, neoplasms were more frequent in Chagas disease patients. Regarding reactivation, the emergence of parasitemia appeared to occur similarly in different stages, but Chagas disease clinical reactivation was more frequent during the first stage in five patients compared to one patient during the second stage ($p < 0.002$). The dose of cyclosporine was reduced, and this change in clinical management was associated with a lower rate of reactivation (2.3 per patient vs 0.25 in the second group ($p < 0.05$) but rejection was similar in both protocols [6].

In 2005, cardiac transplantation had become the treatment of choice for advanced cardiomyopathy due to Chagas disease. At that time, and as previously mentioned, the dose of AZA vs MMF was compared, and its impact on reactivation during the period 1987–2004 was evaluated. One group received AZA and the other MMF at the standard doses in sequential periods, along with equal doses of prednisone and cyclosporine. The reactivation episodes were less frequent in the AZA group ($p < 0.0001$) vs the MMF group, and the rejection rates were similar ($p = 0.88$), suggesting that the more intense immunosuppression by MMF increased the risk of Chagas disease reactivation [28].

1.4.2 Kidney Transplantation

In September 1988, Argentina's INCUCAI (National Institute for the Coordination of Ablation and Implant) liberalized the criteria for organ donation due to the critical situation of patients with terminal illnesses on the waiting list. A decade later, a first study based on kidney transplantation and Chagas disease in seronegative or seroreactive recipients for *T. cruzi* was published in Argentina. An strict monitoring was carried out to detect chronic Chagas disease reactivation or transmission by grafts from infected living or deceased donors during follow-up. During the 1989–1996 period, Chagas disease was detected in 17.2% (41/238) of kidney transplant patients at a high-complexity hospital in Argentina (General Hospital Dr. Cosme Argerich). In 21.7% (5/23) of the cases, reactivation was diagnosed in chronic Chagas disease patients within a median time of 63 days (range IQ 47.5–483.5) by patent parasitemia in three and skin lesions in two patients. Transmission of *T. cruzi* was diagnosed in 18.7% (3/16) of patients by patent parasitemia using the Strout test. After these outcomes, two of them persisted with indeterminate Chagas disease, while parasitemia in the third patient turned prompt negative as a result of benznidazole treatment with no seroconversion. There was no mortality associated with Chagas disease [29].

More recently, in 2013, a study at a reference center from Formosa, an endemic province from Argentina, reported that 73 kidney transplants were performed from 2007 to 2013 and that 8% (6/73) of patients with no history of Chagas disease received *T. cruzi*-infected grafts. Prophylactic treatment for Chagas disease was not administered. Patients were screened once per week for the first 3 months after transplantation and once per month thereafter. In case of infection, benznidazole at a dose of 5 mg/kg/day would be given for 60–90 days. The immunosuppressive regimen included tacrolimus, sodium mycophenolate, and corticosteroids at standard doses. Four patients received induction therapy with thymoglobulin or basiliximab. A living donor infected with *T. cruzi* received benznidazole 5 mg/kg/day during 30 days before donation, with no adverse events, a recommendation by an Argentinian guidelines in cases of *T. cruzi* donors, to decrease the potential inocula of parasites [18]. Uninfected recipients may develop *T. cruzi* infection from the graft of infected donors [30]. In a report from the United States, the infection rate among kidney recipients was 13% (2/15). Those patients were monitored closely, diagnosed early, and treated in the course of the infection. This prevention in the interaction of immunosuppression and Chagas disease was able to modulate the morbidity associated with infected kidney transplantation and Chagas disease [31].

1.4.3 Liver Transplantation

In 2005, the follow-up of a non-Chagas recipient was reported, describing a 33-year-old woman in a clinical emergency status due to fulminant autoimmune hepatitis who had an orthotopic liver transplantation in 1998 at a high-complexity reference center from Argentina (Italian Hospital, Buenos Aires). The donor was a 58-year-old female, born in an endemic Argentine province for Chagas disease (a rural area in Catamarca), who had died of a hemorrhagic stroke. Her serologic tests for cytomegalovirus, Chagas disease, and toxoplasmosis were positive. Potential risk of Chagas disease transmission was communicated to her family, who accepted the risk of *T. cruzi* infection. Primary immunosuppression was performed with mycophenolate mofetil, steroids, and tacrolimus. During posttransplant follow-up, the patient showed moderate rejection at day 7 and treated with a 1 g bolus of methylprednisolone for 3 days, with favorable response. A magnetic resonance imaging of the brain was performed due to persisting coma, showing diffuse severe brain damage presumably caused by hypoxia or toxicity. At day 42, a new magnetic resonance imaging demonstrated improvement of brain lesions but severe cortical atrophy probable sequelae of pre-transplant encephalopathy. Table 1 shows a summary of serologic and parasitological data observed after transplant during the monitoring period. At day 76, ELISA testing for *T. cruzi* antigen showed results very close to cutoff, and at day 84 primary *T. cruzi* infection was diagnosed according to positive parasitemia by the Strout test and seroconversion by ELISA and IFA. No clinical signs nor laboratory abnormalities were observed during the primary infection. Preemptive therapy with benznidazole (Radanil, Roche Laboratory, Buenos Aires, Argentina) was administered at 5 mg/kg/day for 60 days. Parasitemia decreased rapidly, becoming negative 11 days after therapy began. During and after parasiticidal therapy, antibody titers detected by ELISA and IFA decreased, showing a variable result becoming consistently negative at day 196, about 3 months after the acute episode (Table 2). At day 357, there was no laboratory and/or clinical evidence that the patient had suffered an acute episode of Chagas disease [32].

A Brazilian experience was published in 2007, showing the outcome of orthotopic liver transplantation in six patients with severe liver disease, with livers from donors serologically positive for Chagas disease. Transplantations were performed from November 2000 to January 2005, and the patients received prophylactic treatment with benznidazole for 60 days, according to the Brazilian Consensus in

Table 1 Post-transplant immunosuppressive therapy

Drugs used in immunosuppressive maintenance therapy	
Steroids	
Calcineurin inhibitors	Ciclosporin Tacrolimus
Antimetabolites	Mycophenolate mofetil (MMP) Mycophenolate sodium (MMS) Azathioprine (AZA)
Drugs used for acute rejection and/or induction of immunosuppression	
Antithymocyte globulins	Antithymocyte globulins
Anti-CD3 murine antibodies	OKT3
Monoclonal antibodies	Alemtuzumab Rituximab Bortezomib Eculizumab
IL-2 receptor antagonists	Basiliximab Daclizumab

Table 2 Transmission of *Trypanosoma cruzi* infection from a donor to a non-Chagas recipient submitted to liver transplantation. Synthesis of parasitological and serological data during follow-up

Day post Tx	Strout test	ELISA	IHA	IFA
1	Neg	Neg	Neg	Neg
13	Neg	Neg	Neg	Neg
76	Neg	0.168	Neg	Neg
84	Pos Benz	0.207	Neg	1/32
95	Neg	0.219	Neg	1/64
117	Neg	0.186	Neg	1/32
188	Neg	0.245	Neg	1/32
196	Neg	0.152	Neg	Neg
357	Neg	0.061	Neg	Neg

References: Tx, transplant; Benz, benznidazole; Pos, positive; Neg, negative; ELISA ≥0.20 (optical density), Pos; indirect hemagglutination (IHA) and immunofluorescent assay (IFA) ≥ 1/32, Pos. Barcan L, Luna C, Clara L, Sinagra A, Valledor A, de Rissio AM, Gadano A, Martın Garcia M, Santibañes E de, and Riarte A. Liver Transpl 2005; 11: 1112–6

Chagas Disease. Patients received immunosuppression by cyclosporine, tacrolimus, MMF, and steroids. There was no development of acute Chagas disease in the six patients, but one patient died of pulmonary tuberculosis and another one of septicemia [33].

More recently, in 2011, a report from a non-endemic area was published [31]. Liver transplantation was performed in two patients who received graft from *T. cruzi*-infected donors. The serology of both recipients for *T. cruzi* was nonreactive. After transplant, both patients received primary prophylaxis with benznidazole (300 mg/day) for 60 days. Immunosuppression was achieved by administering tacrolimus and MMF, and eventually everolimus was added due to toxicity of tacrolimus. Patients have a good evolution, but reactive serology, with negative qPCR after transplant, raises doubts about possible *T. cruzi* transmission by the donor controlled with trypanocidal therapy [34].

A new report was published in Argentina in 2012 about a prospective protocol using livers from *T. cruzi*-infected donors for non-Chagas recipients with no administration of prophylactic therapy. During a 13-month period, 24% (9/37) liver transplants were performed within this protocol. After transplant, each recipient was sequentially and strictly monitored for infection transmission using the Strout test and promptly treated with benznidazole if this occurred. During follow-up, two patients died without Chagas infection, and *T. cruzi* transmission rate was 22% (2/9). Patients developed *T. cruzi* infection with no clinical symptoms. The median follow-up time for the seven living patients was 15 months (range, 13–20). At present, they are all symptom-free with excellent allograft function and no evidence of Chagas disease [35]. These published data and others show that *T. cruzi*-infected donors are a promising source of liver grafts that would reduce the high mortality in the liver waiting lists in the America.

1.4.4 Bone Marrow Transplantation

The evolution of patients who underwent bone marrow transplantation (BMT) between 1989 and 2002 in four reference centers in Argentina was reported in 2005 [13]. All Chagas disease patients under immunosuppression for underlying diseases, and therefore at risk of reactivation, were screened using the direct Strout test before BMT. As in other types of transplants and Chagas disease, the recipients were checked on a weekly basis using Strout test and serological methods during the first 2 months, every 15 days during the third month, and on monthly controls from then on. Monitoring took place throughout the immunosuppressive period after allogenic (Allo) BMT and during 60 days in autologous BMT (ABMT). However, the Chagas disease protocol returned to a weekly frequency in every instance of clinical suspicion that an infectious episode was in course and until the etiological cause was diagnosed. Basal diseases in these patients were chronic leukemias, Hodgkin and non-Hodgkin lymphomas, and solid tumors. The incidence rate of Chagas disease was 1.9% (25/1328), and all patients were in the indeterminate phase. In ABMT, parasitemia was detected only in two patients with non-Hodgkin lymphoma by blood cultures 4 days before BM infusion during the running neutropenic period, with no signs or symptoms of Chagas disease reactivation. Preemptive therapy with benznidazole was started and administered during 30 days. The investigation of *T. cruzi* parasites in the infused bone marrow was negative. During weekly controls, parasitemia was always negative using Strout test. After treatment, there were no signs of relapsing *T. cruzi* parasitemia. In the second patient after chemotherapy for the underlying disease, at −20 days of ABMT, parasitemia was detected using SM, with no clinical signs. He received preemptive therapy with benznidazole until bone marrow infusion day, with remission of parasitemia. After ABMT, *T. cruzi* parasitemia did not relapse nor did the recipient show any clinical signs of reactivation. In nine Allo-BMT patients, parasitemia reactivated in four patients in a median of 110.5 days (IQ range 37.25–163.5). One of these patients also showed skin lesions. Treatment with benznidazole abrogates reactivation. Three non-Chagas disease recipients received Allo-BMT from *T. cruzi*-infected donors. Two living donors received prophylaxis during 30 days, and the third donor was not treated due to a BMT emergency in the recipient [13, 15].

1.5 *Clinical Management of Immunosuppressed Patients and Chagas Disease*

It is known that immunosuppression associated with hematological diseases, solid neoplasms, autoimmune diseases, or organ transplantation can modify the natural evolution of Chagas disease. Reactivation of chronic Chagas disease, a new acute form of the disease, characterized by exacerbation of parasitemia or clinical manifestation may cause high morbidity and mortality in the posttransplant setting [19].

Potential transplant candidates and donors born in endemic areas for Chagas disease, born to mothers infected by *T. cruzi*, who have traveled rural areas of endemic countries or who have received multiple blood transfusions must access before transplantation to serological diagnosis according to international recommendations [16]. At present organs or tissues as the kidney, liver, and bone marrow from infected donors by *T. cruzi* have shown the feasibility of their use in transplantation [29–35]. Any recipient of organs or tissues, in cases where Chagas disease is involved in the recipient and/or the donor, must be subjected to close and rigorous monitoring in order to detect early reactivation or infection, so as to have the ability to administer preemptive or therapeutic treatment that enables the recovery of patients undergoing an acute process. Regarding the living *T. cruzi*-infected donor, it is advisable to administer trypanocidal treatment in order to reduce the potential inoculum of parasites transmitted by the infected graft. The first-line drug of choice is benznidazole and the second-line drug is nifurtimox. Occasionally allopurinol was used in some transplant scenarios. There is an agreement that cardiac transplant recipients should not receive *T. cruzi*-infected graft. This is an empirical concept in the sense of not providing *T. cruzi* parasites different from the receptor that could impact the new heart, an target organ of *T. cruzi*.

For other organs not currently used in transplantation, it is necessary to proceed with caution and consider the intensity of the immunosuppression and be able to provide a strict monitoring in the follow-up. Not all organ transplantations from *T. cruzi*-infected donors result in transmission to the recipient. If the donor status is known and appropriate posttransplant monitoring ensured, clinical outcomes can be controlled in transplant recipients.

There are controversial data regarding primary prophylaxis in Chagas disease recipients. Back in 2000, a Brazilian report did not recommend giving primary prophylaxis with benznidazole in the face of the existing evidence that steroids reactivated parasitemia in those patients [36]. In 1996, the first work in cardiac transplantation performed primary prophylaxis prior to transplantation, and this did not prevent the reactivation of Chagas disease [27]. A recent study recommends that benznidazole or nifurtimox should be started before immunosuppression to reduce the risk of reactivation in Chagas disease recipients [5]. Others argue that prophylactic treatment early after transplantation is not beneficial to prevent reactivation or transmission from the *T. cruzi*-infected donors although the data come from small cohorts of patients. There are also no data from randomized clinical trials to provide evidence levels that allow greater certainty in this issue; however considering that patients with hematological or end-stage liver and kidney diseases may be potentially vulnerable to drug of benznidazole or nifurtimox, at present primary prophylaxis does not seem to be beneficial [4, 19, 37].

Our recommendations stress the need for systematic monitoring of recipients by polymerase chain reaction, qPCR, and Strout test in the peripheral blood or buffy coat or other parasitological tests in reference laboratories and advance planning for immediate antitrypanosomal treatment if recipient infection or reactivation is detected. Data on management and outcomes of all cases should be collected to generate more evidence for future guidelines.

A clear concept emerges from all the studies: only strict post-transplant monitoring enables early diagnosis of *T. cruzi* reactivation or infection episodes, and parasiticide treatment applied in these clinical events decreases the morbidity and mortality associated with immunosuppression and Chagas disease. It is also important to emphasize that bioethics and informed consent should govern all immunosuppression and Chagas disease protocols.

2 HIV and Chagas Disease

2.1 Brief Epidemiological and Historical Review of HIV/AIDS Infection

According to the latest UNAIDS—Joint United Nations Programme on HIV and AIDS—report, the HIV/AIDS pandemic has caused about 35 million deaths, and 36.7 million individuals currently live with HIV. In 2016, of all people living with HIV around the world, only 53% had access to treatment, and 1.8 million people became newly infected with HIV. In Latin America, 1.8 million people were living with HIV, and 36,000 people died of AIDS-related illnesses in 2016. There were an estimated 97,000 new HIV infections per year in the region, and this number did not change between 2010 and 2016 [38].

2.1.1 The First Cases of the Syndrome and the HIV/AIDS Pandemic

The first five cases of previously healthy, homosexual patients with manifestations of opportunistic infections (OIs), such as extensive *Candida albicans* mucosal lesions, CMV, and *Pneumocystis jiroveci* pneumonia, were published in June 1981. At that time, these patients were found to show severe cellular immunosuppression similar to that of patients with lymphoma [39, 40]. These OIs were initially linked to homosexual men, later to intravenous drug users and recipients of blood products, and ultimately to the general population. This disease was called acquired immunodeficiency syndrome (AIDS) from the very beginning. Due to its rapid spread and lethality worldwide, the impact of AIDS quickly encouraged intensive research activities.

2.1.2 Isolation of the Etiological Agent, Diagnosis, and Pathogenesis of AIDS

The prematurely called HTLV-III retrovirus was isolated in 1983, and currently it is known as HIV-1 [41]. In 1984, different assays were developed for the diagnosis of antibodies against HIV using ELISA and Western blot tests to differentiate individuals with and without HIV/AIDS infection [42]. In the same year, the CD4 cell surface molecule was identified as the main HIV receptor [43–45]. This finding provided insight into the essential substrate of immunosuppression in AIDS, which is a steady decrease of the CD4+ T lymphocyte cells circulating in peripheral blood, and reinforced the rationale for monitoring CD4+ cell counts in the clinical follow-up of HIV patients. CD4+ cell counts are part of the clinical monitoring of these patients even nowadays [46].

Later on, in 1993, plasma viral levels could be measured thanks to polymerase chain reaction (PCR). The viral dynamics could be described during infection. At the early phase of HIV infection, high RNA copies per mL of plasma were detected [47]. After this initial burst of viral replication in peripheral blood, viremia in plasma reduces, but HIV disease remains active or latent in lymphoid tissues and other "sanctuaries" [48]. As these new insights were learned, these advances in molecular biology also benefited patients. HIV-infected individuals have a test that allows making a very early diagnosis before seroconversion occurs, which takes between 2 and 6 weeks, and the viral load has become another predictor of disease progression and a surrogate marker for antiretroviral drugs efficacy. This surrogate marker is the most used outcome in clinical trials for new drugs and therapeutic regimens. Today, the determination of viral load has become a routine diagnosis, and maintaining an undetectable viral load, below the limit of detection of HIV RNA 20–75 copies/mL, is one of the standards of care for antiretroviral therapy [49, 50].

Patients on antiretroviral therapy with good adherence, who have repeated detectable viral load in peripheral blood, are considered a treatment failure. In these cases, virus resistance has to be ruled out using more sophisticated assays, such as genotypic drug-resistance testing. These assays determined drug-resistance mutations in patient isolated viral genomes by PCR amplification of known mutations that are associated with antiviral resistance [51].

2.1.3 First Medical Interventions: Management of Opportunistic Infections

OIs were the principal cause of morbidity and mortality in the HIV population in the 1980s. At the time of the first report of AIDS, clinicians could only treat OIs with limited success. These OIs are more frequent and more severe because of immunosuppression in HIV-infected people. In the 1990s, before the HAART (highly active antiretroviral therapy) era, the strategies used to treat and prevent OIs, such as chemoprophylaxis, contributed to improved quality of life and improved survival (e.g., *Pneumocystis jiroveci* pneumonia prophylaxis alone increased survival to 0.26 person/year when HIV disease remained untreated) [52]. Subsequently, the widespread use of HAART has had the most profound influence on reducing OI-related mortality in HIV-infected persons [53–58].

At present, despite the global availability of HAART, OIs continue to be a cause of higher AIDS-related morbidity and mortality. The "burden of OIs" occurs mostly for the following reasons: because people are unaware of their HIV infection, because they are aware of their HIV infection but have no access to HAART due to psychosocial or economic factors, and because patients on HAART are unsuccessful due to poor adherence, unfavorable pharmacokinetics, or still unexplained biological factors [59–62].

OIs can have global and regional distribution. In Latin America or countries with Latin American migration flows, chronic *T. cruzi* infection may manifest as a disease reactivation. Chagas disease has been considered an AIDS-defining OI since 2004 by the Brazilian Society of Tropical Medicine [63]. The reactivation of *T. cruzi* in the HIV/AIDS model has serious clinical manifestations such as meningoencephalitis and myocarditis with high morbidity and mortality.

2.1.4 The Intervention That Changed the History of AIDS: Highly Active Antiretroviral Therapy

In a pandemic situation with a high-mortality viral infection such as AIDS, all the information obtained from research was applied to the affected population. Even today, the scientific work in the field of HIV is constantly providing new knowledge that leads to continuous changes in diagnosis and treatment, which are quickly applied to clinical care when evidence quality is high. The strategies for the development of knowledge include the initiative of the National Institute of Allergy and Infectious Diseases (NIAID) and National Institutes of Health (NIH), which founded the AIDS Clinical Trials Group (ACTG) in 1986. The ACTG has supported clinical research with high-quality standards [64]. Originally starting with 14 AIDS treatment evaluation units in the United States and growing to 62 sites at present in the United States, Canada, India, Thailand, South Africa, Kenya, Tanzania, Haiti, Malawi, Botswana, Brazil, Uganda, and Peru, the ACTG is the largest network of clinical trial therapeutic units in the world. The first step in HIV antiviral therapy was taken in 1987, when the ACTG Executive Committee conducted the ACTG019 study, a clinical trial showing that azidothymidine (AZT, also known as zidovudine) decreased OIs and mortality in AIDS patients. Then, administration of treatment was considered before HIV infection progressed to an advanced stage of AIDS with OIs [65, 66]. AZT, which had been originally synthesized as an anticancer treatment, was repositioned because it was also found to block the reverse transcription step of the HIV-1 life cycle [67]. However, viral resistance quickly developed, and therefore, new drugs had to be developed on the basis of insights into the HIV replication cycle and how to target it. After that, in 1994, the ACTG 076/ANRS 024 showed that ante- and intrapartum AZT treatment to the pregnant mother significantly reduced the risk of HIV acquisition by the newborn. This proved that antivirals could be effective in the prevention of mother-to-child HIV transmission [68].

In 1995, David Ho proposed that "when safe and effective treatment regimens are defined, the time will be right to test the concept of ablative treatment in HIV-1 infection. To stack the deck in favor of success, we should exert maximal antiviral pressure (using the optimal regimen) on the virus when it is most homogeneous—during the initial phase of infection. We should bear in mind the lessons learned from the treatment of tuberculosis and childhood leukemia, in which monotherapy resulted in transient responses that were quickly followed by relapses. It was aggressive combination chemotherapy at the outset that led to cures. Optimistically, we can hope that such an approach will become possible in patients infected with HIV-1" [69].

Indeed, the introduction of a combination of several drugs to limit the development of resistance was a major breakthrough. Studies showed that the introduction of a protease inhibitor (PI) or non-nucleoside reverse transcriptase inhibitors (NNRTIs) alongside two nucleoside reverse transcriptase inhibitors (NRTIs) in combination with antiretroviral therapy markedly reduced morbidity and mortality

[70–75]. In the late 1990s, early initial trials tested combinations of drugs aimed at controlling or preventing resistance, leading to the highly active antiretroviral therapy, or HAART, which is used today.

The availability of HAART and its universal distribution, including developing countries, is considered to be one of the greatest advances in modern medicine, despite concerns about toxicity, adherence, and development of viral resistance. These concerns cannot deny the great reduction in morbidity and mortality associated with HIV infection, resulting from the widespread introduction, in the mid-1990s, of triple therapy, which had a great impact in terms of time and quality of survival. The availability of safe and effective HAART for millions of individuals in places with limited resources strengthened its impact on the reduction of morbidity and mortality related to HIV infection.

2.1.5 HAART Improvement, Better Immune Profile, Decrease of Opportunistic Diseases

Clinical research continues to improve the therapeutic options available, with the aim of successfully controlling viral replication with minimal side effects and manageable treatment regimens [76].

HAART has steadily improved since the advent of potent combination therapy in 1996. It has dramatically reduced HIV-associated morbidity and mortality and has transformed HIV infection into a manageable chronic condition. In addition, HAART is highly effective in preventing HIV transmission [77]. However, only 55% of people with HIV in the United States have suppressed viral loads, mostly resulting from undiagnosed HIV infection and failure to link or retain diagnosed patients in care [78].

2.1.6 Continuous Training on New Knowledge and Strategies

Although HAART improves the quality of life of HIV patients enormously, additional strategies continue to be developed as prophylaxis. There are two very valuable studies that were successful in the prevention of newly infected people in the population with risk behavior.

In 2010, the initiative for pre-exposure prophylaxis trial probed that a once-daily oral pill combination of two antiretrovirals—emtricitabine/tenofovir disoproxil fumarate—decreased HIV acquisition by 44% in men or transgender women who have sex with men. Again, risk reduction strongly correlated with adherence [79].

In 2011, a study applied the concept of using effective HAART to make the viral load undetectable, thereby limiting infection via sexual transmission to an HIV-negative partner. The study showed one of the most significant findings in HIV prevention in serodiscordant couples, couples in which one partner is HIV-positive and the other is HIV-negative, 96% reduction in the transmission rate [77].

There is still a need of clinical research in generating evidence to address unanswered questions related to the optimal safety and efficacy of HAART, and the development of new protocols and clinical research with high-quality standards is still encouraged [51].

Despite a declining incidence rate, the AIDS epidemic remains uncontrolled, particularly in the most-at-risk populations. Secondly, although treatment provides successful viral control, HIV remains persistent and has not been eradicated, and testing and treatment remain insufficiently implemented for many people in resource-limited settings. Thirdly, HAART does not fully restore the immune system and, consequently, health. In addition, it is associated with many side effects, and it remains a lifelong commitment. Furthermore, AIDS is associated with chronic inflammation and immune activation, which probably results in immunosenescence and aging, as well as coinfections that frequently lead to complications. Finally, an effective vaccine against HIV is still elusive [80–83].

The short and intense evolution of knowledge about HIV infection has been, and continues to be, a constant challenge. Collaborative work with high standards of quality and ethical commitment has been promoted and established as a global humanitarian need. All this scientific information, at times overwhelming, needs to be used quickly and consensually with the commitment of scientific societies and governmental and nongovernmental organizations, as well as the participation of patients. Healthcare guidelines must be constantly updated, and health providers can quickly adapt to paradigm changes and improve healthcare practice for the benefit of the patient. There has been a radical change in morbidity and mortality, as well as the quality of life and biopsychosocial factors, of HIV/AIDS patients, but we still need to move forward and improve our knowledge and medical practice [84].

2.2 *Current Epidemiological Situation of HIV and* T. cruzi *Coinfection in the World and Argentina*

The evolution of knowledge about HIV infection in nearly the last 40 years was described above in order to contextualize the implications of a simultaneous *T. cruzi* and HIV infection. The relevance of knowing about the simultaneous infection with both microorganisms lies mainly, as in any model of severe immunocompromise, in the risk of *T. cruzi* reactivation.

Coinfected patients, especially those severely immunocompromised, may have *T. cruzi* reactivation, which is considered an AIDS-defining OI, and, as such, it has high mortality. Early diagnosis and treatment are central to lower mortality in small cohorts of patients with AIDS and reactivation of Chagas disease. Conditions in Latin America are fit for HIV and *T. cruzi* coinfection.

HIV is a pandemic, and 1.8 million people are estimated to be infected in the region. The infection with *T. cruzi* is estimated to be affecting six million people, and it is mostly concentrated in Latin America because of the biological cycle of *T. cruzi* and the vector that constitutes the main route of transmission and spread of the parasitosis [38, 85].

Both infections circulate in different population niches—the retrovirus in big cities and the parasite in rural areas—but the peasant urbanization that began in the mid-twentieth century continued over the years, and more and more rural people infected with *T. cruzi* went to live in large cities [86]. In addition, both infections share certain transmission mechanisms (mainly blood transfusion and congenital transmission), which is why HIV/*T. cruzi* coinfection is not uncommon. This coinfection is not limited to Latin America, since Chagas disease has been detected in people around the world. The HIV/*T. cruzi* coinfection has already been reported in Spain, Italy, the United Kingdom, and the United States [2, 87–92]. The two countries with the highest recorded prevalence of Chagas disease in the world are Argentina and Brazil. In Argentina, there are approximately 1,500,000 people infected with *T. cruzi* and 122,000 are infected with HIV [86, 93]. Assuming independence from risk factors, coinfected people in Argentina would be approximately 4575. Under HIV care guidelines in Argentina, a screening for *T. cruzi* infection is required in the initial control of HIV-infected individuals, but there is little current data on prevalence of both simultaneous infections in Argentina [94]. There are cohort publications that report 2.9–8% prevalence of *T. cruzi* infection in patients infected with HIV [95–98].

In Brazil, a reference center for HIV/AIDS patients verified that the frequency of coinfection with *T. cruzi* was of 1.3% [99]. Based on this study, it is estimated that, out of all HIV-infected individuals, approximately 8354 people have the coinfection in Brazil [100, 101]. However, the frequency of this association remains unestablished, since the overall prevalence of Chagas disease in Brazil has decreased dramatically due to vector and transfusion control. This reduced the prevalence in people under 14 years old to less than 0.13% at the end of the last century [102].

The overlapping of HIV and *T. cruzi* infection does not only occur in endemic areas but also in non-endemic areas, such as Europe, where the diagnosis may be even more difficult due to low diagnostic suspicion [100]. In Spain, 14.8% of all new diagnoses of HIV infection during 2004–2008 involved Latin American people. The confirmed prevalence of *T. cruzi* infection in Spain is 1.9%, and 3.9% of Latin American immigrants are HIV-positive [101, 103, 104].

2.3 *Coinfection with HIV and* T. cruzi

As mentioned above, HIV/AIDS care guidelines in the two countries with the highest number of Chagas cases in the world (Argentina and Brazil) require a screening for concomitant infections, including Chagas disease, during the first HIV visit [94, 105]. However, compliance with this requirement is inconsistent, even in reference centers. Only half of the patients in a HIV/AIDS reference center in Argentina were screened for *T. cruzi*, and the diagnosis test for Chagas disease was requested only in 3.2% of the cases in Brazil [97, 106]. This data from the reference center demonstrates extremely poor adherence to protocols by healthcare providers. Therefore, it is necessary to emphasize the importance of diagnosing Chagas disease at any time, particularly in people affected by HIV/AIDS in our region.

2.3.1 Chronic Infection with *T. cruzi*

As discussed above, chronic *T. cruzi* infection should be ruled out in HIV-infected patients from Latin America and in HIV patients with epidemiological risk for *T. cruzi* infection, even in non-endemic countries. The diagnosis of chronic *T. cruzi* infection (with or without HIV/AIDS infection) is based on the positive result of at least two different serological tests (ELISA, indirect immunofluorescence or indirect hemagglutination, etc.) that detect specific antibodies in patient serum according to PAHO recommendations [107]. However, serology could be negative in immunosuppressed patients. The serological tests for *T. cruzi* in HIV/AIDS patients can have low sensitivity, and false-negative results have been detected [108]. This phenomenon was observed in HIV/AIDS cohorts from Brazil where 21.2% of patients were discordant or nonreactive, but positive parasitological data was drawn from xenodiagnosis/blood culture or direct exam. Similar observations occur in Argentina on few cohorts of patients with Chagas disease reactivation who were diagnosed by visualization of trypomastigotes in blood, CSF, cardiac and pleural effusion, and *T. cruzi* amastigotes in tissue biopsies, and some of them, between 2.5% and 14%, have nonreactive serologies [109–112].

The importance of knowing about Chagas disease serology first of all is being aware that a patient can have or develop myocardiopathy or digestive manifestations by Chagas disease, in addition to the HIV/AIDS infection. Therefore, together with lifelong follow-up for the HIV infection, as well as other potential, associated infections (e.g., HCV, HBV, HPV, etc.), and HAART, routine follow-up for early diagnosis of heart disease and digestive disorders is important, as recommended for any individual suffering from Chagas disease. Early diagnosis of any heart or digestive illness will allow for intervention to control it, thus reducing the risk of sudden death and improving the patient's quality of life. There is limited data about heart or digestive complaint progression related to chronic Chagas disease in HIV patients. According to a review of literature published in 2015, out of 120 cases of chronic coinfection, 51 presented with no complaints, 37.5 suffered heart disease, five had digestive disorders, and 6.5 had a combination of both [108]. Therefore, follow-up recommendations are similar to those for HIV-negative individuals [100]. Another unique characteristic of coinfection is that vertical transmission of Chagas rate is higher in HIV-infected women than in immunocompetent mothers. One study described a rate of congenital *T. cruzi* transmission as high as 75%. Concomitant vertical transmission of both organisms has also been described, and symptoms, especially neurologic symptoms, are more likely to develop on the newborn infected [113–115].

In addition to diagnosis and clinical examination for heart disease or digestive disorder, prescription of trypanocidal therapy should also be assessed. During this phase, it is considered an interesting strategy, which will be discussed in the treatment section of this chapter.

2.3.2 Reactivation of Chagas Disease

The risk of *T. cruzi* reactivation in patients with HIV ranges from 11% to 35% according to limited information [99, 110, 116]. Even in the absence of symptoms, patients with chronic Chagas disease who are HIV coinfected have significantly higher levels of *T. cruzi* parasitemia than their immunocompetent counterparts [117, 118]. Most cases of clinically apparent reactivation occur in patients with low CD4 T lymphocyte cell counts, other opportunistic infections, or both [110, 111, 116, 119]. Almost all cases of symptomatic Chagas disease in HIV-coinfected patients represent a reactivation. Reactivation is not universal, even among those with very low CD4 lymphocyte cell counts and with other immunosuppression conditions, such as neoplasia, and/or those who are on systemic glucocorticoids (unpublished data).

Reactivation of Chagas disease may present with severe manifestations, atypical clinical forms, and high associated morbidity and mortality. The clinical manifestations and differential diagnosis of *T. cruzi* infection in HIV/AIDS hosts are similar to many other OIs.

Reactivation of Chagas disease is defined as the occurrence of any form of acute-phase clinical manifestations in chronic-phase carrier individuals, mainly with confirmed parasitemia in direct parasitological exams [116]. It is also defined as the situation in which *T. cruzi* is found in any organic liquid or any tissue in areas where the parasite is not usually found [108].

Chagas disease reactivation is an AIDS-defining event. The first cases where reactivation of Chagas disease was suspected in the context of HIV/AIDS were reported in Brazil in 1988. The first published cases with direct parasite view in LCR occurred in Brazil in 1989 and with amastigote nests in brain necropsy in Argentina in 1990 [120–122].

The most common manifestation of *T. cruzi* reactivation is meningoencephalitis, with or without brain abscesses (also denominated *chagomas*) [99, 111, 116, 117, 123]. The differential diagnosis for AIDS patients with central nervous system (CNS) symptoms and mass lesions on imaging may be with other OIs, such as neurotoxoplasmosis, lymphoma, progressive multifocal leucoencephalopathy, and tuberculosis. CNS symptoms without mass lesion should rule out other OIs that usually manifest with meningitis such as cryptococcosis, tuberculosis, CMV, VZV, HHV-6, HSV, and HIV itself. With so many options for differential diagnosis, the suspicion of Chagas disease reactivation and its diagnosis is a real challenge for physicians and health system effectors.

The second organ that is most frequently affected by Chagas reactivation in HIV-infected patients is the heart and pericardium. An acute myocarditis and/or pericarditis could develop, sometimes superimposed on pre-existing chronic Chagas heart disease [123]. Patients may present with new arrhythmias, pericardial effusion, and hemodynamic imbalance [119].

Cases of pleural effusion, cervicitis, peritonitis, chronic diarrhea, or skin involvement have been occasionally reported. Furthermore, although reactivation is the most characteristic type of involvement, asymptomatic and pauci-symptomatic forms with parasites in peripheral blood smears have also been found [115, 124–127].

Mortality is higher (79–100%) when the CNS is involved, and it is around 19% when only the heart is affected [109, 111, 116, 128, 129].

Diagnosis of Chagas Disease Reactivation

Necrosis and hemorrhage in tissue and abundant parasites can be observed in patients with meningoencephalitis and neuroimaging lesions. It has been suggested that Chagas disease mainly affects the white matter and subcortical region, unlike toxoplasmic encephalitis, in which the affected areas are the basal ganglia and the cortex. However, radiology—CT or MRI brain scan—does not confirm the diagnosis. This must be based on parasitological tests [130, 131]. Radiology reveals single or multiple tumor-like lesions, often showing a mass effect and edema, and ring uptake of contrast. The absence of lesions on a computed tomography scan does not rule out CNS involvement, because it may be normal in 17% of patients [110, 111].

A conclusive diagnosis of reactivation is established by identification of the parasite or its products in tissue, such as brain biopsy, CSF or blood [116]. In the absence of reactivation, circulating parasites are rarely detected microscopically in immunocompetent patients with chronic Chagas disease or HIV-coinfected patients [130]. If observed in an HIV-infected patient, circulating parasites suggest reactivation, and immediate treatment is necessary.

In Chagas disease reactivation, about 75% of diagnoses are established by detection of the trypomastigotes using direct microscopy in blood, CSF, or other body fluids (pleural, ascitic, and pericardial fluid). Histopathological diagnosis based on biopsy specimens or autopsy findings was often the result of many other differential diagnoses, late diagnosis, or rapid disease progression [111, 112, 116, 122, 124–126].

One of the methods to detect trypomastigotes in blood is the Strout test; it is a concentration method with higher sensitivity than a drop smear. In this test 5–10 mL of venous blood without heparin is collected. After spontaneous retraction of the coagulum, the serum is separated and centrifuged at low speed (90–160 g). The supernatant is then transferred to another tube and centrifuged (600–900 g). The sediment is observed under a light microscope for motile *T. cruzi* parasites [132].

A negative result for *T. cruzi* in the blood using direct techniques should not rule out a diagnosis of reactivation. In these patients, the parasite must be sought in other body fluids (CSF, peritoneal and pericardial fluid, or pleural effusion), if possible, or through biopsy, especially if other concomitant opportunistic infections are suspected. The CSF in coinfected patients with CNS involvement contains the parasite in more than 80% of cases [111, 116, 133]. If encephalitis is suspected and there are no contraindications for lumbar puncture, CSF examination provides a diagnostic opportunity.

Systemic corticosteroids for the treatment of other opportunistic diseases, such as *Pneumocystis jiroveci* pneumonia, or even as anti-edema for CNS mass lesions, can trigger a reactivation [111]. Therefore, the possibility of reactivation should be suspected as an intercurrent disease in coinfected patients receiving corticosteroids [134].

As serology can be negative in cases of reactivation in HIV individuals with risk factors for *T. cruzi* infection, even with no reactive serology for *T. cruzi*, Chagas disease reactivation should be suspected, and close follow-up is recommended to allow early detection of possible reactivation episodes.

Parasite detection by PCR or blood culture should not be considered evidence of reactivation, since a positive result can be yielded during the chronic phase, even in immunocompetent patients. However, real-time PCR can detect increased parasitemia in serial blood specimens before it can be detected using direct parasitological methods or before the patient shows clinical symptoms. Consequently, early testing could be useful for preemptive therapy, an early treatment strategy usually used in transplant patients before the onset of symptoms, in coinfected patients [118, 135, 136].

Limited data exists on PCR of CSF, but it could have high sensitivity for the diagnosis of reactivation in the CNS [24, 137].

It is clear that the gold standard for the diagnosis of Chagas disease reactivation is the visualization of trypomastigotes in the blood or any corporal fluid or the observation of amastigotes nests in a tissue biopsy with acute inflammatory infiltrate. However, this could not be achieved on many occasions. Therefore, strong suspicion with the help of *T. cruzi* PCR should lead to a probable diagnosis of Chagas disease reactivation, allowing for prompt treatment with antitrypanosomal drugs.

Treatment

Reactivation in an HIV-coinfected patient should be treated immediately with a standard course of trypanocidal treatment and optimization of antiretroviral therapy. Mortality is high even with medication, although early initiation of parasiticides will improve the prognosis [99, 111, 116, 138–141]. A delay in the diagnosis of reactivation is not unusual. This occurs because other differential diagnoses are much more frequent than Chagas disease reactivation itself and *T. cruzi* reactivation diagnosis depends on suspicion and the ability of the microbiologist team.

Benznidazole (5–8 mg/kg/day for 30–60 days) is the most commonly recommended initial treatment. Nifurtimox (8–10 mg/kg/day administered for 90–120 days) is an alternative. The duration of therapy with either of these agents has not been studied in HIV-coinfected patients [134, 142]. Both trypanocidal drugs are associated with significant toxicities. Data on tolerability in HIV-coinfected patients is scarce, although up to 50% of patients can have adverse reactions [111]. Benznidazole causes rash, increased liver enzymes, granulocytopenia, and peripheral neuropathy. Nifurtimox causes anorexia, abdominal pain, weight loss, restlessness, tremors, and peripheral neuropathy. The adverse effects of both drugs wane when the drugs are discontinued, except for peripheral neuropathy, which can persist for several months.

Even though there is little information on posaconazole, itraconazole, fluconazole, and ketoconazole, these drugs have been used in some cases where trypanocidal drugs were not an option because of toxicity [140, 143]. Nevertheless, information is too scarce to make recommendations for their use. However, if therapy fails or severe adverse effects are observed with benznidazole or nifurtimox, posaconazole is currently an alternative for immunosuppressed patients as the best option in the azole family but very inferior to benznidazole [144].

Pregnancy. Two cases of Chagas disease treatment with benznidazole during pregnancy have been reported. One report was about an acute infection with continued treatment during the first weeks of a subsequently diagnosed pregnancy, with normal infant outcome. The other one was about treatment in a severely immunosuppressed HIV-infected woman with Chagas disease encephalitis during the third trimester of pregnancy. The infant was small for gestational age but otherwise healthy and without evidence of *T. cruzi* infection [145, 146].

2.4 Prevention of Chagas Disease Reactivation

2.4.1 HAART

Initiation or optimization of HAART may prevent Chagas disease reactivation in chronic HIV-coinfected patients and recurrence after a reactivation event. If a patient has CNS Chagas disease, HAART should be started as soon as acute symptoms are stable [134]. Once HAART is started in a *T. cruzi*-coinfected patient, the theoretical possibility of immune restoration syndrome should be considered. Even though there is limited information, one report describes immune reconstitution inflammatory syndrome with skin manifestation and no CNS events [127].

Drug Interactions: HAART and Trypanocidal Drugs

There is no available data on the interaction between benznidazole and nifurtimox and antiretrovirals, but for some authors, little potential for pharmacokinetic interaction is expected because trypanocidal drugs undergo NADPH-dependent nitro reduction transformations and have a different metabolic pathway compared with nearly all antiretrovirals. However, these combinations require caution when coadministered [147].

Drug Interactions: HAART and Heart Disease Drugs

It is well known that many antiretroviral drugs modulate cytochrome P450 pathways, the most common drug-metabolizer liver enzymes, so it is important to check drug interactions between HAART and other therapies. For example, amiodarone is a very common anti-arrhythmogenic drug used in Chagas disease to treat ventricular arrhythmia, but amiodarone interacts with the protease inhibitor (PI) family, so if a patient needs amiodarone, the existence of PI in his/her HAART regimen should be checked. If PI is present, this should probably be switched to an integrase inhibitor antiretroviral.

2.4.2 Secondary Prophylaxis

A few number of cases of patients with CNS reactivation suggest that those who respond to a treatment regimen may benefit from subsequent secondary prophylaxis with a lower dose of benznidazole (5 mg/kg or 200 mg thrice weekly). The recommended doses and duration of prophylaxis, according to expert opinions, are based on other OIs and consist in maintaining the secondary prophylaxis until achieve of virologic control and CD4 counts of 200–250 cells/mm^3 in at least two determinations [148, 149].

2.4.3 Primary Prophylaxis

Even the primary prophylaxis, as it is usually called in others OIs where patients have CD4 lymphocytes count below 200, 100, or 50 cells/ml, consists in continued treatment with lower doses of standard treatment form some OIs. The standard treatment for Chagas disease, for 60 days, may produce parasitological clearance and possibly reduce the risk of reactivation, so this could be called and function as a primary prevention measure for reactivation.

2.4.4 Preemptive Therapy

As it is suggested that parasitemia may increase prior to reactivation and clinical manifestations, a preemptive therapy strategy could apply. Like in the model of chronic infection due to CMV or Chagas disease in transplant patients, periodic follow-up could be performed using quantitative real-time PCR techniques, and when the parasitemia increases, trypanocidal therapy may be started before clinical symptoms onset and before trypanosomes can be detected in blood or fluids [118, 136]. However, it is not a validated strategy yet as data is scarce [135].

2.5 *Conclusions Present and Future*

Even though *T. cruzi* and HIV coinfection has been described since 1988, it has not been studied thoroughly. Mortality in Chagas reactivation is high, and it is usually associated with low CD4 lymphocyte counts, other OIs, or both. In spite of currently available HIV knowledge and tools, poor immune status can occur because patients are unaware of their HIV infection, because they are aware of their HIV infection but are not on HAART for various reasons, or because they are on HAART but with poor adherence, virologic failure, or other biological interferences without an effective control of HIV infection.

The only two drugs used in Chagas disease treatment—benznidazole and nifurtimox—are effective in reducing parasitemia, are mandatory in any case of Chagas disease reactivation, and also raise the possibility of administering regimens that could lower the risk of *T. cruzi* infection and reactivation. However, in spite of all these possibilities, the unknown of concomitant infection HIV-*T. cruzi* due to poor adherence to existing clinical practice guidelines limits the possibilities of better attention to those patients.

Clearly, in the event of suspected Chagas disease reactivation, it is important to proceed with all the necessary interventions and exams to make the etiological diagnosis and start treatment, and if suspicion is still high but the diagnosis cannot be confirmed, empirical trypanocidal therapy may also be started because of the high mortality rate.

References

1. Kuschnir E, Sgammini H, Castro R, Evequoz C, Ledesma R, Brunetto J. Evaluation of cardiac function by radioisotopic angiography, in patients with chronic Chagas cardiopathy. Arq Bras Cardiol. 1985;45(4):249–56.
2. Schmunis GA, Yadon ZE. Chagas disease: a Latin American health problem becoming a world health problem. Acta Trop. 2010;115(1–2):14–21.
3. Fiorelli AI, Santos RHB, Oliveira JL Jr, Lourenço-Filho DD, Dias RR, Oliveira AS, et al. Heart transplantation in 107 cases of Chagas' disease. Transplant Proc. 2011;43(1):220–4.
4. Martinez-Perez A, Norman FF, Monge-Maillo B, Perez-Molina JA, Lopez-Velez R. An approach to the management of Trypanosoma cruzi infection (Chagas' disease) in immunocompromised patients. Expert Rev Anti-Infect Ther. 2014;12(3):357–73.
5. Pinazo MJ, Espinosa G, Cortes-Lletget C, Posada EJ, Aldasoro E, Oliveira I, et al. Immunosuppression and Chagas disease: a management challenge. PLoS Negl Trop Dis. 2013;7(1):e1965. https://doi.org/10.1371/journal.pntd.0001965.
6. Bacal F, Bocchi EA. Trasplante cardíaco para la enfermedad de Chagas. Rev Insuf Cardiac. 2008;3(2):88–90.
7. Jordán R, Valledor A. Guías de recomendaciones de prevención de infecciones en pacientes que reciben modificadores de la respuesta biológica. Rev Arg Reumatol. 2014;25(2):08–26.
8. González-Ramos J, Alonso-Pacheco ML, Mora-Rillo M, Herranz-Pinto P. Need to screen for Chagas disease and Strongyloides infestation in non-endemic countries prior to treatment with biologics. Actas Dermosifiliogr. 2017;108(4):373–5. https://doi.org/10.1016/j.ad.2016.10.013.30.
9. Navarrete-Dechent C, Majerson D, Torres M, Armijo D, Patel M, Menter A, et al. Use of tumor necrosis factor alpha (TNF α) antagonists in a patient with psoriasis and Chagas disease. Ann Bras Dermatol. 2015;90(3 Suppl 1):S171–4. https://doi.org/10.1590/abd1806-4841.20153538.
10. Garzón MI, Sánchez AG, Goy MC, Alvarellos T, Zarate AH, Basquiera AL, et al. Reactivation of Chagas disease in a patient with follicular lymphoma diagnosed by means of quantitative real-time polymerase chain reaction. Open Forum Infect Dis. 2015;30(2):ofv060. https://doi.org/10.1093/ofid/ofv060.
11. Fontes Rezende RE, Lescano MA, Zambelli Ramalho LN, de Castro Figueiredo JF, Oliveira Dantas R, Garzella Meneghelli U, et al. Reactivation of Chagas' disease in a patient with non-Hodgkin's lymphoma: gastric, oesophageal and laryngeal involvement. Trans R Soc Trop Med Hyg. 2006;100(1):74–8.
12. Vicco MH, César LI, Musacchio HM, Bar DO, Marcipar IS, Bottasso OA. Chagas disease reactivation in a patient non-Hodgkin's lymphoma. Rev Clin Esp. 2014;214(7):e83–5. https://doi.org/10.1016/j.rce.2014.04.006.
13. Altclas J, Sinagra A, Dictar M, Luna C, Verón MT, de Rissio AM, et al. Chagas disease in bone marrow transplantation: an approach to preemptive therapy. Bone Marrow Transplant. 2005;36:123–9. https://doi.org/10.1038/sj.bmt.1705006.
14. Dictar M, Sinagra A, Verón MT, Luna C, Dengra C, de Rissio A, et al. Recipients and donors of bone marrow transplants suffering from Chagas' disease: management and preemptive

therapy of parasitemia. Bone Marrow Transplant. 1998;21:391–3. https://doi.org/10.1038/sj.bmt.1701107.
15. Guiang KMU, Cantey P, Montgomery SP, Ailawadhi S, Qvarnstrom Y, Price T, et al. Reactivation of Chagas disease in a bone marrow transplant patient: case report and review of screening and management. Transpl Infect Dis. 2013;15(6):E264–7.
16. WHO. Consultation on international biological reference. Geneva: WHO; 2007.
17. Strout R. A method for concentrating hemoflagellate. J Parasitol. 1962;48:100–1.
18. Ministerio de Salud de la Nación. Argentina. Guías Para La Atencion Al Paciente Infectado Con Trypanosoma cruzi Enfermedad De Chagas. Agosto de 2012. Revisión Noviembre 2011 – Julio 2012 (Resolución Ministerial 1337/14).
19. Chin-Honga PV, Schwartz BS, Bern C, Montgomery SP, Kontak S, Kubak B, et al. Screening and treatment of Chagas disease in organ transplant recipients in the United States: recommendations from the Chagas in transplant working group. Am J Transplant. 2011;11:672–80.
20. The Chagas' Disease Argentine Collaborative Transplant Consortium et al. Chagas' disease and solid organ transplantation. Transplant Proc. 2010;42:3354–9.
21. Schijman AG, Vigliano C, Burgos J, Favaloro R, Perrone S, Laguens R, et al. Early diagnosis of recurrence of Trypanosoma cruzi infection by polymerase chain reaction after heart transplantation of a chronic chagas heart disease patient. Heart Lung Transplant. 2000;19(11):1114–7.
22. Diez M, Favaloro L, Bertolotti A, Burgos JM, Vigliano C, Lastra MP, et al. Usefulness of PCR strategies for early diagnosis of Chagas' disease reactivation and treatment follow-up in heart transplantation. Am J Transplant. 2007;7:1633–40.
23. Schijman AG, Bisio M, Orellana L, Sued M, Duffy T, Mejia Jaramillo AM, et al. International study to evaluate PCR methods for detection of Trypanosoma cruzi DNA in blood samples from Chagas disease patients. PLoS Negl Trop Dis. 2011;5(1):e931. https://doi.org/10.1371/journal.pntd.0000931.
24. Qvarnstrom Y, Schijman AG, Veron V, Aznar C, Steurer F, da Silva AJ. Sensitive and specific detection of Trypanosoma cruzi DNA in clinical specimens using a multi-target real-time PCR approach. PLoS Negl Trop Dis. 2012;6(7):e1689. https://doi.org/10.1371/journal.pntd.0001689.
25. Cura CI, Lattes R, Nagel C, Gimenez MJ, Blanes M, Calabuig E, et al. Early molecular diagnosis after transplantation with donors from Trypanosoma cruzi-infected donors. Am J Transplant. 2013;13(12):3253–61.
26. Ramírez JC, Cura CI, da Cruz MO, Lages-Silva E, Juiz N, Velázquez E, et al. Analytical validation of quantitative real-time PCR methods for quantification of Trypanosoma cruzi DNA in blood samples from Chagas disease patients. J Mol Diagn. 2015;17(5):605–15. https://doi.org/10.1016/j.jmoldx.2015.04.010.
27. Bocchi EA, Bellotti G, Mocelin AO, Uip D, Bacal F, Higuchi ML, et al. Heart transplantation for chronic Chagas' heart disease. Ann Thorac Surg. 1996;61:1727–33.
28. Bacal F, Pereira Silva C, Bocchi EA, Vieira Pires P, Moreira LFP, Issa S, et al. Mychophenolate mofetil increased Chagas disease reactivation in heart transplanted patients: comparison between two different protocols. Am J Transplant. 2005;5:2017–21.
29. Riarte A, Luna C, Sabatiello R, Sinagra A, Schiavelli R, de Rissio A, et al. Chagas' disease in patients with kidney transplants: 7 years of experience 1989–1996. Clin Infect Dis. 1999;29:561–7.
30. Cicora F, Escurra V, Silguero S, Gonzalez IM, Roberti JE. Use of kidneys from Trypanosoma cruzi infected donors in naive transplant recipients without prophylactic therapy: the experience in a high-risk area. Transplantation. 2014;97(1):e3–4. https://doi.org/10.1097/01.TP.0000437673.86339.82.
31. Huprikar S, Bosserman E, Patel G, Moore A, Pinney S, Anyanwu A, et al. Donor-derived Trypanosoma cruzi infection in solid organ recipients in the United States, 2001-2011. Am J Transplant. 2013;3:2418–25.
32. Barcan L, Luna C, Clara L, Sinagra A, Valledor A, de Rissio AM, et al. Liver transplantation from deceased donors serologically positive for Chagas disease. Liver Transpl. 2005;11:1112–6.
33. Albuquerque LACD, Gonzalez AM, Filho HLVN, Copstein JLM, Larrea FIS, Mansero JMP, et al. Liver transplantation from deceased donors serologically positive for Chagas disease. Am J Transplant. 2007;7:680–4.

34. Salvador F, Len O, Molina I, Sulleiro E, Sauleda S, Bilbao I, et al. Safety of liver transplantation with Chagas disease-seropositive donors for seronegative recipients. Liver Transpl. 2011;17:1304–8.
35. McCormack L, Quinonez E, Goldaracena N, Anders M, Rodrıguez V, Orozco F, et al. Liver transplantation using Chagas-infected donors in uninfected recipients: a single-center experience without prophylactic therapy. Am J Transplant. 2012;12:2832–7.
36. Nishioka SA. Benznidazol na quimioprofilaxia primária da reativação de doença de Chagas em chagásicos crônicos em uso de corticosteróides em dose imunodepressoras: há evidência suficiente para a recomendação do seu uso? Rev Soc Bras Med Trop. 2000;33(1):83–5.
37. Lattes R, Lasala MB. Chagas disease in the immunosuppressed patient. Clin Microbiol Infect. 2014;20(4):300–9. https://doi.org/10.1111/1469-0691.12585.
38. UNAIDS. World AIDS Day report, 2017. Geneva: UNAIDS; 2017. http://www.unaids.org/en/resources/fact-sheet. Accessed 18 Feb 2018
39. CDC. Pneumocystis pneumonia—Los Angeles. MMWR. 1981;30(21). https://www.cdc.gov/mmwr/hiv_aids20.html
40. Gottlieb MS, Schroff R, Schanker HM, Weisman JD, Fan PT, Wolf RA, et al. Pneumocystis carinii pneumonia and mucosal candidiasis in previously healthy homosexual men: evidence of a new acquired cellular immunodeficiency. N Engl J Med. 1981;305(24):1425–31.
41. Barré-Sinoussi F, Chermann JC, Rey F, Nugeyre MT, Chamaret S, Gruest J, et al. Isolation of a T-lymphotropic retrovirus from a patient at risk for acquired immune deficiency syndrome (AIDS). Science. 1983;220(4599):868–71.
42. Brun-Vezinet F, Rouzioux C, Barre-Sinoussi F, Klatzmann D, Saimot AG, Rozenbaum W, et al. Detection of IgG antibodies to lymphadenopathy-associated virus in patients with AIDS or lymphadenopathy syndrome. Lancet. 1984;1(8389):1253–6.
43. Dalgleish AG, Beverley PC, Clapham PR, Crawford DH, Greaves MF, Weiss RA. The CD4 (T4) antigen is an essential component of the receptor for the AIDS retrovirus. Nature. 1984;312(5996):763–7.
44. Klatzmann D, Champagne E, Chamaret S, Gruest J, Guetard D, Hercend T, et al. T lymphocyte T4 molecule behaves as the receptor for human retrovirus LAV. Nature. 1984;312(5996):767–8.
45. Maddon PJ, Dalgleish AG, McDougal JS, Clapham PR, Weiss RA, Axel R. The T4 gene encodes the AIDS virus receptor and is expressed in the immune system and the brain. Cell. 1986;47(3):333–48.
46. Fahey JL, Prince H, Weaver M, Groopman J, Visscher B, Schwartz K, et al. Quantitative changes in T helper or T suppressor/cytotoxic lymphocyte subsets that distinguish acquired immune deficiency syndrome from other immune subset disorders. Am J Med. 1984;76(1):95–100.
47. Piatak M Jr, Saag MS, Yang LC, Clark SJ, Kappes JC, Luk KC, et al. High levels of HIV-1 in plasma during all stages of infection determined by competitive PCR. Science. 1993;259(5102):1749–54.
48. Pantaleo G, Graziosi C, Demarest JF, Butini L, Montroni M, Fox CH, et al. HIV infection is active and progressive in lymphoid tissue during the clinically latent stage of disease. Nature. 1993;362(6418):355–8.
49. Mellors JW, Rinaldo CR Jr, Gupta P, White RM, Todd JA, Kingsley LA. Prognosis in HIV-1 infection predicted by the quantity of virus in plasma. Science. 1996;272(5265):1167–70.
50. Murray JS, Elashoff MR, Iacono-Connors LC, Cvetkovich TA, Struble KA. The use of plasma HIV RNA as a study endpoint in efficacy trials of antiretroviral drugs. AIDS. 1999;13(7):797–804.
51. Panel on Antiretroviral Guidelines for Adults and Adolescents. Guidelines for the use of antiretroviral agents in adults and adolescents living with HIV. Department of Health and Human Services. http://www.aidsinfo.nih.gov/ContentFiles/AdultandAdolescent. Accessed 18 Feb 2018.
52. Walensky RP, Paltiel AD, Losina E, Mercincavage LM, Schackman BR, Sax PE, et al. The survival benefits of AIDS treatment in the United States. J Infect Dis. 2006;194(1):11–9.
53. Palella FJ Jr, Delaney KM, Moorman AC, Loveless MO, Fuhrer J, Satten GA, et al. Declining morbidity and mortality among patients with advanced human immunodeficiency virus infection. HIV outpatient Study Investigators. N Engl J Med. 1998;338(13):853–60.

54. Detels R, Muñoz A, McFarlane G, Kingsley LA, Margolick JB, Giorgi J, et al. Effectiveness of potent antiretroviral therapy on time to AIDS and death in men with known HIV infection duration. Multicenter AIDS Cohort Study Investigators. JAMA. 1998;280(17):1497–503.
55. Jones JL, Hanson DL, Dworkin MS, Alderton DL, Fleming PL, Kaplan JE, et al. Surveillance for AIDS-defining opportunistic illnesses, 1992–1997. MMWR CDC Surveill Summ. 1999;48(2):1–22.
56. Mocroft A, Vella S, Benfield TL, Chiesi A, Miller V, Gargalianos P, et al. Changing patterns of mortality across Europe in patients infected with HIV-1. EuroSIDA Study Group. Lancet. 1998;352(9142):1725–30.
57. AD MN, Hanson DL, Jones JL, Dworkin MS, Ward JW. Effects of antiretroviral therapy and opportunistic illness primary chemoprophylaxis on survival after AIDS diagnosis Adult/Adolescent Spectrum of Disease Group. AIDS. 1999;13(13):1687–95.
58. Dore GJ, Li Y, McDonald A, Ree H, Kaldor JM, National HIV Surveillance Committee. Impact of highly active antiretroviral therapy on individual AIDS-defining illness incidence and survival in Australia. J Acquir Immune Defic Syndr. 2002;29(4):388–95.
59. Centers for Disease Control and Prevention. Vital signs: HIV prevention through care and treatment—United States. MMWR Morb Mortal Wkly Rep. 2011;60(47):1618–23.
60. Campsmith ML, Rhodes PH, Hall HI, Green TA. Undiagnosed HIV prevalence among adults and adolescents in the United States at the end of 2006. J Acquir Immune Defic Syndr. 2010;53(5):619–24. https://doi.org/10.1097/QAI.0b013e3181bf1c45.
61. Perbost I, Malafronte B, Pradier C, Santo LD, Dunais B, Counillon E, et al. In the era of highly active antiretroviral therapy, why are HIV-infected patients still admitted to hospital for an inaugural opportunistic infection? HIV Med. 2005;6(4):232–9.
62. Palacios R, Hidalgo A, Reina C, de la Torre M, Márquez M, Santos J. Effect of antiretroviral therapy on admissions of HIV-infected patients to an intensive care unit. HIV Med. 2006;7(3):193–6.
63. Ramos AN Jr. Inclusion of Chagas' disease reactivation as a condition for AIDS case definition to epidemiological surveillance in Brazil [Article in Portuguese]. Rev Soc Bras Med Trop. 2004;37(2):192–3.
64. Rubin R. Collaboration and conflict: looking back at the 30-year history of the AIDS clinical trials group. JAMA. 2015;314(24):2604–6. https://doi.org/10.1001/jama.2015.14848.
65. Fischl MA, Richman DD, Grieco MH, Gottlieb MS, Volberding PA, Laskin OL, et al. The efficacy of azidothymidine (AZT) in the treatment of patients with AIDS and AIDS-related complex. A double-blind, placebo-controlled trial. N Engl J Med. 1987;317(4):185–91.
66. Volberding PA, Lagakos SW, Koch MA, Pettinelli C, Myers MW, Booth DK, et al. Zidovudine in asymptomatic human immunodeficiency virus infection. A controlled trial in persons with fewer than 500 CD4-positive cells per cubic millimeter. The AIDS Clinical Trials Group of the National Institute of Allergy and Infectious Diseases. N Engl J Med. 1990;322(14):941–9.
67. Mitsuya H, Weinhold KJ, Furman PA, St Clair MH, Lehrman SN, Gallo RC, et al. 3′ Azido 3′ deoxythymidine (BW A509U): an antiviral agent that inhibits the infectivity and cytopathic effect of human T lymphotropic virus type III/lymphadenopathy-associated virus in vitro. Proc Natl Acad Sci U S A. 1985;82(20):7096–100.
68. Connor EM, Sperling RS, Gelber R, Kiselev P, Scott G, O'Sullivan MJ, et al. Reduction of maternal-infant transmission of human immunodeficiency virus type 1 with zidovudine treatment. N Engl J Med. 1994;331(18):1173–80.
69. Ho DD. Time to hit HIV, early and hard. N Engl J Med. 1995;333(7):450–1.
70. Hammer SM, Squires KE, Hughes MD, Grimes JM, Demeter LM, Currier JS, et al. A controlled trial of two nucleoside analogues plus indinavir in persons with human immunodeficiency virus infection and CD4 cell counts of 200 per cubic millimeter or less. N Engl J Med. 1997;337(11):725–33.
71. Ho DD, Neumann AU, Perelson AS, Chen W, Leonard JM, Markowitz M. Rapid turnover of plasma virions and CD4 lymphocytes in HIV 1 infection. Nature. 1995;373(6510):123–6.
72. Collier AC, Coombs RW, Schoenfeld DA, Bassett RL, Timpone J, Baruch A, et al. Treatment of human immunodeficiency virus infection with saquinavir, zidovudine and zalcitabine. N Engl J Med. 1996;334(16):1011–7.

73. Staszewski S, Morales-Ramirez J, Tashima KT, Rachlis A, Skiest D, Stanford J, et al. Efavirenz plus zidovudine and lamivudine, efavirenz plus indinavir plus zidovudine and lamivudine in the treatment of HIV-1 infection in adults. N Engl J Med. 1999;341(25):1865–73.
74. The Antiretroviral Therapy Cohort Collaboration. Life expectancy of individuals on combination antiretroviral therapy in high-income countries: a collaborative analysis of 14 cohort studies. Lancet. 2008;372(9635):293–9.
75. Egger M, Hirschel B, Francioli P, Sudre P, Wirz M, Flepp M, et al. Impact of new antiretroviral combination therapies in HIV infected patients in Switzerland: prospective multicentre study. BMJ. 1997;315(7117):1194–9.
76. Barré-Sinoussi F, Ross AL, Delfraissy JF. Past, present and future: 30 years of HIV research. Nat Rev Microbiol. 2013;11(12):877–83.
77. Cohen MS, Chen YQ, McCauley M, Gamble T, Hosseinipour MC, Kumarasamy N, et al. Prevention of HIV-1 infection with early antiretroviral therapy. N Engl J Med. 2011;365(6):493–505.
78. Centers for Disease Control and Prevention. Monitoring selected national HIV prevention and care objectives by using HIV surveillance data—United States and 6 dependent areas, 2014. HIV surveillance supplemental report 21. Atlanta, GA: Centers for Disease Control and Prevention. p. 2016. https://www.cdc.gov/hiv/pdf/library/reports/surveillance/cdc-hiv-surveillance-supplemental-reportvol-21-4.pdf
79. Van Damme L, Corneli A, Ahmed K, Agot K, Lombaard J, Kapiga S, et al. Preexposure prophylaxis for HIV infection among African women. N Engl J Med. 2012;367(5):411–22.
80. Chun TW, Finzi D, Margolick J, Chadwick K, Schwartz D, Siliciano RF. In vivo fate of HIV 1 infected T cells: quantitative analysis of the transition to stable latency. Nat Med. 1995;1(12):1284–90.
81. Chun TW, Stuyver L, Mizell SB, Ehler LA, Mican JA, Baseler M, et al. Presence of an inducible HIV 1 latent reservoir during highly active antiretroviral therapy. Proc Natl Acad Sci U S A. 1997;94(24):13193–7.
82. Finzi D, Hermankova M, Pierson T, Carruth LM, Buck C, Chaisson RE, et al. Identification of a reservoir for HIV 1 in patients on highly active antiretroviral therapy. Science. 1997;278(5341):1295–300.
83. Wong JK, Hezareh M, Günthard HF, Havlir DV, Ignacio CC, Spina CA, et al. Recovery of replication-competent HIV despite prolonged suppression of plasma viremia. Science. 1997;278(5341):1291–5.
84. van der Kuy AC, Bakker M, Jurriaans S, Back NK, Pasternak AO, Cornelissen M, et al. Translational HIV-1 research: from routine diagnostics to new virology insights in Amsterdam, the Netherlands during 1983–2013. Retrovirology. 2013;10:93. https://doi.org/10.1186/1742-4690-10-93.
85. World Health Organization. Chagas disease in Latin America: an epidemiological update based on 2010 estimates. Wkly Epidemiol Rec. 2015;90(6):33–43.
86. Coura JR, Dias JC. Epidemiology, control and surveillance of Chagas disease: 100 years after its discovery. Mem Inst Oswaldo Cruz. 2009;104(Suppl 1):31–40.
87. Schmunis GA. Epidemiology of Chagas disease in non-endemic countries: the role of international migration. Mem Inst Oswaldo Cruz. 2007;102(Suppl 1):75–85.
88. Balasso V, Almeida E, Molina I, Campis M, et al. A coinfeccao T cruzi/HIV em regioes nao endemicas para doenca de chagas. Libro: Epidemiologia e clinica da coinfeccao Trypanosoma cruzi e virus da imunofediciencia adquirida. Campinas, SP: Editorial Unicamp; 2015. p. 215–36.
89. Yasukawa K, Patel SM, Flash CA, Stager CE, Goodman JC, Woc-Colburn L. Trypanosoma cruzi meningoencephalitis in a patient with acquired immunodeficiency syndrome. Am J Trop Med Hyg. 2014;91(1):84–5.
90. Pinto A, Pett S, Jackson Y. Identifying Chagas disease in Australia: an emerging challenge for general practitioners. Aust Fam Physician. 2014;43(7):440–2.
91. Antinori S, Galimberti L, Bianco R, Grande R, Galli M, Corbellino M. Chagas disease in Europe: a review for the internist in the globalized world. Eur J Intern Med. 2017;43:6–15. https://doi.org/10.1016/j.ejim.2017.05.001.

92. Satoh F, Tachibana H, Hasegawa I, Osawa M. Case report sudden death caused by chronic Chagas disease in a non-endemic country: autopsy report. Pathol Int. 2010;60(3):235–40. https://doi.org/10.1111/j.1440-1827.2009.02503.x.
93. Dirección de Sida y ETS, Ministerio de Salud de la Nación, Argentina. Boletín sobre el VIH, SIDA e ITS en la Argentina n°34. 2017.
94. Benetucci J, Bissio E, Bologna R, Bruno M, et al. Guía para el manejo de los pacientes adultos con infección por VIH 2013. http://www.msal.gob.ar/images/stories/bes/graficos/0000000109cnt-2013-05_guia-manejo-pacientes-adultos.pdf
95. Benchetrit A, Andreani G, Avila MM, Rossi D, de Rissio AM, Weissenbacher M, et al. High HIV-Trypanosoma cruzi coinfection levels in vulnerable populations in Buenos Aires, Argentina. AIDS Res Hum Retroviruses. 2017;33(4):330–1.
96. Diez M, Nocito I, Frade RA, Cappello S, et al. Evidencia serologica de infecciones por citomegalovirus, hepatitis b y c, vírus de Ebstein Barr, Toxoplasma gondii, Trypanosoma cruzi y Treponema pallidum en infectados por HIV. Medicina (B Aires). 2001;61(4):376–80.
97. Dolcini G, Ambrosioni J, Andreani G, Pando MA, Martínez Peralta L, Benetucci J. Prevalencia de la coinfeccion virus e la inmunodeficiencia humana (VIH)-Trypanosoma cruzi e impacto del abuso de drogas inyectables en un centro de salud de la ciudad de Buenos Aires. Rev Argent Microbiol. 2008;40(3):164–6.
98. Arias L. Enfermedad de chagas: coinfección con VIH. Congreso Argentino de Parasitología 2012. Rev Arg Parasitol. 2012;1:117–8.
99. Almeida EA, Lima JN, Lages-Silva E, Guariento ME, Aoki FH, Torres-Morales AE, et al. Chagas' disease and HIV co-infection in patients without effective antiretroviral therapy: prevalence, clinical presentation and natural history. Trans R Soc Trop Med Hyg. 2010;104(7):447–52.
100. Pérez-Molina JA, Rodríguez-Guardado A, Soriano A, Pinazo MJ, Carrilero B, García-Rodríguez M, et al. Guidelines on the treatment of chronic coinfection by Trypanosoma cruzi and HIV Outside Endemic Areas. HIV Clin Trials. 2011;12(6):287–98.
101. Llenas-García J, Hernando A, Fiorante S, Maseda D, Matarranz M, Salto E, et al. Chagas disease screening among HIV-positive Latin American immigrants: an emerging problem. Eur J Clin Microbiol Infect Dis. 2012;31(8):1991–7.
102. Pinto Dias JC, Coura JR. Epidemiologia da doenca de Chagas. libro: epidemiologia e clinica da coinfeccao Trypanosoma cruzi e virus da imunofediciencia adquirida. Campinas, SP: Editorial Unicamp; 2015.
103. Salvador F, Molina I, Sulleiro E, Burgos J, Curran A, Van den Eynde E, et al. Tropical diseases screening in immigrant patients with human immunodeficiency virus infection in Spain. Am J Trop Med Hyg. 2013;88(6):1196–202.
104. Rodríguez-Guardado A, Asensi Alvarez V, Rodríguez Perez M, Mejuto Alvarez P, Flores-Chavez M, Alonso González P, et al. Screening for Chagas' disease in HIV-positive immigrants from endemic areas. Epidemiol Infect. 2011;139(4):539–43.
105. Departamento de Vigilância Prevenção e Controle das IST, do HIV/Aids e das Hepatites Virais (DIAHV), Ministerio da Saude, Brasil. Protocolo Clínico e Diretrizes Terapêuticas para Manejo da Infecção pelo HIV em Adultos. http://www.aids.gov.br/pt-br/pub/2013/protocolo-clinico-e-diretrizes-terapeuticas-para-manejo-da-infeccao-pelo-hiv-em-adultos
106. Stauffert D, da Silveira MF, Mesenburg MA, Gaspar T, Manta AB, Bicca GL, et al. Serological diagnosis of Chagas disease in HIV-infected patients. Rev Soc Bras Med Trop. 2015;48(3):331–3.
107. PAHO/WHO. Communicable diseases and health analysis neglected infectious diseases: Chagas disease. http://www.paho.org/hq/index.phpoption=com_content&view=article&id=5856&Itemid=41506&lang=es
108. Almeida E, Ramos A Jr, Correa D, Shikanai Yasuda MA. Coinfeccao T. cruzi/HIV/AIDS: revisão da literatura. Epidemiologia da doenca de Chagas. Epidemiologia e clinica da coinfeccao Trypanosoma cruzi e virus da imunodeficiencia adquirida. Campinas, SP: Editorial Unicamp; 2015.
109. Almeida EA, Ramos AN Jr, Correia D, Shikanai-Yasuda MA. Co infection Trypanosoma cruzi/HIV: systematic review (1980–2010). Rev Soc Bras Med Trop. 2011;44(6):762–70.
110. Benchetrit AG, Fernández M, Bava AJ, Corti M, Porteiro N, Martínez PL. Clinical and epidemiological features of chronic Trypanosoma cruzi infection in patients with HIV/AIDS in Buenos Aires, Argentina. Int J Infect Dis. 2018;67:118–21. https://doi.org/10.1016/j.ijid.2017.11.027.

111. Cordova E, Boschi A, Ambrosioni J, Cudos C, Corti M. Reactivation of Chagas disease with central nervous system involvement in HIV-infected patients in Argentina, 1992–2007. Int J Infect Dis. 2008;12(6):587–92.
112. Menghi RC, de NC G, Angeleri GA, Rocher CA, Palaoro IL. Trypanosoma cruzi: trypomastigote forms in pleural fluid of a patient with AIDS. Rev Chil Infectol. 2011;28(6):597–8.
113. Nisida IV, Amato Neto V, Braz LM, Duarte MI, Umezawa ES. A survey of congenital Chagas' disease, carried out at three health institutions in Sao Paulo City, Brazil. Rev Inst Med Trop Sao Paulo. 1999;41(5):305–11.
114. Scapellato PG, Bottaro EG, Rodriguez-Brieschke MT. Mother-child transmission of Chagas disease: could coinfection with human immunodeficiency virus increase the risk? Rev Soc Bras Med Trop. 2009;42(2):107–9.
115. Freilij H, Altcheh J, Muchinik G. Perinatal human immunodeficiency virus infection and congenital Chagas' disease. Pediatr Infect Dis J. 1995;14(2):161–2.
116. Sartori AM, Ibrahim KY, Nunes Westphalen EV, Braz LM, Oliveira OC Jr, Gakiya E, et al. Manifestations of Chagas disease (American trypanosomiasis) in patients with HIV/AIDS. Ann Trop Med Parasitol. 2007;101(1):31–50.
117. Sartori AM, Caiaffa-Filho HH, Bezerra RC, Guilherme SC, Lopes MH, Shikanai-Yasuda MA. Exacerbation of HIV viral load simultaneous with asymptomatic reactivation of chronic Chagas' disease. Am J Trop Med Hyg. 2002;67(5):521–3.
118. de Freitas VL, da Silva SC, Sartori AM, Bezerra RC, Westphalen EV, Molina TD. Real-time PCR in HIV/Trypanosoma cruzi coinfection with and without Chagas disease reactivation: association with HIV viral load and CD4 level. PLoS Negl Trop Dis. 2011;5(8):e1277. https://doi.org/10.1371/journal.pntd.0001277.
119. Sartori AM, Shikanai-Yasuda MA, Amato Neto V, Lopes MH. Follow-up of 18 patients with human immunodeficiency virus infection and chronic Chagas' disease, with reactivation of Chagas' disease causing cardiac disease in three patients. Clin Infect Dis. 1998;26(1):177–9.
120. Spina-França A, Livramento JA, Machado LR, Yasuda N. Trypanosoma cruzi antibodies in the cerebrospinal fluid: a search using complement fixation and immunofluorescence reactions. Arq Neuropsiquiatr. 1988;46(4):374–8.
121. Livramento JA, Machado LR, Spina-França A. Cerebrospinal fluid abnormalities in 170 cases of AIDS. Arq Neuropsiquiatr. 1989;47(3):326–31.
122. Del Castillo M, Mendoza G, Oviedo J, Perez Bianco RP, Anselmo AE, Silva M. AIDS and Chagas' disease with central nervous system tumor-like lesion. Am J Med. 1990;88(6):693–4.
123. Karp CL, Auwaerter PG. Coinfection with HIV and tropical infectious diseases. I Protozoal pathogens. Clin Infect Dis. 2007;45(9):1208–13.
124. Concetti H, Retegui M, Pérez G, Pérez H. Chagas disease of the cervix uteri in a patient with acquired immunodeficiency syndrome. Hum Pathol. 2000;31(1):120–2.
125. Iliovich E, Lopez R, Kum M, Uzandizaga G. Spontaneous chagasic peritonitis in a patient with AIDS. Medicina (B Aires). 1998;58(5 Pt 1):507–8.
126. Oelemann W, Velásquez JN, Carnevale S, Besasso H, Teixeira MG, Peralta JM. Intestinal Chagas disease in patients with AIDS. AIDS. 2000;14(8):1072–3.
127. Sartori AM, Sotto MN, Braz LM, Oliveira Júnior Oda C, Patzina RA, Barone AA, et al. Reactivation of Chagas disease manifested by skin lesions in a patient with AIDS. Trans R Soc Trop Med Hyg. 1999;93(6):631–2.
128. Diazgranados CA, Saavedra-Trujillo CH, Mantilla M, Valderrama SL, Alquichire C, Franco-Paredes C. Chagasic encephalitis in HIV patients: common presentation of an evolving epidemiological and clinical association. Lancet Infect Dis. 2009;9(5):324–30.
129. Lazo J, Meneses AC, Rocha A, Ferreira MS, Marquez JO, Chapadeiro E, et al. Chagasic meningoencephalitis in the immunodeficient. Arq Neuropsiquiatr. 1998;56(1):93–7.
130. Rocha A, de Meneses AC, da Silva AM, Ferreira MS, Nishioka SA, Burgarelli MK, et al. Pathology of patients with Chagas disease and acquired immunodeficiency syndrome. Am J Trop Med Hyg. 1994;50(3):261–8.
131. Lazo JE, Meneses AC, Rocha A, Frenkel JK, Marquez JO, Chapadeiro E, et al. Toxoplasmic and chagasic meningoencephalitis in patients with human immunodeficiency virus infection: anatomopathologic and tomographic differential diagnosis. Rev Soc Bras Med Trop. 1998;31(2):163–71.

132. Strout RG. A method for concentrating hemoflagellates. J Parasitol. 1962;48:100.
133. Lages-Silva E, Ramirez LE, Silva-Vergara ML, Chiari E. Chagasic meningoencephalitis in a patient with acquired immunodeficiency syndrome: diagnosis, follow-up, and genetic characterization of Trypanosoma cruzi. Clin Infect Dis. 2002;34(1):118–23.
134. Panel on Opportunistic Infections in HIV-Infected Adults and Adolescents. Guidelines for the prevention and treatment of opportunistic infections in HIV-infected adults and adolescents: recommendations from the Centers for Disease Control and Prevention, the National Institutes of Health, and the HIV Medicine Association of the Infectious Diseases Society of America. http://aidsinfo.nih.gov/contentfiles/lvguidelines/adult_oi.pdf. Accessed 18 Feb 2018.
135. Castro-Sesquen YE, Gilman RH, Mejia C, Clark DE, Choi J, Reimer-McAtee MJ, et al. Use of a Chagas Urine Nanoparticle Test (Chunap) to correlate with parasitemia levels in T. cruzi/HIV co-infected patients. PLoS Negl Trop Dis. 2016;10(2):e0004407. https://doi.org/10.1371/journal.pntd.0004407.
136. Cura CI, Lattes R, Nagel C, Gimenez MJ, Blanes M, Calabuig E, et al. Early molecular diagnosis of acute Chagas disease after transplantation with organs from Trypanosoma cruzi-infected donors. Am J Transplant. 2013;13(12):3253–61. https://doi.org/10.1111/ajt.12487.
137. Simioli F, Sánchez-Cunto M, Velázquez E, Lloveras S, Orduna T. Chagas disease in the central nervous system in patient infected with HIV: diagnostic and therapeutic difficulties. Rev Chil Infectol. 2017;34(1):62–6. https://doi.org/10.4067/S0716-10182017000100009.
138. Ferreira MS, Nishioka SA, Silvestre MT, Borges AS, Nunes-Araújo FR, Rocha A. Reactivation of Chagas' disease in patients with AIDS: report of three new cases and review of the literature. Clin Infect Dis. 1997;25(6):1397–400.
139. Ferreira MS, Nishioka S, Rocha A, Silva AM, Ferreira RG, Olivier W, et al. Acute fatal Trypanosoma cruzi meningoencephalitis in a human immunodeficiency virus-positive hemophiliac patient. Am J Trop Med Hyg. 1991;45(6):723–7.
140. Solari A, Saavedra H, Sepúlveda C, Oddó D, Acuña G, Labarca J, et al. Successful treatment of Trypanosoma cruzi encephalitis in a patient with hemophilia and AIDS. Clin Infect Dis. 1993;16(2):255–9.
141. Valerga M, Bases O, Martin M, Papucci T. Multifocal encephalitis in an AIDS patient. Enferm Infecc Microbiol Clin. 2005;23(9):569–70.
142. Panel de expertos de Grupo de Estudio del SIDA. Prevention of opportunistic infections in HIV-infected adolescents and adults guidelines. Recommendations of GESIDA/National AIDS Plan AIDS Study Group (GESIDA) and National AIDS Plan. Enferm Infecc Microbiol Clin. 2008;26(7):437–64.
143. de Almeida EA, Silva EL, Guariento ME, Aoki FH, Pedro RJ. Aetiological treatment with itraconazole or ketoconazole in individuals with Trypanosoma cruzi/HIV coinfection. Ann Trop Med Parasitol. 2009;103(6):471–6.
144. Pinazo MJ, Espinosa G, Gállego M, López-Chejade PL, Urbina JA, Gascón J. Successful treatment with posaconazole of a patient with chronic Chagas disease and systemic lupus erythematosus. Am J Trop Med Hyg. 2010;82(4):583–7.
145. Corrêa VR, Barbosa FG, Melo Junior CA, D'Albuquerque e Castro LF, Andrade HF Jr, Nascimento N. Uneventful benznidazole treatment of acute Chagas disease during pregnancy: a case report. Rev Soc Bras Med Trop. 2014;47(3):397–400.
146. Bisio M, Altcheh J, Lattner J, Moscatelli G, Fink V, Burgos JM, et al. Benznidazole treatment of chagasic encephalitis in pregnant woman with AIDS. Emerg Infect Dis. 2013;19(9):1490–2.
147. Seden K, Khoo S, Back D, Prevatt N, Lamorde M, Byakika-Kibwika P, et al. Drug-drug interactions between antiretrovirals and drugs used in the management of neglected tropical diseases: important considerations in the WHO 2020 Roadmap and London Declaration on Neglected Tropical Diseases. AIDS. 2013;27(5):675–86. https://doi.org/10.1097/QAD.0b013e32835ca9b4.
148. Almeida EA, Silva EL, Guariento ME, Souza ML, Aoki FH, Pedro RJ. Fatal evolution of Chagas' disease/AIDS co-infection: diagnostic difficulties between myocarditis reactivation and chronic chagasic myocardiopathy. Rev Soc Bras Med Trop. 2009;42(2):199–202.
149. Bern C, Montgomery SP, Herwaldt BL, Rassi A Jr, Marin-Neto JA, Dantas RO, et al. Evaluation and treatment of Chagas disease in the United States: a systematic review. JAMA. 2007;298(18):2171–81.

Part VI
Treatment

Clinical Pharmacology of Drugs for the Treatment of Chagas Disease

Facundo Garcia-Bournissen

Abstract Chagas disease is a parasitic disease endemic to Latin America that, if left untreated, can lead to severe long-term organ damage, particularly in the heart and gastrointestinal tract. Treatment with benznidazole or nifurtimox has been shown to be effective in the acute phase and during childhood, and limited evidence suggest a beneficial effect later in life too. However, both drugs require prolonged treatments (30–60 days) and are associated to significant adverse events that increase in severity and prevalence with age. Prompt pediatric diagnosis and treatment are vital for an effective use of these medications.

1 Introduction

1.1 Chagas Disease

Chagas disease, caused by infection with the parasite *Trypanosoma cruzi*, affects 8–15 million people in Latin America, leading to an estimated 10,000 annual deaths, mostly due to cardiac complications [1].

T. cruzi is transmitted mainly by an arthropod vector of the triatomine family, by congenital transmission, transfusions, organ transplants, or oral ingestion of contaminated foods. Congenital transmission in particular has recently become an important route of infection after significant decreases in zoonotic transmission by

F. Garcia-Bournissen (✉)
Servicio de Parasitología y Enfermedad de Chagas, Hospital de Niños "Ricardo Gutiérrez", Buenos Aires, Argentina

Instituto Multidisciplinario de Investigación en Patologías Pediátricas (IMIPP-GCBA), Consejo Nacional de Investigaciones Científicas y Técnicas (CONICET), Buenos Aires, Argentina

J. M. Altcheh, H. Freilij (eds.), *Chagas Disease*, Birkhäuser Advances in Infectious Diseases, https://doi.org/10.1007/978-3-030-00054-7_14

improved vector control in many areas of South America. Congenital transmission is also the main reason for acute Chagas disease in non-endemic countries, such as North America and Europe [2–4].

The acute *T. cruzi* infection takes place mostly in children either by parasite inoculation after contact with the vector in endemic areas or by congenital transmission from a chronically infected mother. The acute phase of the infection is usually asymptomatic but can be associated with fever, lymphadenopathies, and malaise and rarely (less than 5% of acute infections) to systemic disseminated infection and acute target organ failure (e.g., acute myocarditis, encephalitis, acute megacolon, etc.). The acute phase is followed by a chronic, mostly silent, infection that leads to progressive and irreversible damage in the heart and other organs in about 30% of infected patients, decades after the acute infection [5, 6].

1.2 Pharmacological Treatment for Chagas Disease

In spite of a relative abundance of preclinical molecular candidates and potential repositionable drugs [7–10], there are only two drugs currently in clinical use for the treatment of Chagas disease: benznidazole and nifurtimox. Clinical trials with azole medications such as posaconazole and the ravuconazole prodrug E1224 have so far yielded disappointing results [11], and evidence to support the efficacy and safety of drugs previously used in clinical cases such as allopurinol and itraconazole [7] is still deeply lacking [12, 13].

Benznidazole (Rochagan©, Radanil©, Abarax©) and nifurtimox (Lampit©) are nitroheteroerocyclic drugs developed over four decades ago by Roche and Bayer, respectively. Both drugs have in vitro and in vivo activity against *Trypanosoma* and *Leishmania* parasites. The parasitic activity of these drugs is believed to be secondary to the production of reactive metabolites in the parasite that lead to alkylation and oxidative damage of vital parasite structures such as DNA and RNA [14, 15]. Benznidazole and nifurtimox are considered prodrugs from the pharmacological point of view, as they need to be activated in the parasite in order to exert their effect. Activation is most probably carried out by the mitochondrial enzyme nitroreductase type 1 [16]. Protective *T. cruzi* mechanisms against nifurtimox or benznidazole action include detoxifying molecules such as trypanothione [17, 18], a vital part of the free radical scavenging cycle that is recycled by the enzyme trypanothione disulfide reductase.

Even though pharmacological treatment of Chagas disease is highly effective during the acute phase of the infection and in the chronic stage in children and young adults, available medications need to be administered for 30–60 days and are associated to toxicity, especially in adults [19–24]. Prevalence and degree of adverse drug reactions, which has been often used as an excuse to avoid treating patients, has been questioned recently [21], and current guidelines emphasize the need to provide all Chagas disease patients, especially children, with appropriate pharmacological treatment [25–28].

1.3 *Benznidazole*

1.3.1 Brief Recent History

Benznidazole (N-benzyl-2-nitro-1-imidazole-1-acetamide; CAS Number 22994-85-0) was developed by Roche (Ro 07-1051) [29]. There have been three producers of benznidazole so far: Roche (the original producer), Lafepe (public pharmaceutical company of the state of Pernambuco, Brazil), and Chemo (formerly Elea, an Argentine pharmaceutical company).

Roche manufactured and distributed the drug (as Radanil© or Rochagan©) from 1967 until the early 2000s, when production was discontinued mainly due to economic reasons [29]. However, encouraged by pressure from scientific and medical organizations, Roche eventually transferred benznidazole production technology and remaining stocks to Lafepe, a Pernambuco State (Brazil) pharmaceutical company, which committed to re-establish production and supply. Initially, Lafepe reportedly started producing benznidazole tablets (100 mg) using active product that had been produced by Roche and using the original formulation; these tablets continued to be distributed in many countries by Roche as Radanil©. Lafepe also developed a pediatric tablet (12.5 mg) that was tested in clinical pediatric study in Argentina (sponsored by Drugs for Neglected Diseases initiative; clinicaltrials.gov #NCT01549236). Following successful clinical trials, and pharmaceutical development, the pediatric formulation was approved by ANVISA (Brazilian Food and Drug Administration) in 2011. Unfortunately, to date, neither the 100 mg tablet nor the pediatric formulations produced by Lafepe are available in most countries. This lack of availability has been blamed on active product shortages associated with a diverse range of production problems [30, 31].

By the end of 2011, a third company, the Argentine pharmaceutical Elea (part of the Chemo group), disclosed advanced plans to develop benznidazole formulations. This company, one of the larger drug producers in Argentina, is associated with the Mundo Sano foundation, a charity that focuses on the treatment of neglected diseases, including Chagas disease, in underserviced areas of Argentina and neighboring countries. Chemo officially announced that it had produced the first batch of benznidazole tablets on March 2012 and that it had also received approval from the ANMAT (Argentine Food and Drug Administration) to start distributing the tablets for treatment in two formulations, a 100 mg tablet and a 50 mg tablet, prepared similar to the original Roche (and Lafepe) product. In spite of the early phase of development of the product, ANMAT gave authorization to Chemo to distribute the tablets due to a lack of alternatives and the public health importance of Chagas disease. Since then, Chemo has disclosed that it plans to complete the remaining regulatory steps commonly required to obtain marketing authorization for the drug, particularly in association to Chemo intentions to obtain access to the US market with benznidazole. FDA gave Chemo approval for their benznidazole application in August 2017 [30, 31].

1.3.2 Benznidazole Pharmacology

Benznidazole is a liposoluble drug with low solubility in water (400 mg/L), moderate solubility in ethanol, and no charge at physiological pH. Benznidazole has demonstrated efficacy against in vitro *T. cruzi* strains and in several in vivo animal models [10, 32–34].

Benznidazole is considered a prodrug, requiring activation by a parasite type 1 nitroreductase that reduces benznidazole and initiates a cascade of reactions leading to the formation of highly reactive drug metabolites. These metabolites bind to parasite macromolecules, disrupting *T. cruzi* metabolism and other vital functions, leading to parasite death [35].

Benznidazole is quickly absorbed from the human intestine after oral administration (Ka = 1.14/h), with a relatively rapid plasmatic peak (within 2–4 h of drug intake) [27, 36]. Absolute bioavailability in humans has never been formally estimated due to the absence of an intravenous formulation apt for human use; data from animal studies suggest that the drug is abundantly absorbed from the gut but seems to undergo first-pass elimination [37, 38]. The mechanisms associated to this pre-systemic elimination have not been investigated in detail to date, but limited evidence points to predominantly hepatic biotransformation and enterohepatic recirculation, possibly with some degree of enteric metabolism as well. However, very little research has been conducted in this area.

Benznidazole pharmacokinetics follows a one-compartment model with first-order absorption and elimination [27, 37]. Benznidazole distributes widely into tissues, including the central nervous system (CNS) [39], with a volume of distribution (V) of 0.6 L/Kg in adults and 0.7 L/Kg in children [27]. The drug reaches CNS concentrations close to 70% of those observed in plasma, which has allowed successful treatment of Chagas CNS infections in immunosuppressed patients with Chagas encephalitis and other CNS involvements [40]. Plasma protein binding of benznidazole is approximately 50% and is thus not expected to lead to significant interactions with other drugs [32].

Clearance of benznidazole is mainly by biotransformation (>80%) [38, 41, 42], currently believed to take place mostly in the liver probably by members of the cytochrome P450 (CYP) family and/or tissue nitroreductases. However, few studies to date have explored the details of the metabolic pathways responsible for benznidazole elimination. Approximately 6–20% of the drug can be found unchanged in urine, with differences depending on age of the patient (e.g., children seem to eliminate more unchanged drug in urine compared to adults); the rest of the drug has been observed as reduced and conjugated metabolites (Rocco D, Perez-Montilla C, Altcheh A, Garcia-Bournissen F, unpublished).

Mean half-life of benznidazole is 13 h in adults [36, 37] and significantly shorter in children, as observed in two prospective clinical trials (approximately 6 h for 2- to 7-year-old children) [27]. The significant differences present in clearance and half-lives between children and adults imply a large difference in average steady-state concentrations of the drug (i.e., children have approximately half the average steady-state benznidazole concentrations compared to adults). These differences do not, however, seem to impact efficacy of the drug as all children treated in the prospective clinical trial responded to the treatment despite lower plasma concentrations of the drug.

Observed differences in pharmacokinetics of benznidazole in children versus adults were put in evidence using a population pharmacokinetic design with sparse sampling in children 2–12 years of age with Chagas disease [27]. The results obtained from this study were compared to adult data obtained from early clinical studies performed by Roche [37]. Comparison among these data showed a progressive decrease in the clearance rate of benznidazole with increasing age (i.e., the older the patient, the slower the drug was eliminated). However, this information cannot identify the actual reason for these differences in drug elimination, which could be due to a slower drug metabolism in adults, to impaired drug absorption in younger children, or to other, yet undiscovered, mechanisms. Presently, research in the area is actively testing these hypotheses. Additionally, pharmacokinetics and response in teenagers and young adults have never been studied, and the assumption that it would be in between children and adults has never been confirmed. Currently, several research studies are expected to address these deficiencies, but several years may pass before specific knowledge obtained from this underserved population is known and applied into clinical practice.

The most commonly used BNZ dosing regimen, reported in the majority of the case series and clinical trials published to date [11, 19, 21–24, 27, 43–59], ranges from 5–8 mg/kg/day in two daily doses, for 30–60 days. No studies have been published to date formally studying or comparing alternative dosing schedules, even if some evidence points toward possible efficacy of lower doses for adults and teenagers, as well as a potential role for less frequent dosing, such as perhaps alternative-day dosing or even weekly dosing [27, 60]. It should be mentioned, though, that a small cohort study of adult patients treated with 5 mg/kg/day in two divided doses every 5 days showed good parasitological response of the patients but did not seem to reduce the incidence of adverse drug reactions [60].

Treatment of asymptomatic and early chronic-phase Chagas disease in adult patients with benznidazole has not been studied appropriately in a randomized double-blind study, but there is moderate evidence of therapeutic benefit obtained from cohort and historical controlled trials [47, 49]. Unfortunately, long-term follow-up of a sizable cohort of patients treated with benznidazole has not been carried out to date which precludes conclusions on the actual impact of treatment on cardiac and gastrointestinal complications in the long run. On the other hand, treatment of chronic-stage patients who already have developed advanced cardiomyopathy (NYHA class I, II, or III) has been shown by the BENEFIT study to be unlikely to produce a significant impact on prognosis or disease progression [59, 61]. BENEFIT study conclusions and methodology, however, have been recently challenged by one of the participating investigators [62].

The most commonly observed adverse drug reactions (ADRs) associated to benznidazole use include rash and pruritus (usually after 7–12 days of treatment), headache, myalgia, and gastrointestinal discomfort (in the first days of the treatment). Other, less common, ADRs include drug-associated hepatitis, leucopenia, peripheral neuropathy, and, in rare occasions, severe drug hypersensitivity (Stevens-Johnson syndrome and drug reaction with systemic symptoms). Even though the incidence of ADRs has not been formally compared in a head-to-head study enrolling children and adults, ADRs seem rare and almost universally mild in children and appear to gradually increase in frequency and severity after 7 years of age [20, 21, 23, 45, 56, 63]. ADRs are a frequent cause of treatment interruption and discontinuation in adults.

The underlying biological mechanisms for the observed ADRs have not been studied in depth to date, but the immune system seems to play an important role, particularly in the case of cutaneous rashes and hypersensitivity reactions. It is interesting to point out that the timing for the moderate cutaneous reactions (7–12 days after onset of treatment) mimics the time course of similar reactions associated to other unrelated medications such as fluoroquinolones and lamotrigine, suggesting common underlying immunological mechanisms. The observation of rare severe adverse reactions such as Stevens-Johnson syndrome and drug reactions with eosinophilia and systemic symptoms (DRESS) [56, 63], which have also been associated to those other medications, adds support for a common immunological trigger for these reactions and possibly a pharmacogenetic predisposition. However, the actual nature of these reactions remains currently unknown, and studies of potential pharmacogenomic markers are lacking.

Benznidazole has never been formally studied during pregnancy, but it is considered by many authors to be incompatible with pregnancy mainly due to the lack of safety data. Benznidazole has been shown in some in vitro tests to have the potential for mutagenesis, but these preclinical tests are markedly inaccurate for the prediction of teratogenic potential of drugs. At this point, it is impossible to say much about reproductive safety of this drug, other than the fact that there have been no reports of malformations or any other pregnancy complications even though it is very likely that some women were accidentally exposed to it in the first trimester (i.e., received treatment before realizing that they were pregnant). Also, some reports exist of treatment during late-stage pregnancy in emergency situations that did not result in any complications for the baby (but may have saved the mother's life) [40]. The main recommendation therefore remains to avoid benznidazole during the first trimester of pregnancy and throughout pregnancy whenever possible until further information becomes available. However, in case of an emergency (e.g., a pregnant woman with Chagas encephalitis), treatment should not be withheld with the excuse of an unproven teratogenic risk.

Benznidazole has been considered a contraindication during lactation since its initial development due to lack of data on safety and potential accumulation into breastmilk due to its lipophilicity. However, recent prospective studies and pharmacokinetic evaluations suggest that the risk of exposure to benznidazole from breastmilk for a breastfed baby is negligible, and lactation should not be considered a contraindication for Chagas disease treatment in those circumstances when treatment cannot be postponed [26].

1.4 Nifurtimox

1.4.1 Brief Recent History

(For a more comprehensive review of supply issues with nifurtimox, please see Jannin [29].)

Nifurtimox was manufactured by Bayer as Lampit© since the 1970s. However, its development started in the early 1960s, as Bayer-2502. Near the end of the 1990s, Bayer discontinued the production of Lampit© due to a perceived lack of demand and low profitability. However, as a consequence of clinical trials showing that the drug was highly effective when used in combination with eflornithine against sleeping sickness [64], and significant pressure by medical organizations such as Medecins Sans Frontieres, Bayer restarted nifurtimox production after constructing a model pharmaceutical plant in El Salvador and committed to donating all its production through the World Health Organization (WHO) for the treatment of sleeping sickness and Chagas disease. Since then, country-level access to the drug has depended on individual states' agreements and negotiations with WHO and Bayer and local bureaucratic and political decisions. Availability currently seems erratic in many South American countries due to these factors.

1.4.2 Nifurtimox Clinical Pharmacology

Many aspects of nifurtimox clinical pharmacology mirror those of benznidazole (see Table 1), in particular the lack of specific knowledge on many aspects of its pharmacokinetics, actual effectiveness, and metabolism. However, nifurtimox is currently undergoing extensive redevelopment by Bayer, most likely in order to apply for approval in Europe and North America, including several new clinical trials in children to confirm nifurtimox effectiveness and safety in this population (clinicaltrials.gov; https://clinicaltrials.gov/ct2/show/NCT02625974) [13].

Similar to benznidazole, nifurtimox is a hydrophobic, highly liposoluble, drug which distributes widely to tissues, including the central nervous system. Animal

Table 1 Comparison of pharmacokinetics and adverse events between benznidazole and nifurtimox

	Benznidazole	Nifurtimox
Absorption	Fast (peak ~3 h)	Fast (peak 2–4 h)
Clearance	Mostly metabolized Half-life 12 h in adults (6 h in children)	Mostly metabolized Half-life 3 h (similar in children, based on limited data)
Distribution	Widely to all tissues, included CNS	Widely to all tissues, included CNS
Adverse events	Rash and pruritus, headache, myalgia, and gastrointestinal discomfort. Rarely severe cutaneous and systemic hypersensitivity	Headache, anorexia, irritability, sleepiness, and other CNS signs and symptoms of myalgia. Rarely severe cutaneous and systemic hypersensitivity

studies have shown that absorption of the drug from the gut is rapid (Ka 0.8/h) and virtually complete but that it undergoes significant first-pass metabolism (much higher than benznidazole), leading to a small fraction of orally administered nifurtimox reaching the systemic circulation [65, 66]. Nifurtimox bioavailability in humans is not known due to the absence of an intravenous formulation but, as mentioned above, is expected to be relatively low based on animal studies and observed drug concentrations in humans.

Oral administration of nifurtimox reaches peak plasma concentrations after approximately 2–4 h [67–69]. The half-life of the drug is relatively short (approximately 3 h in adults, and similar in children, based on very limited data [70]). Elimination of nifurtimox is mostly hepatic and accounts for virtually all the clearance of the drug (i.e., unchanged elimination in urine is less than 1%) [69, 71]. Active metabolites have been suggested by isolated (animal) liver experiments [72], but this aspect has never been studied in humans. Data from animal studies also suggests that CYP enzymes are responsible for metabolism of the drug, but no human data is publicly available to date to confirm this suspicion [73]. Similar to benznidazole, nifurtimox plasma protein binding is approximately 50% and not expected to play a significant role in drug-drug interactions [74]. The observed (apparent) volume of distribution is high (V/F = 760 L), suggesting both an extensive distribution into tissues and also a significant pre-systemic elimination (i.e., a limited bioavailability), such as that observed in animal studies [65, 66]. Nifurtimox readily enters the CNS, which is a useful property both for the treatment of *T. cruzi* CNS infections and for the management of African trypanosomiasis. Nifurtimox is a substrate for the BCRP transporter [75, 76].

Presently neither the optimal dose nor the optimal duration of treatment with nifurtimox for Chagas disease is well defined. Initially, treatments tended to be long (90–120 days) [77] but were subsequently reduced to mimic benznidazole treatment spans (approximately 60 days) [7, 78–80]. Commonly used doses range from 8 to 15 mg/kg/day divided in three daily administrations. However, alternative dosing strategies, or doses, have not been tested in clinical trials, and therefore it is hard to conclude that this is the most appropriate (or safer) dosing schedule.

The most commonly observed ADRs are anorexia and weight loss, irritability, sleepiness, and other central nervous system signs and symptoms [80, 81]. Nifurtimox use is also associated with rash, pruritus, and drug-associated hepatitis but less frequently than benznidazole. Depression, peripheral neuropathy, and psychiatric symptoms have also been reported, less commonly. Similar to benznidazole, nifurtimox-associated ADRs seem much more common and severe in adults and are usually mild in children, including neonates [78, 82]. Notably, a case report suggests that patients who develop a severe drug reaction to benznidazole may still be treated safely with nifurtimox [83].

Similar to benznidazole, nifurtimox is considered contraindicated during pregnancy and lactation. Unfortunately, virtually no data is available on the safety of this drug during the first trimester of pregnancy, and therefore it is still advisable to avoid its use at this stage. However, published and ongoing studies on nifurtimox transfer into breastmilk strongly suggest that the drug is safe during breastfeeding, and treatment, if necessary, of a lactating mother should not be discouraged if needed [75].

Patients taking either nifurtimox or benznidazole are discouraged from drinking alcohol during the treatment [21]. The basis for this is the occasional observation that patients may develop disulfiram-like reactions (i.e., flushing, facial redness, hypotension, diarrhea, etc.) when concomitantly taking either benznidazole or nifurtimox and alcohol. Also, both drugs are structurally similar to disulfiram and other nitro drugs, and it is conceivable that they may inhibit aldehyde dehydrogenase and thus trigger the reaction. The actual prevalence of the reaction is unknown, but it is nonetheless advisable to avoid ingesting alcohol during the treatment.

1.4.3 Biomarker-Related Uncertainties

Appropriate evaluation of drug response requires the existence of unequivocal markers of therapeutic benefit. For example, therapeutic benefit of anticonvulsants is measured in the decrease or disappearance of seizures in the treated patient. Similarly, resolution of acute infections after antibiotic treatment is measured by disappearance of the infectious agent and resolution of the symptoms of inflammation. Unfortunately, certification of drug response in chronic diseases, such as Chagas disease, requires the prospective of large cohorts of treated patients followed for many decades, a task that is beyond the possibilities of most research groups at this time.

In the case of chronic Chagas disease, the evaluation of therapeutic drug response is complicated by several facts. First of all, not all patients develop cardiac or gastrointestinal involvement (approximately 30% are expected to develop organ damage decades after infection). Also, no biomarker has been shown to date to accurately correlate with the risk of target organ dysfunction, and the biomarkers commonly used to guide therapy do not fulfill the criteria required for a biomarker (i.e., none of them have been shown to correlate with actual outcome in the form of organ involvement). This situation is further complicated by the fact that several of these so-called biomarkers, anti-*T. cruzi* antibodies in particular, have been traditionally used to define success or failure of drug treatments (i.e., patients were considered to be "cured" when their anti-*T. cruzi* antibodies became undetectable, and they were considered to have failed treatment when anti-*T. cruzi* antibodies remained positive after a certain amount of time). However, these considerations fail to take into account how the immune system actually works (e.g., the longer the antigen, *T. cruzi* in this case, persists in the body, the more prolonged and sustained will be the immune response, including antibody production, against the antigen even if the parasite is not present anymore). Failure to incorporate these considerations into the evaluation of pharmacological therapy against Chagas disease has led to confusing perceptions of effectiveness (or lack thereof) of the available drugs, particularly in the chronic stage.

At this point, both nifurtimox and benznidazole are considered effective (or not) by most researchers solely based on the use of two main biomarker criteria after treatment: disappearance of the parasite from circulation (in those cases where this can be ascertained) and disappearance of anti-*T. cruzi*-specific antibodies from circulation. As mentioned before, both criteria have problems that have opened the door to significant discussion and discord, particularly because both criteria are subject to variability that depends on many things other than the treatment (e.g., age of the patient, reinfections, methods used to quantify the biomarker, interlaboratory variability, etc.) that have been poorly investigated, at best.

Unfortunately, the development of new and better biomarkers will require a benchmark to which compare them. Chagas disease research has yet to produce one.

2 Conclusion

Both nifurtimox and benznidazole are relatively safe drugs that are widely considered effective in the treatment of acute Chagas disease and in the early chronic stage (e.g., pediatric Chagas disease and early adult). Adverse events are usually mild, and less likely in children, and in general should not be an excuse for treating the disease. Many aspects of the pharmacotherapy of Chagas disease are still in the research area and will require new perspectives on the development and use of appropriate cure criteria.

References

1. Rassi A Jr, Rassi A, Marin-Neto JA. Chagas disease. Lancet. 2010;375(9723):1388–402. PubMed PMID: 20399979.
2. Schmunis GA. Epidemiology of Chagas disease in non-endemic countries: the role of international migration. Mem Inst Oswaldo Cruz. 2007;102(Suppl 1):75–85. PubMed PMID: 17891282.
3. Schmunis GA, Yadon ZE. Chagas disease: a Latin American health problem becoming a world health problem. Acta Trop. 2010;115(1–2):14–21. PubMed PMID: 19932071.
4. Carlier Y. Globalization of Chagas disease (American trypanosomiasis): the situation in Europe and Belgium. Bull Mem l'Acad R Med Belg. 2011;166(10–12):347–55. discussion 56–7. PubMed PMID: 23082500.
5. Teixeira AR, Nitz N, Guimaro MC, Gomes C, Santos-Buch CA. Chagas disease. Postgrad Med J. 2006;82(974):788–98. Pubmed Central PMCID: 2653922.
6. Teixeira AR, Nascimento RJ, Sturm NR. Evolution and pathology in Chagas disease—a review. Mem Inst Oswaldo Cruz. 2006;101(5):463–91. PubMed PMID: 17072450.
7. Apt W. Current and developing therapeutic agents in the treatment of Chagas disease. Drug Des Devel Ther. 2010;4:243–53. PubMed PMID: 20957215. Pubmed Central PMCID: 2948934.
8. Stoppani AOM. Quimioterapia de la enfermedad de Chagas. Medicina (B Aires). 1999;59(Suppl 2):147–65.
9. Buckner FS, Bahia MT, Suryadevara PK, White KL, Shackleford DM, Chennamaneni NK, et al. Pharmacological characterization, structural studies, and in vivo activities of anti-Chagas disease lead compounds derived from tipifarnib. Antimicrob Agents Chemother. 2012;56(9):4914–21. PubMed PMID: 22777048. Pubmed Central PMCID: 3421879.
10. Chatelain E. Chagas disease drug discovery: toward a new era. J Biomol Screen. 2015;20(1):22–35. PubMed PMID: 25245987.
11. Molina I, Gomez i, Prat J, Salvador F, Trevino B, Sulleiro E, Serre N, et al. Randomized trial of posaconazole and benznidazole for chronic Chagas' disease. N Engl J Med. 2014;370(20):1899–908. PubMed PMID: 24827034.
12. Perez-Molina JA, Molina I. Chagas disease. Lancet. 2018;391(10115):82–94. PubMed PMID: 28673423.

13. Sales Junior PA, Molina I, Fonseca Murta SM, Sanchez-Montalva A, Salvador F, Correa-Oliveira R, et al. Experimental and clinical treatment of Chagas disease: a review. Am J Trop Med Hyg. 2017;97(5):1289–303. PubMed PMID: 29016289.
14. Wilkinson SR, Bot C, Kelly JM, Hall BS. Trypanocidal activity of nitroaromatic prodrugs: current treatments and future perspectives. Curr Top Med Chem. 2011;11(16):2072–84. PubMed PMID: 21619510.
15. Hall BS, Bot C, Wilkinson SR. Nifurtimox activation by trypanosomal type I nitroreductases generates cytotoxic nitrile metabolites. J Biol Chem. 2011;286(15):13088–95. Pubmed Central PMCID: 3075655.
16. Wilkinson SR, Taylor MC, Horn D, Kelly JM, Cheeseman I. A mechanism for cross-resistance to nifurtimox and benznidazole in trypanosomes. Proc Natl Acad Sci U S A. 2008;105(13):5022–7. PubMed PMID: 18367671. Pubmed Central PMCID: 2278226.
17. Vazquez K, Paulino M, Salas CO, Zarate-Ramos JJ, Vera B, Rivera G. Trypanothione reductase: a target for the development of anti-Trypanosoma cruzi drugs. Mini Rev Med Chem. 2017;17(11):939–46. PubMed PMID: 28302040.
18. Paulino M, Iribarne F, Dubin M, Aguilera-Morales S, Tapia O, Stoppani AO. The chemotherapy of Chagas' disease: an overview. Mini Rev Med Chem. 2005;5(5):499–519. PubMed PMID: 15892691.
19. de Andrade AL, Zicker F, de Oliveira RM, Almeida Silva S, Luquetti A, Travassos LR, et al. Randomised trial of efficacy of benznidazole in treatment of early Trypanosoma cruzi infection. Lancet. 1996;348(9039):1407–13. PubMed PMID: 8937280.
20. Viotti R, Alarcon de Noya B, Araujo-Jorge T, Grijalva MJ, Guhl F, Lopez MC, et al. Towards a paradigm shift in the treatment of chronic Chagas disease. Antimicrob Agents Chemother. 2014;58(2):635–9. PubMed PMID: 24247135. Pubmed Central PMCID: 3910900.
21. Viotti R, Vigliano C, Lococo B, Alvarez MG, Petti M, Bertocchi G, et al. Side effects of benznidazole as treatment in chronic Chagas disease: fears and realities. Expert Rev Anti-Infect Ther. 2009;7(2):157–63. PubMed PMID: 19254164.
22. Sosa Estani S, Segura EL, Ruiz AM, Velazquez E, Porcel BM, Yampotis C. Efficacy of chemotherapy with benznidazole in children in the indeterminate phase of Chagas' disease. Am J Trop Med Hygiene. 1998;59(4):526–9. PubMed PMID: 9790423.
23. Altcheh J, Moscatelli G, Moroni S, Garcia-Bournissen F, Freilij H. Adverse events after the use of benznidazole in infants and children with Chagas disease. Pediatrics. 2011;127(1):e212–8. PubMed PMID: 21173000.
24. Marin-Neto JA, Rassi Júnior A, Mattos AC, Avezum Júnior A, Rassi A. The BENEFIT trial: testing the hypothesis that trypanocidal therapy is beneficial for patients with chronic Chagas heart disease. Mem Inst Oswaldo Cruz. 2009;104(Suppl 1):319–24.
25. Fernández ML, Marson ME, Ramirez JC, Mastrantonio G, Schijman AG, Altcheh J, et al. Pharmacokinetic and pharmacodynamic responses in adult patients with Chagas disease treated with a new formulation of benznidazole. Mem Inst Oswaldo Cruz. 2016;111(3):218–21.
26. Garcia-Bournissen F, Moroni S, Marson ME, Moscatelli G, Mastrantonio G, Bisio M, et al. Limited infant exposure to benznidazole through breast milk during maternal treatment for Chagas disease. Arch Dis Child. 2015;100(1):90–4. PubMed PMID: 25210104.
27. Altcheh J, Moscatelli G, Mastrantonio G, Moroni S, Giglio N, Marson ME, et al. Population pharmacokinetic study of benznidazole in pediatric Chagas disease suggests efficacy despite lower plasma concentrations than in adults. PLoS Negl Trop Dis. 2014;8(5):e2907. PubMed PMID: 24853169. Pubmed Central PMCID: 4031103.
28. Garcia-Bournissen F, Altcheh J, Giglio N, Mastrantonio G, Della Vedova CO, Koren G. Pediatric clinical pharmacology studies in Chagas disease: focus on Argentina. Paediatr Drugs. 2009;11(1):33–7. PubMed PMID: 19127950.
29. Jannin J, Villa L. An overview of Chagas disease treatment. Mem Inst Oswaldo Cruz. 2007;102(Suppl 1):95–7. PubMed PMID: 17906803.
30. Pinheiro E, Brum-Soares L, Reis R, Cubides J-C. Chagas disease: review of needs, neglect, and obstacles to treatment access in Latin America. Rev Soc Bras Med Trop. 2017;50(3):296–300.
31. Alpern JD, Lopez-Velez R, Stauffer WM. Access to benznidazole for Chagas disease in the United States-cautious optimism? PLoS Negl Trop Dis. 2017;11(9):e0005794. PubMed PMID: 28910299. Pubmed Central PMCID: 5598921.

32. Richle RW, Raaflaub J. Difference of effective antitrypanosomal dosages of benznidazole in mice and man. Chemotherapeutic and pharmacokinetic results. Acta Trop. 1980;37(3):257–61. PubMed PMID: 6106364.
33. Chatelain E, Ioset JR. Phenotypic screening approaches for Chagas disease drug discovery. Expert opinion on drug discovery. Expert Opin Drug Discov. 2018;13:141–53. PubMed PMID: 29235363.
34. Romanha AJ, Castro SL, Soeiro MNC, Lannes-Vieira J, Ribeiro I, Talvani A, et al. In vitro and in vivo experimental models for drug screening and development for Chagas disease. Mem Inst Oswaldo Cruz. 2010;105(2):233–8.
35. Hall BS, Wilkinson SR. Activation of benznidazole by trypanosomal type I nitroreductases results in glyoxal formation. Antimicrob Agents Chemother. 2012;56(1):115–23. PubMed PMID: 22037852. Pubmed Central PMCID: 3256028.
36. Roberts JT, Bleehen NM. Benznidazole with CCNU: a clinical phase I toxicity study. Int J Radiat Oncol Biol Phys. 1985;11(2):331–4. PubMed PMID: 3972652.
37. Raaflaub J. Multiple-dose kinetics of the trypanosomicide benznidazole in man. Arzneimittelforschung. 1980;30(12):2192–4. PubMed PMID: 6783051.
38. Workman P, White RA, Walton MI, Owen LN, Twentyman PR. Preclinical pharmacokinetics of benznidazole. Br J Cancer. 1984;50(3):291–303. PubMed PMID: 6466543. Pubmed Central PMCID: 1976805.
39. Roberts JT, Bleehen NM, Lee FY, Workman P, Walton MI. A phase I study of the combination of benznidazole and CCNU in man. Int J Radiat Oncol Biol Phys. 1984;10(9):1745–8. PubMed PMID: 6480457.
40. Bisio M, Altcheh J, Lattner J, Moscatelli G, Fink V, Burgos JM, et al. Benznidazole treatment of chagasic encephalitis in pregnant woman with AIDS. Emerg Infect Dis. 2013;19(9):1490–2. PubMed PMID: 23965334. Pubmed Central PMCID: 3810932.
41. Walton MI, Workman P. Nitroimidazole bioreductive metabolism. Quantitation and characterisation of mouse tissue benznidazole nitroreductases in vivo and in vitro. Biochem Pharmacol. 1987;36(6):887–96. PubMed PMID: 3105539.
42. Workman P, Walton MI, Lee FY. Benznidazole: nitroreduction and inhibition of cytochrome P-450 in chemosensitization of tumour response to cytotoxic drugs. Biochem Pharmacol. 1986;35(1):117–9. PubMed PMID: 3940522.
43. Streiger ML, del Barco ML, Fabbro DL, Arias ED, Amicone NA. [Longitudinal study and specific chemotherapy in children with chronic Chagas' disease, residing in a low endemicity area of Argentina]. Estudo longitudinal e quimioterapia especifica em criancas, com doenca de Chagas cronica, residentes em area de baixa endemicidade da Republica Argentina. Rev Soc Bras Med Trop. 2004;37(5):365–75.
44. Alvarez MG, Vigliano C, Lococo B, Bertocchi G, Viotti R. Prevention of congenital Chagas disease by Benznidazole treatment in reproductive-age women. An observational study. Acta Trop. 2017;174:149–52. PubMed PMID: 28720492.
45. Bertocchi GL, Vigliano CA, Lococo BG, Petti MA, Viotti RJ. Clinical characteristics and outcome of 107 adult patients with chronic Chagas disease and parasitological cure criteria. Trans R Soc Trop Med Hyg. 2013;107(6):372–6. PubMed PMID: 23612468.
46. Viotti R, Vigliano C, Lococo B, Bertocchi G, Alvarez M, Laucella S, et al. Tratamiento antiparasitario en la enfermedad de Chagas. Enferm Emerg. 2008;10(Suppl 1):10–3.
47. Viotti R, Vigliano C, Lococo B, Bertocchi G, Petti M, Alvarez MG, et al. Long-term cardiac outcomes of treating chronic Chagas disease with benznidazole versus no treatment: a nonrandomized trial. Ann Intern Med. 2006;144(10):724–34. PubMed PMID: 16702588.
48. Sosa-Estani S, Armenti A, Araujo G, Viotti R, Lococo B, Ruiz Vera B, et al. Tratamiento de Chagas con benznidazol y ácido tióctico. Medicina (Buenos Aires). 2004;64(1):1–6.
49. Viotti R, Vigliano C, Armenti H, Segura E. Treatment of chronic Chagas' disease with benznidazole: clinical and serologic evolution of patients with long-term follow-up. Am Heart J. 1994;127(1):151–62. PubMed PMID: 8273735.
50. Morillo CA, Waskin H, Sosa-Estani S, Del Carmen Bangher M, Cuneo C, Milesi R, et al. Benznidazole and posaconazole in eliminating parasites in asymptomatic *T. cruzi* carriers: the STOP-CHAGAS trial. J Am Coll Cardiol. 2017;69(8):939–47. PubMed PMID: 28231946.

51. Sguassero Y, Cuesta CB, Roberts KN, Hicks E, Comande D, Ciapponi A, et al. Course of chronic Trypanosoma cruzi Infection after treatment based on parasitological and serological tests: a systematic review of follow-up studies. PLoS One. 2015;10(10):e0139363. PubMed PMID: 26436678. Pubmed Central PMCID: 4593559.
52. Carlier Y, Torrico F, Sosa-Estani S, Russomando G, Luquetti A, Freilij H, et al. Congenital Chagas disease: recommendations for diagnosis, treatment and control of newborns, siblings and pregnant women. PLoS Negl Trop Dis. 2011;5(10):e1250. PubMed PMID: 22039554. Pubmed Central PMCID: 3201907.
53. Coura JR, de Abreu LL, Willcox HP, Petana W. [Comparative controlled study on the use of benznidazole, nifurtimox and placebo, in the chronic form of Chagas' disease, in a field area with interrupted transmission. I. Preliminary evaluation]. Estudo comparativo controloado com emprego de benznidazole, nifurtimox e placebo, na forma cronica da doenca de Chagas; em uma area de campo com transmissao interrompida. I. Avalicao preliminar. Rev Soc Bras Med Trop. 1997;30(2):139–44.
54. Gallerano RR, Sosa RR. Estudio de intervención en la evolución natural de la enfermedad de Chagas. Evaluación del tratamiento antiparasitario específico. Estudio retrospectivo-prospectivo de terapéutica antiparasitaria. Rev Fac Cien Med Univ Nac Cordoba. 2000;57(2):135–62.
55. de Castro AM, Luquetti AO, Rassi A, Chiari E, Galvao LM. Detection of parasitemia profiles by blood culture after treatment of human chronic Trypanosoma cruzi infection. Parasitol Res. 2006;99(4):379–83. PubMed PMID: 16570199.
56. Yun O, Lima MA, Ellman T, Chambi W, Castillo S, Flevaud L, et al. Feasibility, drug safety, and effectiveness of etiological treatment programs for Chagas disease in Honduras, Guatemala, and Bolivia: 10-year experience of Medecins Sans Frontieres. PLoS Negl Trop Dis. 2009;3(7):e488. PubMed PMID: 19582142. Pubmed Central PMCID: 2700957.
57. Molina I, Salvador F, Sanchez-Montalva A, Artaza MA, Moreno R, Perin L, et al. Pharmacokinetics of benznidazole in healthy volunteers and implications in future clinical trials. Antimicrob Agents Chemother. 2017;61(4):e01912–6. PubMed PMID: 28167552. Pubmed Central PMCID: 5365666.
58. Soy D, Aldasoro E, Guerrero L, Posada E, Serret N, Mejia T, et al. Population pharmacokinetics of benznidazole in adult patients with Chagas disease. Antimicrob Agents Chemother. 2015;59(6):3342–9. PubMed PMID: 25824212. Pubmed Central PMCID: 4432184.
59. Morillo CA, Marin-Neto JA, Avezum A, Sosa-Estani S, Rassi A Jr, Rosas F, et al. Randomized trial of benznidazole for chronic Chagas' cardiomyopathy. N Engl J Med. 2015;373(14):1295–306. PubMed PMID: 26323937.
60. Alvarez MG, Hernandez Y, Bertocchi G, Fernandez M, Lococo B, Ramirez JC, et al. New scheme of intermittent benznidazole administration in patients chronically infected with Trypanosoma cruzi: a pilot short-term follow-up study with adult patients. Antimicrob Agents Chemother. 2016;60(2):833–7. PubMed PMID: 26596935. Pubmed Central PMCID: 4750658.
61. Morillo CA, Marin-Neto JA, Avezum A. Benznidazole for chronic Chagas' cardiomyopathy. N Engl J Med. 2016;374(2):189–90. PubMed PMID: 26760092.
62. Rassi A Jr, Marin Neto JA, Rassi A. Chronic Chagas cardiomyopathy: a review of the main pathogenic mechanisms and the efficacy of aetiological treatment following the BENznidazole Evaluation for Interrupting Trypanosomiasis (BENEFIT) trial. Mem Inst Oswaldo Cruz. 2017;112(3):224–35.
63. Sperandio da Silva GM, Mediano MFF, Hasslocher-Moreno AM, Holanda MT, Silvestre de Sousa A, Sangenis LHC, et al. Benznidazole treatment safety: the Medecins Sans Frontieres experience in a large cohort of Bolivian patients with Chagas' disease. J Antimicrob Chemother. 2017;72(9):2596–601. PubMed PMID: 28645201.
64. Yun O, Priotto G, Tong J, Flevaud L, Chappuis F. NECT is next: implementing the new drug combination therapy for Trypanosoma brucei gambiense sleeping sickness. PLoS Negl Trop Dis. 2010;4(5):e720. PubMed PMID: 20520803. Pubmed Central PMCID: 2876135.
65. Duhm B, Maul W, Medenwald H, Patzschke K, Wegner LA. Investigations on the pharmacokinetics of nifurtimox-35 S in the rat and dog. Arzneimittelforschung. 1972;22(9):1617–24. PubMed PMID: 4630484.

66. Haberkorn A, Gonnert R. Animal experimental investigation into the activity of nifurtimox against Trypanosoma cruzi. Arzneimittelforschung. 1972;22(9):1570–82. PubMed PMID: 4630483.
67. Gonzalez-Martin G, Thambo S, Paulos C, Vasquez I, Paredes J. The pharmacokinetics of nifurtimox in chronic renal failure. Eur J Clin Pharmacol. 1992;42(6):671–3. PubMed PMID: 1623911.
68. Paulos C, Paredes J, Vasquez I, Kunze G, Gonzalez-Martin G. High-performance liquid chromatographic determination of nifurtimox in human serum. J Chromatogr. 1988;433:359–62. PubMed PMID: 3235567.
69. Paulos C, Paredes J, Vasquez I, Thambo S, Arancibia A, Gonzalez-Martin G. Pharmacokinetics of a nitrofuran compound, nifurtimox, in healthy volunteers. Int J Clin Pharmacol Ther Toxicol. 1989;27(9):454–7. PubMed PMID: 2807618.
70. Saulnier Sholler GL, Bergendahl GM, Brard L, Singh AP, Heath BW, Bingham PM, et al. A phase 1 study of nifurtimox in patients with relapsed/refractory neuroblastoma. J Pediatr Hematol Oncol. 2011;33(1):25–30. PubMed PMID: 21063221.
71. Medenwald H, Brandau K, Schlossmann K. Quantitative determination of nifurtimox in body fluids of rat, dog and man. Arzneimittel-Forschung. 1972;22(9):1613–7. PubMed PMID: 4678720.
72. Gonzalez-Martin G, Paulos C, Guevara A, Ponce G. Disposition of nifurtimox and metabolite activity against Trypanosoma cruzi using rat isolated perfused liver. J Pharm Pharmacol. 1994;46(5):356–9. PubMed PMID: 8083806.
73. Letelier ME, Izquierdo P, Godoy L, Lepe AM, Faundez M. Liver microsomal biotransformation of nitro-aryl drugs: mechanism for potential oxidative stress induction. J Appl Toxicol. 2004;24(6):519–25. PubMed PMID: 15558828.
74. Masana M, de Toranzo EG, Castro JA. Studies on nifurtimox nitroreductase activity in liver and other rat tissues. Arch Int Pharmacodynam Ther. 1984;270(1):4–10. PubMed PMID: 6497502.
75. Garcia-Bournissen F, Altcheh J, Panchaud A, Ito S. Is use of nifurtimox for the treatment of Chagas disease compatible with breast feeding? A population pharmacokinetics analysis. Arch Dis Child. 2010;95(3):224–8. PubMed PMID: 19948512.
76. Watson CP, Dogruel M, Mihoreanu L, Begley DJ, Weksler BB, Couraud PO, et al. The transport of nifurtimox, an anti-trypanosomal drug, in an in vitro model of the human blood-brain barrier: evidence for involvement of breast cancer resistance protein. Brain Res. 2012;1436:111–21. PubMed PMID: 22200378. Pubmed Central PMCID: 3281990.
77. Cancado JA, Salgado AA, Marra UD, Alvares JM, Machado JR. [Clinical therapeutic trial in chronic Chagas' disease using nifurtimox in 3 schedules of long duration]. Ensaio terapeutico cl'inico na doenca de Chagas cronica com o nifurtimox em tres esquemas de dura coa prolongada. Rev Inst Med Trop Sao Paulo. 1975;17(2):111–27.
78. Freilij H, Altcheh J. Congenital Chagas' disease: diagnostic and clinical aspects. Clin Infect Dis. 1995;21(3):551–5. PubMed PMID: 8527542.
79. Coura JR. [Current prospects of specific treatment of Chagas' disease] Perspectivas actuales del tratamiento especifico de la enfermedad de Chaga. Bol Chil Parasitol. 1996;51(3–4):69–75. PubMed PMID: 9302778.
80. Jackson Y, Alirol E, Getaz L, Wolff H, Combescure C, Chappuis F. Tolerance and safety of nifurtimox in patients with chronic Chagas disease. Clin Infect Dis. 2010;51(10):e69–75. PubMed PMID: 20932171.
81. Forsyth CJ, Hernandez S, Olmedo W, Abuhamidah A, Traina MI, Sanchez DR, et al. Safety profile of nifurtimox for treatment of Chagas disease in the United States. Clin Infect Dis. 2016;63(8):1056–62. PubMed PMID: 27432838. Pubmed Central PMCID: 5036918.
82. Alarcon de Noya B, Ruiz-Guevara R, Noya O, Castro J, Ossenkopp J, Diaz-Bello Z, et al. Long-term comparative pharmacovigilance of orally transmitted Chagas disease: first report. Expert Rev Anti-Infect Ther. 2017;15(3):319–25. PubMed PMID: 28132566.
83. Perez-Molina JA, Sojo-Dorado J, Norman F, Monge-Maillo B, Diaz-Menendez M, Albajar-Vinas P, et al. Nifurtimox therapy for Chagas disease does not cause hypersensitivity reactions in patients with such previous adverse reactions during benznidazole treatment. Acta Trop. 2013;127(2):101–4. PubMed PMID: 23583863.

In Vivo Drug Testing for Experimental *Trypanosoma cruzi* Infection

Julián Ernesto Nicolás Gulin

Abstract Experimental animals have contributed to basic and translational research on Chagas disease. Although many species and models have been used depending on the main experimental objective, there is a lack of uniformity and harmonization in preclinical in vivo studies. This chapter focuses on the description of relevant *T. cruzi*-infected animal models for drug discovery, and results are discussed in a translational research perspective, while some key concepts are presented in order to choose and establish a suitable animal model of *T. cruzi* infection to assess new chemotherapies' efficacy. Also, some strategies related to the 3Rs principles (replacement, reduction, and refinement) are proposed to apply on Chagas disease research to achieve scientific aims while ensuring animal well-being.

Animal models play a key role in various aspects of Chagas disease experimental research, helping researchers elucidate several issues related to disease physiopathology, immunology, diagnosis, drug evaluation, and vaccine development [1].

Dr. Carlos Chagas himself sent the first *T. cruzi* isolates to the Oswaldo Cruz Institute in Brazil to experimentally infect different animal species, including dogs, cats, monkeys, rabbits, guinea pigs, and other rodents, in order to reproduce the different phases of the disease [2].

The Chagas Disease Committee of the Training and Research Program on Parasitic Diseases suggested that a suitable animal model for the study of Chagas disease should meet the following characteristics [3]:

- Allow parasite isolation throughout the course of infection.
- Present positive serology as a marker of active infection.
- Exhibit the various and distinctive clinical presentations from chronic stage.
- Develop myocarditis, myositis, and other characteristic alterations of the disease.

J. E. N. Gulin (✉)
Servicio de Parasitología y Enfermedad de Chagas, Hospital de Niños "Ricardo Gutiérrez", Buenos Aires, Argentina

Instituto Multidisciplinario de Investigación en Patologías Pediátricas (IMIPP-GCBA), Consejo Nacional de Investigaciones Científicas y Técnicas (CONICET), Buenos Aires, Argentina
e-mail: jgulin@fvet.uba.ar

J. M. Altcheh, H. Freilij (eds.), *Chagas Disease*, Birkhäuser Advances in Infectious Diseases, https://doi.org/10.1007/978-3-030-00054-7_15

- Induce an immune response against host tissues.
- Be easy to maintain with affordable costs.

Considering the complex pathophysiology developed in Chagas disease, there may not be a single animal model that reproduces all aspects of the disease, and the choice of an animal model over another will depend on the working hypothesis, research objectives, and variables of interest, among other factors.

Several animal species have been employed as models for experimental *T. cruzi* infection: domestic dog (*Canis lupus familiaris*), rabbits (*Oryctolagus cuniculus*), hamsters (*Mesocricetus auratus*), guinea pigs (*Cavia porcellus*), rats (*Rattus norvegicus*), mice (*Mus musculus*), and primates from the Old and the New World (*Macaca mulatta* and *Cebus* and *Saimiri* and *Callithrix* spp., respectively).

The course of *T. cruzi* infection varies widely among laboratory animals, depending upon the host and parasite strains used, the route of inoculation, and the size of the inoculums [2, 4]. Each animal species develops different pathophysiological responses to *T. cruzi* infection, different from those that occur in the human host in some cases. For drug discovery purposes, mice, rats, and dogs are the most employed animals in recent publications.

The dog was suggested as a suitable model for experimental infection since it reproduces the acute and chronic phases of the disease. Early works reported that the development of the acute stage of the infection depends mainly on the parasite's characteristics and the host immunological modulation [5].

Subsequently, the canine model was employed to determine the therapeutic response to BZ, supporting the usefulness of this model for evaluation of trypanocidal activity of new compounds [6]. Later, the efficacy of the antifungal ravuconazole was assessed in experimentally infected dogs. The drug was able to eliminate bloodstream trypomastigotes, but not all the parasites lodged in the tissues of the host [7]. This response predicted the effect on subsequent clinical proof-of-concept trials carried out with the ravuconazole prodrug E1224 in patients in the indeterminate phase of Chagas disease [8, 9].

Rats have been considered more refractory to experimental infection with *T. cruzi* as they develop low levels of parasitemia and have low mortality rates during the acute phase, even in newly weaned pups [10]. However, these characteristics would turn as an appropriate to develop models of chronic infection.

Likewise, the presence of trypomastigotes in amniotic fluid of infected Wistar rats at 10 days of gestation was reported, suggesting the viability of congenital transmission [11].

Potential therapeutic effects of hormones have been investigated in this models [12–14] and also the immunomodulatory effect of a tumor necrosis factor α-antagonist [15]. Despite some advantages of the rat model over the mouse such as greater docility, ease of handling, maintenance, and the possibility of obtaining more blood in each sample and applying imaging techniques easily (e.g., echocardiogram, electrocardiogram), this experimental model has not been widely chosen as the first step in drug discovery.

Mice (*Mus musculus*) models are the most frequently used for the study of the experimental course *T. cruzi* infection and for the assessment of compound efficacy, although it is not exempt from questioning [16]. The murine model of *T. cruzi* infection has been most extensively studied maybe due to its ease of handling, housing, and low cost added to the wide variety of available strains and transgenic mice and the extensively studied immune mechanisms [17–19].

Depending on the combination of *T. cruzi* and mouse strain, the results may vary from early death with high parasitemia and high tissue parasitism to transient parasitemia, no death in the acute phase, and the development of chronic disease often (but not always) manifested as a cardiomyopathy and electrocardiographic alterations [19]. Similarly, vertical infection in mice is also possible, with a transmission rate to the offspring of 4% [10, 20].

The final result obtained in any *T. cruzi* infection in an animal model will depend on factors related both to the host (species, strain, sex, age) and the parasite (strain, stage, source, inoculum, route of infection). In this sense, a higher morbi-mortality was observed in C57BL/6 mice than in BALB/c both infected with the same inoculum of *T. cruzi* Tulahuén strain by the subcutaneous route [21].

The main objective of in vivo drug screening is to assess the trypanocidal effect of the compound candidate [17]. Although it may not resemble human pathology, the murine acute *T. cruzi* infection is the most accepted animal model as the first approach to the evaluation of new drugs. The trypanocidal ability of a compound is unequivocally tested by the effect on parasitemia levels and mortality prevention [4, 22].

On the other hand, the vast diversity of mice models and the lack of unified criteria to standardize some key characteristics in order to assess drug efficacy properly prevent comparison of results obtained by independent researchers.

Some steps were made toward the standardization for the development and drug evaluation for experimental Chagas disease, in a symposium held in Rio de Janeiro in 2008, where some criteria for in vitro and in vivo compound screening were unified. As concluding remarks, the animal model currently accepted for drug discovery is a murine acute model infected with *T. cruzi* strains with different BZ response, and the current cure criteria are established by the parasitological sterilization paradigm determined by conventional PCR after some immunosuppression cycles [22].

As described above, many aspects of Chagas disease have been studied in several animal species, which have similarities and differences both among themselves and with human pathology. It is clear that the choice of the animal model is conditioned by multiple variables and that although there is literature that attempts to standardize animal models to systematically evaluate new compounds, there is a lack of consensus to unify criteria that contribute to the comparison of results and to obtain solid conclusions.

As in the follow-up of infected patients, there is no current cure criterion, so the validation of surrogate markers could contribute to improving the predictability and transfer of results obtained in animal models to clinical studies [8, 18].

1 Applying the 3Rs Principles in Experimental *Trypanosoma cruzi* Infection

The animal use for research purposes is a privilege that carries individual and scientific obligations and responsibilities to ensure animal welfare to the greatest extent possible [23]. Since Russell and Burch formulated the principles of the 3Rs (replacement, reduction, and refinement) in animal experimentation in 1959, many strategies were conceived and updated to address these guidelines. Applying these principles in Chagas disease research is ethically desired and legally required in some cases.

Absolute replacement (i.e., replacing animals with computer simulation or reliable in vitro systems that can predict drug efficacy in human host) cannot prevent, to date, the use of animals as a previous step to clinical trials in drug discovery, but these strategies could be really useful to rapidly discard inactive compounds without passing through an animal model, and, inversely, there is no enough rationale to evaluate drugs that were not previously evaluated in vitro regarding the activity against clinically relevant *T. cruzi* stages (amastigotes and/or trypomastigotes) and the effect on host cell viability.

On the other hand, relative replacement (i.e., using invertebrates instead of vertebrates) may contribute to understand underlying pathogenic mechanism as seen in infected zebrafish (*Danio rerio*) or as a toxicology model as seen in zebrafish embryos, but its application in assessing drug efficacy is not developed yet [24, 25].

Once it has been determined that an animal model must be used to continue with compound development, different strategies can be applied to reduce the number of animals. As a first step, it is essential to carry out an extensive and systematic review in order to update the state of the art on in vivo drug screening for *T. cruzi* infection. This task may help to correctly determine the hypothesis and objectives, to avoid experiment duplication, and to decide which animal species represent the best model and can answer the main question more accurately. It will also help to decide the best methods to record our primary and secondary variables. For a general experiment planning, recent guidelines can be taking account [26].

Nowadays, it is mandatory to present the study protocol to the Institutional Animal Care and Use Committee (IACUC) before the experiments start. Although the formulary may vary in each institution, in all cases, it constitutes a sworn declaration from the principal researcher.

To comply with the reduction principle in animal experimentation, a proper experimental design and a correct sample size calculation are required. Using the smallest animal number that can bring statistically significant results should be the main goal.

Since parasitemia as a continuous variable is usually hard to estimate, with a wide dispersion and usually a distribution different to normal, the Mead's resource equation could be a helpful tool to support the experimental sample size. Readers are invited to refer to specific reviews for further explanation and practical examples [27, 28].

Another factor regarding reduction is the proper selection of strain, mainly in mice and rat models. Inbred, "genetically defined" strains are more stable, more uniform, more repeatable, and better defined than the "genetically undefined" outbred stocks. Experiments employing inbred strains are more powerful with more accurate dose-response relationships and fewer false-negative results than those done using outbred stocks [29]. The less variable response described in inbred strains contributes to reducing the number of animals.

The ARRIVE guidelines are very useful while preparing a draft or dossier, in order to include all the relevant information about the conducted experiments that will contribute to proper results analysis and in enhancing reproducibility and making possible the translation to clinical research [30]. Despite the science reproducibility crisis [31], a recent review of Chagas disease drug development reported a serious lack of information on animal models [32].

Once that animal use is correctly warranted and all strategies to reduce the number of animals were taken into account, refinement must be considered while conducting procedures. Refinement usually refers to any modification of husbandry or experimental procedures to enhance animal well-being and minimize or eliminate pain and distress [33]. Some key points that need to be addressed are summarized below.

Handling and habituation to experimental procedures previous to the essay are good laboratory practices that are inexpensive and less time consuming but may contribute to reducing stress and anxiety during the experiments [34, 35]. In order to increment well-being, rodents should be located in groups, and individual housing should be clearly justified and evaluated by the IACUC.

One of the most common procedures performed in experimental *T. cruzi* infection is blood sampling in order to count bloodstream trypomastigotes. The most used technique in the mouse is the amputation of the tail tip. Although the required volume (less than 10 μL) does not induce main physiologic responses, it could cause a temporary discomfort. For repeated blood sampling, the tail wound can be sealed with a topical hemostatic pencil and remove the generated clot the following times, avoiding successive tail cuts.

For assessing the efficacy of anti-*T. cruzi* novel compounds, Romanha et al. proposed a treatment schedule of 20 consecutive days by the oral route. Substances can be delivered to the gastrointestinal tract by including them in water or food, by oropharyngeal administration of capsules, pills, or fluids, or through gavage [36]. Oral gavage delivers a known quantity of drug in a single administration step but requires physical restraint by trained staff, which is very labor-intensive and is not indicated for long-term and/or repetitive treatments [37]. Disadvantages of gavage dosing include risks of esophageal or stomach damage and inadvertent lung administration [38].

As a refinement strategy, gavage can be replaced by a homemade device consisting of a disposable tip and automatic pipette, which allows selecting the final volume to the adequate minimum volume to dissolve the compound. The administration is performed by gently pressing the hard palate with the tip to stimulate the swallowing reflex. This method offers many advantages over conventional gavage administration, such as less animal manipulating time, less intensive operator training, and, most importantly, reduced risk of upper respiratory and digestive tract damage or aspiration pneumonia in those protocols that require long-term oral administration.

In most *T. cruzi* animal models, treatment response is traditionally followed by counting bloodstream trypomastigotes in regular timepoints. After finishing the treatment period, and if the parasitemia remains below detection limits, the most used methods to determine the parasitological cure are immunosuppression therapy (in order to induce parasite rebound), histopathology analyses to detect amastigotes nests, and PCR (or qPCR) targeting *T. cruzi*-DNA-specific sequences from blood and organs. Nevertheless, all these methods are postmortem although imply animal euthanasia.

Preclinical imaging techniques have arisen as an alternative and refined method for treatment follow-up. It improves the animal well-being, reduces the number of animals (since each animal serves as its own control), and improves the data amount and quality, due to the possibility of having a real-time progression from the infection and the response to drug administration.

The imaging modalities that have been applied to study infectious disease include magnetic resonance imaging (MRI), computed tomography (CT), positron-emission tomography (PET), bioluminescence imaging (BLI), and intravital imaging. Multiple-modality imaging is a promissory option since it permits the evaluation of the same animals by different imaging technologies [1].

Most of the imaging techniques were applied to unravel *T. cruzi* infection pathogenesis, and only MRI and echocardiography were employed to report the efficacy of verapamil on reducing mortality and development of cardiac pathology [19], but in the last few years, BLI models revolutionized the in vivo method for drug screening. Canavaci et al. reported the use of *T. cruzi* strains expressing the firefly luciferase or the tandem tomato fluorescent protein for in vitro and in vivo high-throughput assays for the identification of new drugs [39]. Subsequently, BALB/c mice were infected with transgenic luciferase expressing *T. cruzi* CL Brener clone [40] either intravenously, intraperitoneally, or subcutaneously, and by applying BLI system, it allowed longitudinal monitoring of parasitic load, the activity of benznidazole in both experimental acute and chronic stages, and the identification of trypanostatic effect of posaconazole [41].

In summary, the development of imaging applications in animal models of infectious diseases can quickly move from basic research to the clinic while reducing the number of animals, refining the follow-up procedures, and improving animal well-being [1].

Finally, it is important to consider applying anticipated endpoints, especially when working with very virulent *T. cruzi* strains, which leads to a poor body condition, emaciation, and dehydration in infected non-treated animals or animals treated with noneffective drugs. Human or anticipated endpoints are defined as point(s) at which pain, distress, or discomfort in an experimental animal is prevented, terminated, or relieved [33]. As a refined strategy, mortality must be replaced as a variable in study, considering biochemical, clinical, or behavior alterations prior to the illness that can serve as an anticipated endpoint. Although weight (or percentage of weight variance), corporal temperature, and physical appearance may help to identify animals entering to a fatal disease, it is important to establish early signs or markers that can predict animal severe pain, distress, and death. Noninvasive parameters and ethogram studies of infected animals may contribute to set up anticipated endpoints and avoid mortality as a variable [42, 43].

Acknowledgment Dr. Facundo García-Bournissen for the critical review of the manuscript.

References

1. Jelicks L, Lisanti M, Machado F, Weiss L, Tanowitz H, Desruisseaux M. Imaging of small-animal models of infectious diseases. Am J Pathol. 2013;182(2):296–304.
2. Desquesnes M, De Lana M. Veterinary aspects and experimental studies. In: Telleria J, Tibayrenc M, editors. American Trypanosomiasis: one hundred years of research. 1st ed. Amsterdam: Elsevier Inc; 2010. p. 277–318.
3. WHO. Report of the scientific working group on the development and evaluation of animal models for Chagas disease. Geneva: WHO; 1984.
4. do Canto Cavalheiro M, Leon L. Animal models of Trypanosoma cruzi infection. In: Zak O, Sande MA, editors. Handbook of animal models of infection: experimental models in antimicrobial chemotherapy. 1st ed. London: Academic Press; 1999. p. 801–10.
5. Castro M, Brener Z. Estudo parasitológico e anátomo-patológico da fase aguda da doença de chagas em cães inoculados com duas diferentes cepas do Trypanosoma Cruzi. Rev Soc Bras Med Trop. 1985;18(1):223–9.
6. Guedes P, Veloso V, Tafuri W, Galvão L, Carneiro C, Lana M, et al. The dog as model for chemotherapy of the Chagas' disease. Acta Trop. 2002;84(1):9–17.
7. Diniz LF, Caldas I, Guedes P, Crepalde G, de Lana M, Carneiro C, et al. Effects of ravuconazole treatment on parasite load and immune response in dogs experimentally infected with Trypanosoma cruzi. Antimicrob Agents Chemother. 2010;54(7):2979–86.
8. Chatelain E. Chagas disease drug discovery: toward a new era. J Biomol Screen. 2015;20(1):22–35.
9. Torrico F, Gascón J, Ribeiro I. E 1224—results of proof-of-concept clinical trial in patients with chronic indeterminate Chagas disease. Washington, DC: American Society of Tropical Medicine and Hygiene, 62nd Annual Meeting; 2013. p. 13.
10. Jorge T, Castro S. Doença de Chagas: manual para experimentação animal. (Editora FIOCRUZ, Ed.) (1st ed.). Rio de Janeiro. 2000.
11. Moreno E, Araujo M, Alarcón M, Lugo de Yarbuh A, Araujo S, Borges R. Effects of acute Chagasic infection on gestating Wistar rats. Rev Cient. 2006;16(5):506–16.
12. Frare E, Santello F, Caetano L, Caldeira J, Toldo M, Prado JJ. Growth hormones therapy in immune response against Trypanosoma cruzi. Res Vet Sci. 2010;88(2):273–8. https://doi.org/10.1016/j.rvsc.2009.10.001.
13. Kuehn C, Rodrigues Oliveira L, Santos C, Ferreira D, Alonso Toldo M, de Albuquerque S, do Prado JJ. Melatonin and dehydroepiandrosterone combination: does this treatment exert a synergistic effect during experimental Trypanosoma cruzi infection? J Pineal Res. 2009;47(3):253–9.
14. Santos C, Loria R, Oliveira L, Kuehn C, Toldo M, Albuquerque S, do Prado JJ. Effects of dehydroepiandrosterone-sulfate (DHEA-S) and benznidazole treatments during acute infection of two different Trypanosoma cruzi strains. Immunobiology. 2010;215(12):980–6. https://doi.org/10.1016/j.imbio.2009.11.002.

15. Perez A, Fontanella G, Nocito A, Revelli S, Bottasso O. Short treatment with the tumour necrosis factor-alpha blocker infliximab diminishes chronic chagasic myocarditis in rats without evidence of Trypanosoma cruzi reactivation. Clin Exp Immunol. 2009;157(2):291–9.
16. Costa S. Mouse as a model for Chagas disease: does mouse represent a good model for Chagas disease? Mem Inst Oswaldo Cruz. 1999;94(Suppl. I):269–72.
17. Buckner FS. Experimental chemotherapy and approaches to drug discovery for Trypanosoma cruzi infection. Adv Parasitol. 2011;75:89–119.
18. Chatelain E, Konar N. Translational challenges of animal models in Chagas disease drug development: a review. Drug Des Devel Ther. 2015;19(9):4807–23.
19. Jelicks L, Tanowitz H. Advances in imaging of animal models of Chagas disease. Adv Parasitol. 2011;75:193–208.
20. Solana M, Celentano A, Tekiel V, Jones M, González-Cappa S. *Trypanosoma cruzi*: effect of parasite subpopulation on murine pregnancy outcome. J Parasitol. 2002;88(1):102–6.
21. Roggero E, Perez A, Tamae-Kakazu M, Piazzon I, Nepomnaschy I, Wietzerbin J, et al. Differential susceptibility to acute Trypanosoma cruzi infection in BALB/c and C57BL/6 mice is not associated with a distinct parasite load but cytokine abnormalities. Clin Exp Immunol. 2002;128(3):421–8.
22. Romanha A, Castro S, Soeiro MN, Lannes-Vieira J, Ribeiro I, Talvani A, et al. *In vitro* and *in vivo* experimental models for drug screening and development for Chagas disease. Mem Inst Oswaldo Cruz. 2010;105(2):233–8.
23. ICLAS-CIOMS. International guiding principles for biomedical research involving animals. Geneva: ICLAS-CIOMS; 2012.
24. Akle V, Agudelo-Dueñas N, Molina-Rodriguez MA, Kartchner L, Ruth A, González J, Forero-Shelton M. Establishment of larval Zebrafish as an animal model to investigate Trypanosoma cruzi motility in vivo. J Vis Exp. 2017;127:PMID: 28994774.
25. Buchanan-Kilbey G, Djumpah J, Papadopoulou MV, Bloomer W, Hu L, Wilkinson SR, Ashworth R. Evaluating the developmental toxicity of trypanocidal nitroaromatic compounds on zebrafish. Acta Trop. 2013;128(3):701–5. https://doi.org/10.1016/j.actatropica.2013.07.022.
26. Smith A, Clutton R, Lilley E, Hansen K, Brattelid T. PREPARE: guidelines for planning animal research and testing. Lab Anim. 2017; https://doi.org/10.1177/0023677217724823.
27. Festing MFW, Altman DG. Guidelines for the design and statistical analysis of experiments using laboratory animals. ILAR J. 2002;43(4):244–58.
28. Pathak RR. Small size sampling. Int J Bas Clin Pharmacol. 2012;1(1):43–4.
29. Festing M. Inbred strains should replace outbred stocks in toxicology, safety testing, and drug development. Toxicol Pathol. 2010;38:681–90.
30. Kilkenny C, Browne WJ, Cuthill IC, Emerson M, Altman DG. Improving bioscience research reporting: the ARRIVE guidelines for reporting animal research. PLoS Biol. 2010;8(6):e1000412.
31. Munafò MR, Nosek BA, Bishop DVM, Button KS, Chambers CD, Percie du Sert N, et al. A manifesto for reproducible science. Nat Hum Behav. 2017;1:21. https://doi.org/10.1038/s41562-016-0021.
32. Gulin J, Rocco D, García-Bournissen F. Quality of reporting and adherence to ARRIVE guidelines in animal studies for Chagas disease preclinical drug research: a systematic review. PLoS Negl Trop Dis. 2015;9(11):e0004194.
33. National Research Council. Guide for the care and use of laboratory animals. 8th ed. Washington, DC: The National Academic Press; 2011.
34. Gouveia K, Hurst J. Reducing mouse anxiety during handling: effect of experience with handling tunnels. PLoS One. 2013;8(6):e664.
35. Hurst JL, West RS. Taming anxiety in laboratory mice. Nat Methods. 2010;7:825. https://doi.org/10.1038/nmeth.1500.
36. Morton DDB, Jennings M, Buckwell A, Ewbank R, Godfrey C, Holgate B, et al. Refining procedures for the administration of substances. Lab Anim. 2001;35(1):1–41. https://doi.org/10.1258/0023677011911345.

37. Küster T, Zumkehr B, Hermann C, Theurillat R, Thormann W, Gottstein B, Hemphill A. Voluntary ingestion of antiparasitic drugs emulsified in honey represents an alternative to gavage in mice. J Am Assoc Lab Anim Sci. 2012;51(2):219–23.
38. Zimmer J, Lewis S, Moyer J. Comparison of gavage, water bottle, and a high-moisture diet bolus as dosing methods for quantitative D-xylose administration to B6D2F1 (Mus musculus) mice. Lab Anim. 1993;27(2):164–70. https://doi.org/10.1258/002367793780810423.
39. Canavaci A, Bustamante J, Padilla A, Perez Brandan C, Simpson L, Xu D, et al. In vitro and in vivo high-throughput assays for the testing of anti-Trypanosoma cruzi compounds. PLoS Negl Trop Dis. 2010;4(7):e740.
40. Lewis MD, Fortes Francisco A, Taylor MC, Burrell-Saward H, McLatchie AP, Miles M a, Kelly JM. Bioluminescence imaging of chronic Trypanosoma cruzi infections reveals tissue-specific parasite dynamics and heart disease in the absence of locally persistent infection. Cell Microbiol. 2014;16(May):1285–300. https://doi.org/10.1111/cmi.12297.
41. Francisco AF, Lewis MD, Jayawardhana S, Taylor MC, Kelly JM, Chatelain E, Kelly JM. Limited ability of posaconazole to cure both acute and chronic Trypanosoma cruzi infections revealed by highly sensitive in vivo imaging. Antimicrob Agents Chemother. 2015;59(8):4653–61. https://doi.org/10.1128/AAC.00520-15.
42. Campos J, Hoppe L, Duque T, de Castro S, Oliveira G. Use of non-invasive parameters to evaluate Swiss Webster mice during Trypanosoma cruzi experimental acute infection. J Parasitol. 2016;102(2):280–5.
43. Silva D, Castro S, Alves M, Batista WS, Oliveira G. Acute experimental Trypanosoma cruzi infection: establishing a murine model that utilises non-invasive measurements of disease parameters. Mem Inst Oswaldo Cruz. 2012;107(2):211–6.

Chagas Disease Treatment Efficacy Biomarkers: Myths and Realities

Elizabeth Ruiz-Lancheros, Eric Chatelain, and Momar Ndao

Abstract Chagas disease (CD), caused by *Trypanosoma cruzi*, affects millions of people worldwide. Although CD R&D has made progress during the last decade, clinicians and general practitioners are still facing the same challenge, i.e., the lack of adequate markers of clinical cure, hindering assessment of new drug efficacy in clinical trials and counseling of patients about treatment outcome. To date, no new markers have been validated as surrogates of seroreversion – the only marker of parasitological cure which is itself considered to be a surrogate of clinical benefit. *T. cruzi* DNA detected using PCR cannot currently be considered as a surrogate of seroconversion. Much emphasis has been placed on different *T. cruzi* antigens but no definite proof of correlation between titers, as determined by serology at a given timepoint, and seroreversion has been shown. Thanks to the improvement of analytical methods and the application of new methodologies, the identification of potential new markers is being facilitated, and some of these are progressing. However, there is a long journey from the identification of a potential biomarker to its clinical validation and acceptance by the regulatory authorities that requires a common effort from the entire Chagas community.

E. Ruiz-Lancheros · M. Ndao (✉)
National Reference Centre for Parasitology, Research Institute of the McGill University Health Centre, Montreal, QC, Canada
e-mail: momar.ndao@mcgill.ca

E. Chatelain (✉)
Drugs for Neglected Diseases *initiative* (DND*i*), Geneva, Switzerland
e-mail: echatelain@dndi.org

J. M. Altcheh, H. Freilij (eds.), *Chagas Disease*, Birkhäuser Advances in Infectious Diseases, https://doi.org/10.1007/978-3-030-00054-7_16

1 Introduction

The last decade has seen an increase in the number of clinical trials assessing the potential of new drugs for Chagas disease (CD), focusing specifically on repurposed azoles. One of the main issues facing clinical researchers in the field, however, is the absence of clearly defined markers of clinical cure, due to the complexity and long development time of the disease. This fact, among others, has hampered efforts toward the development of new drugs for CD. The scope of this review is not to present an extensive overview of all the potential markers for assessment of treatment efficacy described so far, as this has already been done [1–3]. Instead, we focus on the current needs and challenges in this specific area, describe new technologies that have been applied to the identification of potential markers of interest and propose the steps that we consider should be taken in order to tackle this important issue.

Indeed, we believe that a concerted joint effort by the CD community is essential in order to gain a better understanding of how to define a biomarker for CD and how then to further develop and validate it, in order to answer this very complex and demanding research question. This is not only necessary to be able to run clinical trials for the registration of new drugs but also so that general practitioners will be able to inform patients about the outcome of their treatment.

2 Chagas Disease Overview

CD, also known as American trypanosomiasis, and its etiological factor, *Trypanosoma cruzi*, were discovered more than a century ago by Carlos Chagas [4]. Since then, CD epidemiology has changed; although still endemic in Latin America, the disease has spread into non-endemic countries due to population migration and has become a global public health issue [5–8]. CD is the most common cause of infectious cardiomyopathy worldwide [9]. Around 6–7 million people are infected worldwide and 10,000 die annually [10, 11]. The disease presents in two main phases: the acute phase, which is asymptomatic and typically undetected and lasts for a couple of months during which the parasite is readily identified through blood examination, and the chronic phase, which can last for decades while the infection is controlled by the immune system and the parasite is hardly detectable. While most infected patients in the chronic phase will remain asymptomatic, a certain proportion—between 10% and 40%—eventually develop symptoms, mainly cardiomyopathies and in certain cases digestive tract megasyndromes or both [12]. The major causes of mortality in these patients are progressive heart failure and sudden death [13].

There are two treatments currently available, benznidazole (Abarax/ELEA and Rochagan/LAFEPE) and nifurtimox (Lampit/Bayer), which are old nitroheterocyclic trypanocidal drugs. Although these drugs have been shown to be efficacious in both phases of the disease, particularly in children, their use is limited due to side effects occurring during treatment and impeded access to medication [14, 15]. There is an urgent need for new and safer drugs for CD.

3 Disease Progression and Treatment Efficacy Assessment: Current Challenges and Future Needs

A major hurdle for the clinical development of new drugs for CD is the absence of an adequate test that can assess successful treatment or show clinical benefit in a timely manner. The definition of cure criteria for CD has been subject to debate; the complicated development and pathology of the disease, coupled with the complexities of the parasite life cycle and its interactions with the host, make it a very difficult task to determine such criteria. Clinical cure criteria are very often discarded as they are considered to be too difficult to achieve and possibly because of a lack of understanding of the slow evolution of the disease from asymptomatic stage to cardiomyopathy and/or megacolon [16]. Another issue is the lack of consensus on the assessment of treatment efficacy and inadequate tools to address it [17].

Although there is no absolute proof in patients that parasitological cure is synonymous with clinical cure, i.e., halting the progression of the disease toward cardiac or gastrointestinal symptoms, there is a consensus that parasite persistence is needed for the development of CD. All current CD drug development efforts are therefore focused on strategies to eliminate *T. cruzi* from the human body. The only way to assess drug treatment efficacy is to use serological tests showing the disappearance of *T. cruzi* antibodies (seroreversion, synonymous with parasitological cure). This is clearly a major challenge, since seroreversion can take decades to occur in treated adults, if it occurs at all. This makes assessment of parasitological cure with the currently available tools in this category of patients complicated, if not impossible, and thus seroreversion is not useful as a clinical endpoint in clinical trials. There is, therefore, a need to identify surrogate markers for the absence of parasites that are quicker and more sensitive than seroreversion. A surrogate endpoint of a clinical trial is a laboratory measurement or a physical sign that is used as a substitute for a clinically meaningful endpoint that measures directly how a patient feels, functions, or survives. Changes induced by a therapy on a surrogate endpoint are expected to reflect changes in a clinically meaningful endpoint [18].

The need for surrogate markers of parasitological cure is further highlighted by the recent FDA approval of benznidazole (BZN) monotherapy exclusively for the treatment of chagasic children between 2 and 12 years of age [19, 20]. In *T. cruzi*-infected children, seroreversion can be observed fairly quickly, within months to a few years following treatment. The FDA approval of BZN for children was based on seroreversion observed in around 50% of children [21, 22].

Another important feature of CD is the fact that not all *T. cruzi*-infected patients will develop the disease—in the literature it is typically stated that around 10–40% of infected patients will develop symptoms of the disease [12]—and that possible host factors for susceptibility are not well understood [23]. It would be useful to identify markers of disease progression that could be used to predict which *T. cruzi*-infected people are likely to develop the disease, as treatment could then be focused on people most at risk.

Efforts to understand this phenomenon and to identify patterns or indicators that could be used to categorize patients at risk of developing the disease have so far not been successful. Preliminary attempts have been made to identify candidate genes associated with the progression of the disease. Results from a genome-wide association study (GWAS) using the well-established REDS-II cohort of Chagas patients suggested that both cardiovascular- and immune-related polymorphism in some genes of interest could be associated with a genetic predisposition to chronic Chagas cardiomyopathy [24]. Another study found an association between HLA haplotype and resistance to chronic Chagas disease [25]. A recent study showed a potential association between variations in the inflammasome, particularly in NLRP1 and CARD11, and chronic Chagas cardiomyopathy [26]. Although all these studies suffer from the low number of patients used in the analysis, this is certainly an area of research that merits further investigation.

4 CD Biomarker Identification for Treatment Efficacy Assessment and Next Steps: The What(s) and How(s)

There is a need to identify a surrogate marker for the absence of parasites that is quicker and possibly more sensitive than seroreversion. Efforts to identify new potential biomarkers, comparing, for example, samples from healthy people and Chagas patients, can lead to a substantial amount of data that is not always easy to interpret and analyze. Even when potential markers of interest are identified, there is still a long way to go to ensure that these putative markers will be useful in practice [27]. This will include analytic (validation of an assay for the marker of choice) and clinical validation of the marker as well as regulatory acceptance.

It becomes very important therefore to define the attributes that should be required for a biomarker, in particular those related to the methodology used for the analysis of a chosen marker and its suitability in the field, its level of sensitivity and selectivity, and the current level of validation according to the types and number of samples tested, to name but a few criteria. The definition of a target product profile (TPP) for a biomarker and its associated test should clarify these points early on in order to avoid focusing work and testing on a marker of interest for which no suitable test could be available for routine analysis. A tentative TPP for Chagas disease assessment of treatment response has been described but seems to be biased toward the use of PCR [28]. A more general biomarker TPP was highlighted by Pinazo et al. in their biomarker systematic review [2].

Another point for consideration is related to the quality and type of samples used to identify and validate biomarkers. Specimens from patients at different stages of the disease and from healthy people are critical tools for this and for the development of better tests. The appropriate detailed information and handling procedure to be followed are defined in local or international guidelines, and standardization of specimen collection methodologies is critical [29]. Indeed, technical aspects such as the anticoagulant used, sample processing time, processing and storage temperatures, and thaw/freeze cycles are all variables that can impact the quality of specimens and their stability over time, thereby having an impact on analysis results [30].

Finally, the design of clinical trials from which the specimens originate and which are used for the identification and validation of potential biomarkers should be considered very carefully.

5 Pros and Cons of Currently Proposed Biomarkers of Cure for CD

Several CD biomarkers have been suggested and used to discriminate between CD-infected and CD-non-infected individuals when assessing chemotherapy in cohort studies with adults or children and to establish disease progression in CD patients. However, very few of them show high sensitivity, have been systematically studied, or could be used to determine treatment efficacy. In addition, not many would pass the quality criteria mentioned above (see Sect. 4). All the CD biomarkers suggested so far have been reviewed by different authors (see [1, 3]). In 2014, a systematic review by Pinazo et al. proposed 25 potential biomarkers for the evaluation of therapeutic efficacy [2]. Requena-Méndez et al. also reviewed some blood-derived biomarkers useful for disease progression and cure [31]. Here we describe the pros and cons of some of these biomarkers and also some promising new markers with the potential to be surrogate endpoints. A summary of candidates is presented in Table 1.

5.1 Parasite DNA Amplification and Antigens for Serological Tests

Parasite detection in blood by PCR and the evaluation of antibodies by serology are the main techniques used to monitor CD treatment response in patients and in CD clinical trials. The evaluation of treatment efficacy is affected by their limitations, which have been recognized as Achilles' heel of clinical trial outcomes. PCR has shown promising results for the assessment of therapy failure; a positive result clearly evidences failure to clear the parasite and thus ineffective treatment [57]. However, a negative PCR does not guarantee the absence of parasite and cannot confirm parasite clearance. False negatives occur due to fluctuations in parasitemia, the isolation of parasite in tissue or organs, and the intrinsic limit of detection of PCR and qPCR techniques [58]. Other distinct factors may contribute to the overall performance of PCR assays: the size of the serum sample for parasite DNA extraction, the sample collection tubes, the different PCR assay conditions, and the algorithm used to classify results can affect the evaluation of the samples, as was demonstrated by Wei et al. [59] for the STOP CHAGAS clinical trial that evaluated posaconazole for the treatment of CD [60]. Nevertheless, PCR is a promising tool that can be easily performed in clinical settings and used for clinical trials; thus, the investment in improving PCR methodologies is worthwhile. The CD community must focus on suitable strategies for parasite DNA extraction in lower sample volumes, the equivalence between blood and tissue parasitemia; the reduction of false

Table 1 Candidate Chagas disease surrogate biomarkers

Biomarker type	Biomarker	Results in Chagas disease	Potential as a test of cure	References
Parasite proteins	Trypomastigotes F2/3 antigenic fraction	Anti F2/3 decreases after BZN treatment and disappears after 4–21 months in children	Negative results earlier than conventional serology but need to be evaluated in adults	[32]
	Immunofluorescence assay of fixed trypomastigotes (ISIFA)	High titers in infected patients and low titers 6 years after treatment when patients were considered cured. High sensitivity and no cross-reactivity with other diseases	ISIFA can differentiate treated from untreated and those with treatment failure, but the assay requires fixed parasites	[33, 34]
	Trypomastigote mucin antigen A&T CL-ELISA	Measure anti-Gal Abs. Titers decrease after BZN treatment in adults and children	The gradual and consistent decrease of Abs after 3–6 years of treatment. Correlation with seroconversion only in adolescents.	[35–37]

Parasite recombinant proteins	Ag13 85 kDa protein with repeats of 5 amino acids	Anti-Ag13 is suitable for CD diagnosis in different populations, and titers decrease and disappear after 3 years posttreatment	Negative conversion occurs quicker compared to other antigens	[38]
	T. cruzi ribosomal acid protein P2β	Levels of Anti-P2β decrease in asymptomatic treated CD patients	Seroconversion only in half of the population 20 years posttreatment	[39]
	Immunodominant antigens KMP11, HSP70, PFR2, Tgp63	A significant drop in reactivity against antigens between 6 and 9 months in BZN-treated CD adults at different stages of the disease. Titers continue to drop after 24 months	Antigens are recognized by complement-dependent IgG1 which could be an advantage to observe rapid seroreversion	[40, 41]
	24 kDa calcium-binding protein (rTc24)	Anti-rTC24 Abs decreases within 6–36 months posttreatment	Good correlation with the CoML test and seroconversion by other antibodies	[42, 43]
	Flagellar calcium-binding protein (F29)	Seroreversion for the F29 antigen occurs between 6 and 48 months after BZN treatment in children	ELISA-F29 establishes seroreversion earlier than conventional serology in adults. But the time to convert could limit its use in clinical trials	[44, 45]
	Multiplex 16 r *T. cruzi* proteins	Decreased response of the panel 36 months after BZN treatment in adults	Strong correlation with conventional serology, but seroreversion was observed only in a subset of treated patients	[46, 47]
	Recombinant complement regulatory protein (rCRP)	Detect Abs complement-dependent as the CoML test. Positive reactions decrease 1–2 years after BZN treatment	Results correlate well with CoML test and do not require live parasites	[48]
	Putative microtubule-associated protein (MAP) antigen3	Selected antigen from a multiplex array of 15 antigens. Results correlate with PCR-positive and PCR-negative results in a cohort study 5 years after BZN treatment	Abs efficiently detects parasite persistence in infected individuals and PCR-negative treated individuals	[49]

(continued)

Table 1 (continued)

Biomarker type	Biomarker	Results in Chagas disease	Potential as a test of cure	References
Host biochemical markers	ApoA1	Downregulated in CD and normal levels after BZN or NFX treatment	Level return to normal after treatment and 3-year follow-up	[50, 51]
	ApoA1 and FBN fragments	Upregulated in CD and downregulated after BZN or NFX treatments	Correlation with seroreversion, early decrease after BZN treatment	
	Lytic antibody complement-mediated lysis (CoML) test	Abs decreases until becoming negative after parasite elimination in BZN and BFX treatments	Negative results can be obtained 1 year after treatment when serology is still positive but requires live trypomastigotes	[42]
Host prothrombotic markers	Prothrombin fragment 1 + 2 (F_{1+2})	A marker of thrombin generation in vivo increases early in CD and decreases after BZN treatment	Consistent decrease to normal levels after treatment and for 3-year follow-up in 96% study population	[36, 52]
	Endogenous thrombin potential (ETP)	Quantifies the ability to generate thrombin when activated through tissue factor addition upregulated in CD, decreases after BZN treatment		
	Soluble platelet selectin (sP-selectin)	Biomarker of in vivo platelet activation decrease during BZN therapy in adults and children	Consistent decrease after treatment, but upregulation does not occur in all CD cases	[21, 36]
Immunological markers	IFN-γ T cells	Three-fold decrease compared with pre-treatment between 1 and 3 years posttreatment	IFN-γ levels correlate with the severity of disease and can be used to monitor disease progression	[53]
	$CD3^+$ T cells	CD3+ T-cell proportion differs between treated and untreated patients and normalizes in cured patients	Despite values normalizing in cured patients, the number of cells cannot be used to predict parasitological cure	[54]
	$IL12^+$ $CD14^+$ cells	BZN-treated children show low levels of $IL12^+$ $CD14^+$ cells and high levels of IL-10 modulated type 1 cytokines profile	Potential to see immunological effects of treatment but needs further study	[55]
	$CD4^+$ LIR^+ T cells	Decrease of $CD4^+$ LIR^+ T cells after treatment between 2 and 6 months and for at least 2 years		[56]

negatives, as well as the validation and standardization of PCR assays; and the correlation of PCR readouts with seroreversion.

Conventional serology with different parasite antigens is commonly used for CD diagnosis and to evaluate antibody titers against *T. cruzi* after chemotherapy. Due to the long-term persistence of specific antibodies that are detected by serological tests, chronically infected patients must be followed up for several years after treatment until they can be considered cured using seroreversion as a measurement of parasite clearance [33]. The current criteria of cure consist of two nonreactive conventional serological assays with parasite antigens that are commercially available as diagnostic kits in endemic countries. The probability of cure might also be predicted by a decrease in antibody titers for *T. cruzi* over time, but this will depend on the specific antibody and the status of the disease before treatment [57]. However, serodiscordance remains a challenge in Chagas disease diagnosis and raises the question of the reliability of serology tests relying on one specific antigen depending on the region and patient stage [61–63].

Of all the *T. cruzi* antigens published in the literature, only a few have been evaluated in the long term and used to predict treatment efficacy. Antigens obtained directly from *T. cruzi* preparations as the F2/3 antigenic fraction (isolated from trypomastigotes) have been used to assess cure in children with congenital transmission [32]. Anti-F2/3 antibodies become negative 4–21 months after BZN treatment and earlier than conventional serology, suggesting that they may provide an earlier marker of cure. Likewise, Andrade et al. have shown that a chemiluminescent enzyme-linked immunosorbent assay (CL-ELISA) with a trypomastigote mucin antigen (A&T) successfully assesses treatment efficacy in BZN-treated adolescents [35]. When measured as negative A&T CL-ELISA seroconversion, 88.7% of the treated group were cured after 6-year follow-up. Using this assay, the BZN efficacy in children and adolescents by per-protocol analysis and by intention-to-treat approach was 84.7% and 64.7%, respectively. F2/3 and A&T antigens obtained from parasites seem to be candidate surrogate biomarkers, but their use in adult studies needs to be further evaluated since the reduction in titers and seroreversion can take longer. Pinazo et al. have shown that A&T CL-ELISA remains positive for 3 years after treatment in an adult population [36].

Since obtaining pure proteins from the parasite can be laborious, several recombinant proteins have been produced and tested for detecting anti-*T. cruzi* antibodies in ELISAs and immunoblots. Most of the specific antibodies against recombinant proteins are good at discriminating CD individuals from healthy controls and useful for monitoring patients after treatment. Recombinant proteins such as F29 (flagellar calcium-binding protein); P2β (ribosomal acid protein); KMP11, HSP70, PFR2, and Tgp63 (immunodominant antigens); Ag13 (85 kDa protein with repeats of 5 aa); and a multiplex of 16 *T. cruzi* proteins have been used to assess treatment efficacy and to attempt to predict cure (see Table 1).

Anti-F29 decreases quickly after BZN treatment in children and seroconverts in 62.1% of cases after 48 months [44]. Fabbro et al. have shown that Anti-F29 may take up to 14.5 ± 5.7 years to seroconvert after BZN or nifurtimox (NFX) treatment in adults but predicts cure earlier than conventional serology (22 ± 4.9 years) [45].

By contrast, anti-P2β can take more than two decades to seroconvert in treated asymptomatic patients despite a reduction in titers compared to their initial values [39]. Anti-KMP11, HSP70, and PFR2 decrease rapidly a few months (6–9 months) after BZN treatment in more than 70% of CD patients and continue to decrease during the 24 months posttreatment follow-up period [40]. This is perhaps because KMP11 and HSP70 are mainly recognized by IgG1 complement-dependent antibodies [41]. Anti-Ag13 also has shown a constant decrease in titers compared with other antigens and seroconverts in 6/9 patients after 3 years of treatment [38]. Finally, a panel of 16 proteins in a multiplex bead assay has shown a strong correlation with conventional serology tests in a short-term follow-up of 53 BZN-treated patients [46, 47]. Despite the evidence gathered so far, longer follow-up and tests in larger populations are needed to select the best Ag/Abs pairs that can be used to evaluate treatment efficacy in clinical trials, regardless of the type of treatment or the patient's disease status.

In the search for the ideal antigens and antibodies, Zrein et al. used an innovative multiparametric screening technology to identify antibodies that could be used as surrogate biomarkers [49]. After evaluating 15 antigens in a multiplex serology assay, Antibody 3 (Ab3), which recognizes *T. cruzi* putative microtubule-associated protein (MAP) (Antigen 3), showed a strong correlation (92%) with PCR-positive results in treated and untreated CD patients from the SaMi-Trop cohort study [64]. More importantly, Ab3 could discriminate PCR-positive patients from PCR-negative treated patients (AUC 0.74). Ab3 efficiently detected parasite persistence in most of the *T. cruzi*-infected individuals and detected a large number of parasite persistent cases within the PCR-negative group, which shows this assay to be even more informative than PCR [49]. These results suggest that Ab3 could be a good surrogate biomarker; however, Ab3 titers should be evaluated before treatment and in a non-infected population. Validation of other cohort studies, quantitative evaluation of serology, and following titers until seroconversion will determine if parasitological endpoints can be predicted with this antibody. Its usefulness for evaluating treatment efficacy in clinical trials will also depend on a quick change in Ab3 titers after treatment.

Detection of complement-dependent lytic antibodies seems to be an alternative to the detection of specific anti-*T. cruzi* antibodies. These lytic antibodies can be detected either by complement-mediated lysis (CoML) test or by indirect immunofluorescence (IIF). The antibodies appear as soon as 20 days post-infection and disappear as early as 1-year post-chemotherapy [42]. In the 10-year follow-up of a study of CD patients treated with BZN or NFX, patients showed consistently negative CoML test results at 6 to 33 months posttreatment despite positive IIF and conventional serology and thus were considered cured [65]. The disadvantage of this approach is the need for living infective trypomastigotes, which is not practical for clinical trials. One epitope of the lytic antibodies contains a high molecular mass (160 kDa) protein, *T. cruzi* complementary regulatory protein (CRP). Meira et al. found a good correlation between the CoML test and an ELISA using a recombinant CRP [48]. In a cohort study with 31 CD patients, both tests showed the same significant reduction in the number of positive samples over a period of 4 years after treatment [48]. However, evaluation of a bigger population is needed to confirm if this ELISA could replace the CoML test and be used to establish parasite clearance [42]. A recombinant calcium-binding protein (rTc24) has shown a good correlation with CoML test in ELISA and immunoblot, but its potential as a surrogate endpoint has to be further investigated [43].

5.2 *Host Biochemical Molecules*

Since *T. cruzi* is an intracellular parasite that produces a chronic infection, biochemical molecules from the host such as metabolites, proteins, immunomodulators, and cell surface proteins can be affected due to infection and thus may be potential CD biomarkers. In addition, host biochemical molecules can be better surrogate biomarkers than antibodies against *T. cruzi* antigens or DNA amplification techniques, since their evaluation will not depend on the persistence of antibodies or the direct detection of parasites.

In the search for indicators of parasite signature, our group performed the first serum protein analysis of CD patients using mass spectrometry [50]. We used serum fractionation to evaluate both high- and low-abundant serum proteins and surface-enhanced laser desorption ionization time-of-flight mass spectrometry (SELDI-TOF MS) for intact protein analysis [66, 67]. In a panel of 435 sera from Venezuelan asymptomatic CD patients and healthy controls (HC), we identified 18 host proteins that were statistically different between the CD and control populations. To select biomarkers with the greatest discriminatory power, we used a biomarker pattern software to generate candidate decision trees. Five host markers showed high sensitivity (89%) and specificity (100%) and could distinguish asymptomatic CD adult patients from HC. Biomarkers were identified by MS/MS analysis as full-length and fragments of the apolipoprotein-A1 (ApoA1) as well as a fragment of fibronectin (FBN) [50]. It is worth mentioning that our success in detecting and identifying these biomarkers was due to our innovative intact protein approach, also known as "top-down proteomics."

We used the same strategy to search for biomarkers in a Bolivian CD population and to predict cure after treatment with NFX [51]. After comparing the serum proteins of CD vs HC, the same candidate biomarkers were identified, demonstrating the reproducibility of the approach across the South American population, an important factor considering the variety of infective *T. cruzi* strains in patients in this region. In addition, we observed that ApoA1 and FBN fragments were significantly upregulated in chronic or asymptomatic CD subjects compared to HC and 3 years after NFX treatment returned to levels similar to those seen in HC. In contrast, full-length ApoA1 was downregulated in CD individuals compared to HC and returned to normal levels during the follow-up period. All patients were seropositive 3 years after treatment, but using these biomarkers, we were able to predict an overall cure rate of 43.2%. These results suggest that these biomarkers might be useful in assessing treatment efficacy in CD patients and could lead to the development of a test of cure [51]. Following this lead, we have developed a proteomics-based immunoblot that detects ApoA1 and FBN fragments in CD patients. This is a successful translation of proteomics-based studies into accessible tools for bench diagnosis. Results from these studies confirm these fragments as signatures of the parasite and their great potential as surrogate biomarkers.

5.3 Host Prothrombotic and Immune Markers

Host prothrombotic markers have been used to evaluate disease progression in the chronic phase of CD, but they could also be useful for treatment evaluation. In infection, immunothrombosis is activated after the recognition of a pathogen in order to inhibit its dissemination and survival; different infectious agents may cause responses to different degrees. Thromboembolic events are observed in cases of chagasis cardiomyopathy, and an increased risk of peripheral thrombotic phenomena and thrombosis in CD patients without heart failure or structural cardiopathy has been observed. These events can be attributed to the host immune system and to the parasite itself. Of all the biomarkers that can identify a prothrombotic state, markers for clotting activation that have shown the most consistent results are prothrombin fragment 1 + 2 (F_{1+2}) and the endogenous thrombin potential (ETP). These markers are elevated in the early stage of the chronic phase and decrease after therapy with BZN [52]. In a more recent study with 99 individuals, Pinazo et al. observed that F_{1+2} and ETP were abnormally expressed in 77% and 50% of infected patients before treatment but returned to, and remained at, normal levels 6–9 months after treatment in 76% and 96% of cases, respectively [36]. This data suggests these markers could assess short-term response to treatment; however, normal values can be observed in infected patients, and some patients show qPCR-positive results even when ETP values reach baseline after treatment.

Lastly, different cytokines and cell surface markers have been evaluated in CD patients and proposed as immune markers for disease progression. Within them, IFN_Υ T cells, $CD3^+$ T cells, $IL12^+$ $CD14^-$ cells, and $CD4^+$ LIR^+ T cells have been studied in CD-treated or CD-untreated populations. IFN_Υ is one of the main cytokines that regulate Th1 immune responses, and it is critical for innate and adaptive immunity against virus and intracellular parasites. High levels of IFN_Υ in peripheral blood mononuclear cell (PBMC) cultures correlate with severity of CD cardiomyopathy and are probably responsible for the strong Th1 response in CD patients with cardiac disease [68]. Laucella et al. have observed that IFN_Υ T-cell levels decrease after BZN treatment between 1 and 3 years posttreatment and become undetectable in almost 50% of treated patients [53]. Likewise, the proportion of $CD3^+$ T cells differs between treated and untreated patients and normalizes in cured patients without changes in the PBMC phenotype [54]. In patients treated with BZN during the early indeterminate stage, the number of $IL\text{-}12^+CD14^+$ cells decreases, and treatment induces an IL-10-modulated type 1 cytokine profile [55]. On the other hand, chronic CD patients have shown increased numbers of $CD4^+LIR\text{-}1^+$ among total PBMCs, relative to non-infected individuals, and these numbers decreased after BZN treatment [56]. Although these findings suggest that cell types and their markers can be used to assess the influence of treatment, their potential as surrogate biomarkers needs to be further studied.

Nonetheless, immunomarkers can be useful in phase I clinical trials of new drugs. In a mouse immunosuppressed model, the presence of MHC-peptide tetramers, which are specific for $CD8^+$ T cells recognizing a transialidase peptide, was monitored as a biomarker for treatment success [69]. Cured mice show an increased number of T cells displaying a central memory phenotype (T_{CM}) with surface markers CD62L and CD127, these markers are not present in T_{CM} and T effector memory cells (T_{EM}) in non-treated mice, and the phenotype could be used to determine treatment efficacy and cure [70].

6 New Developments: A Hopeful Way Forward?

During the last 15 years, new developments have changed the Chagas biomarker landscape. These developments have included both the emergence of new tools and technologies for the assessment of known/"established" markers (essentially *T. cruzi* antigens or antibodies) and the identification of potential new markers, in particular markers in the host, for the diagnosis of CD and the assessment of drug treatment efficacy. These include, among others, high-throughput technologies to identify RNA aptamers; new analytical devices such as biosensors; new mass spectrometry and NMR technologies allowing comparative analysis of proteins, metabolites, lipids, and mRNA in serum samples from healthy and infected patients (X-omics); as well as FACS and MRI.

6.1 Aptamers

RNA aptamers are short nucleotides that can bind specifically, and with high affinity, to targets in complex protein mixtures, membrane preparations, or whole cells. Their specificity depends on their hydrophobic and ionic interactions with the target as well as on their tertiary structure. Aptamers can be developed in vitro using an iterative procedure known as systematic evolution of ligands by exponential enrichment (SELEX). Without a priori knowing a specific target, this process can select RNA sequences with affinities similar to or lower than those seen with monoclonal antibodies [71–73]. This approach was first explored for CD by Nagarkatti et al. in order to concentrate *T. cruzi* parasites and facilitate their detection by PCR [74]. Using a whole-cell SELEX strategy, they developed serum-stable RNA aptamers that bind to live *T. cruzi* trypomastigotes with an affinity ranging between 8 and 25 nM. The aptamer with the highest affinity, Apt68 (K_d 7.686 ± 1.63 nM), also showed high specificity and did not bind either insect stage epimastigotes of *T. cruzi*, *Leishmania donovani* promastigotes or *Trypanosoma brucei*. The authors also demonstrated that Apt68 was able to bind parasites from different strains, when immobilized in the solid phase and at parasite concentrations as low as 0.33 parasites/ml (five parasites in 15 ml). This approach could be useful for CD diagnosis during the early phase of infection (the window period of PCR detection) or during the chronic phase when there is intermittent parasitemia in the blood [74].

More recently, the same group used the SELEX strategy to select aptamers that bind specifically to TESA (*T. cruzi* excreted-secreted antigens) aiming to develop a new direct non-serological, non-PCR-based assay to detect *T. cruzi* infection. After ten rounds of selection, Apt-L44 showed specific binding to TESA as well as to *T. cruzi* trypomastigote extract from three different strains but not to epimastigotes, host proteins, or *L. donovani* proteins. Using biotinylated Apt-L44 in an enzyme-linked aptamer (ELA) assay, the aptamer showed specific binding to TESA, and higher levels of binding were observed in the serum of *T. cruzi*-infected mice compared to non-infected mice. Additionally, Apt-L44 could detect circulating TESA in mice in both the acute and chronic phases. Apt-L44 ELA assay could be used as a qualitative assay in drug screening to detect *T. cruzi* antigens in infected mice and demonstrate that live parasites are present in the host, even if their direct detection in blood by PCR is negative [75].

Increasing the SELEX rounds and the stringency of the conditions, de Araujo et al. found an aptamer (Apt29) with a higher signal for TESA in infected mice and a higher signal-to-noise ratio compared to Apt-L44 [76]. This aptamer was also able to differentiate infected from non-infected mice and predict treatment failure. In infected and chronically infected mice, the TESA levels detected by Apt-29 ELA were reduced upon BZN treatment. However, levels did not return to those seen in non-infected treated mice, suggesting parasitemia was reduced but parasitological cure was not achieved. These results are in agreement with the detection of parasites in the heart and skeletal muscles by PCR and suggest that the assay can be used to assess treatment efficacy in vivo in murine drug discovery models [76]. However, its ability to predict parasitological cure needs to be confirmed. The ELA assay does not need sophisticated equipment or reagents but can be performed in high-throughput formats, and animals do not need to be sacrificed. In addition, aptamers can be used to evaluate patients' serum and can be coupled to different matrices to increase the detection limit, which makes them a promising tool for CD diagnosis and prognosis. Further validation using human specimens is, however, necessary before drawing conclusions.

6.2 *Biosensors*

The application of new technologies and miniaturization have led to new tools, biosensors, that can be applied to the field of Chagas diagnostics and that can help to detect the presence of *T. cruzi* in serum. A biosensor is an analytical device that converts molecular recognition of a target analyte into a measurable signal via a transducer. Depending on the type of transducer that is employed, they may be electrochemical, acoustic, or optical [77, 78]. To develop a biosensor, a biologically active component needs to be immobilized onto the surface of the transducer; once the target analyte is recognized, a signal response in the sensor is generated, and the signal can be amplified and measured in an electronic system that acquires and records the signal [78]. Biosensors are easy to use, give results in real time and require small sample volumes and short assay times, and have a low energy consumption, which make them excellent tools for point-of-care (PoC) units. Furthermore, they are a sensitive and inexpensive technology platform compared with conventional diagnosis technologies. During the last 15 years, different electrochemical (amperometric or impedimetric) and optical sensors have been designed and tested for the indirect or direct (label-free) detection of *T. cruzi* in serum at acute and chronic stages for CD diagnosis.

In 2011, Pereira et al. optimized an electrochemical immunosensor to quantify IgG *T. cruzi* antibodies in serum patients using *T. cruzi* epimastigote membranes. In order to increase the sensitivity and efficiency of membrane immobilization, a screen-printed carbon electrode (SPCE) was used, and gold nanoparticles (AuNPs) were electrodeposited where the *T. cruzi* antigens were immobilized. The biosensor also uses anti-IgG antibodies coupled to horseradish peroxidase (HRP) and redox reagents to amplify (label) the immunodetection [79]. Their optimized biosensor showed a linear detection of IgG *T. cruzi* antibodies between 11 and 205 ng/mL and a detection limit of 3.065 ng/mL. In addition, the microfluid technology used allowed a fast response and short assay time (26 min). This biosensor is easy to operate and transport, but it still depends on the presence of anti-*T. cruzi* antibodies to predict infection.

More recently, a similar approach was used to develop a biosensor to detect and quantify anti-*T. cruzi* IgM antibodies in newborns and infants and to predict congenital CD [80]. In this case, a SAPA (shed acute-phase antigen) was immobilized in a SPCE together with AuNPs. IgM antibodies appear early in the acute phase of *T. cruzi* infection, and the SAPA has been shown to be a good marker for CD diagnosis by conventional serology. Moreover, anti-SAPA antibodies (IgM or IgG) have been detected in 90% of acute chagasic patients and in 7–10% of chronic patients [81]. This biosensor can distinguish between congenitally infected and non-infected infants when cord blood is tested, the sensitivity is in the ng/mL range (3.03 ng/mL) and the linear response between 10 and 200 ng/mL [80]. This device could facilitate and speed up the unequivocal diagnosis of congenital transmission, since it does not depend on the detection of parasite in newborns or the clearance of maternal antibodies in infants. To continue its validation, a large set of samples from newborns and infants at different ages needs to be tested.

A biosensor to be used in PoC units for serodiagnosis of infectious diseases was fully developed by Cortina et al. [82]. In this device, antigen-coated magnetic beads are used to detect antibodies in serum samples. Immunocapture is amplified using HRP-conjugated secondary antibodies, and the beads magnetically collected are placed on an electrode surface to detect peroxidase activity amperometrically [82]. For CD diagnosis, recombinant proteins of different *T. cruzi* antigens (Ag1, Ag36, SAPA, and TSSA) were used to coat superparamagnetic beads and tested with serum samples from CD patients and HC. Results showed that the magnetic bead-based biosensor discriminates infected from non-infected serum with a minimal overlap and excellent signal-to-noise ratio. It also showed a high level of accuracy in diagnosis, similar to ELISA and IFA, as well as having similar sensitivity and selectivity [82]. Despite the different steps to be performed during the assay, it can be done in PoC units as it uses an eight-channel portable potentiostat powered by a rechargeable battery. The device is not yet commercialized, but there is potential for its use in the diagnosis of CD and other parasitic infections.

All the above biosensors detect *T. cruzi* antigens indirectly by peroxidase activity, which requires multiple steps of incubation with a secondary antibody and redox reagents. In contrast, a free-label biosensor that uses optical transducers needs fewer steps and a shorter assay time and still shows high sensitivity. An optical immunosensor (SPRCruzi) with a surface plasmon resonance (SPR) transducer was recently developed for CD diagnosis [83]. SPR sensors use surface plasmons, which are electromagnetic waves that can be excited by light at gold sensor interfaces, to transduce a biochemical interaction. In brief, the interaction changes the SPR baseline, and real-time measurement of specific analytes in unknown samples flushed over the sensor can be performed; for a review, see [77, 84].

To build SPRCruzi, Luz and collaborators used soluble antigens of *T. cruzi* epimastigotes immobilized on a sensor chip. The biosensor was able to discriminate positive from negative serum, including those infected with other related parasites, and detect antibodies in serum dilutions as high as 1280x. In 2016, the same group tested SPRCruzi with a higher number of positive and negative serum samples and compared their results with conventional serological tests. SPRCruzi showed 100% sensitivity (cutoff $\Delta\theta$SPR 17.2°), 99.6% global accuracy, and a better specificity (97.2%) compared to ELISA [80]. Nonetheless, the use of this device is still limited to laboratories since expensive and heavy SPR equipment is required.

Lastly, a CD nanowire electrical sensor based on field-effect transistor (FET) technology was designed last year by Janissen et al. [85]. In FET sensors, the current-carrying capability of a semiconductor is used, and the sensor response is interpreted as a result of a shift in the threshold voltage of the field-effect structure [77]. In the Janissen et al. device, anti-*T. cruzi* IBMP8-1 antibodies were immobilized on a surface using a biocompatible ethanolamine and poly(ethylene glycol) derivate coating. This biosensor reached detection limits in the femtomolar range (6 fM) for a recombinant IBMP8-1 protein. This limit of detection is 1000-fold lower than ELISA (30 nM), PCR (10 nM), and even electrochemical immunosensors (20 pM). In addition, the assay is fast, taking less than 30 min, and label-free. This highly sensitive biosensor still needs to be tested in human samples and optimized for PoC unit use; but so far it is the only CD biosensor that does not depend on the detection of anti-*T. cruzi* antibodies.

6.3 MicroRNA

MicroRNAs play key roles in intracellular and extracellular protein expression and regulation of biochemical pathways. They may be associated with regulating cellular apoptosis, proliferation, differentiation, metabolism, invasion, and migration. Different studies have also shown that microRNAs may serve crucial functions in the progression of numerous cancers and other diseases and consequently can be used as biomarkers for prognosis of disease progression [86]. For example, microRNA-208a (miR-208a) encoded by the α-myosin heavy chain (MHC) gene has been shown to be involved in pathological cardiac growth, fibrosis, and upregulation of β-MHC expression in human dilated cardiomyopathy (DCM) [87]. It is also an early diagnostic biomarker of acute myocardial infarction (AMI) and can be used for prognosis postinfarction and disease monitoring [88].

Different microRNAs and their mRNA targets appear dysregulated in chronic Chagas disease cardiomyopathy (CCC) patients and CD murine models during acute infection [89, 90]. This year, Linhares-Lacerda et al. suggested microRNA208a could be a potential biomarker of chronic indeterminate CD (CID) [91]. Their results have shown upregulation of miR-208a in serum of CD patients in the indeterminate stage compared to chronically infected cardiac patients with DCM. This suggests the microRNA is participating in the early-onset events responsible for activation of fibrosis and cardiac dysfunction processes in CD. The same microRNA has been reported to be downregulated in the heart muscle of CD patients with CCC or DCM compared to non-infected controls [89] but upregulated in endomyocardial tissues of non-infected DCM patients [87]. Clearly a better understanding and tests in a large number of patients is needed to validate this biomarker. Nevertheless, this study opens the door to looking at microRNA as a possible analyte for CD diagnosis and prognosis.

6.4 *Omics-Based Applications*

Omics-based applications are formidable new technological resources for investigating the status of human diseases and understanding the pathophysiology of disease processes. They can generate enormous amounts of data with high fidelity thanks to recent advances in chromatography, mass spectrometry, and bioinformatics. Furthermore, omics outputs have the advantage of complementarity, enabling cross-corroboration and cross-validation [92]. In the search for biomarkers and tools for CD diagnosis and drug treatment efficacy assessment, our group is using omics applications to detect changes in the proteome, metabolome, and lipidome of CD patients compared to healthy people (Fig. 1). We are using our omics studies to build assays that could be widely employed for diagnosis, prognosis, and evaluation of treatment efficacy of new drugs. To this end, we are focusing primarily on the identification of new host markers following comparative analysis of serum samples issued from patients diagnosed with CD, treated, and followed up several years after treatment and in some cases until they reach seroreversion, the only current surrogate marker for parasitological cure.

In our earlier work involving mass spectrometry serum protein profiling studies, we identified highly sensitive and specific host protein markers (see Sect. 5.2) [50, 51]. We started our proteomics studies using MS SELDI technology for intact serum proteins analysis; in spite of the promising results, the SELDI technology did not allow the direct identification of proteins by tandem MS and had low resolution. Recently, we have used an ultrahigh-resolution quadrupole time-of-flight (UHR-

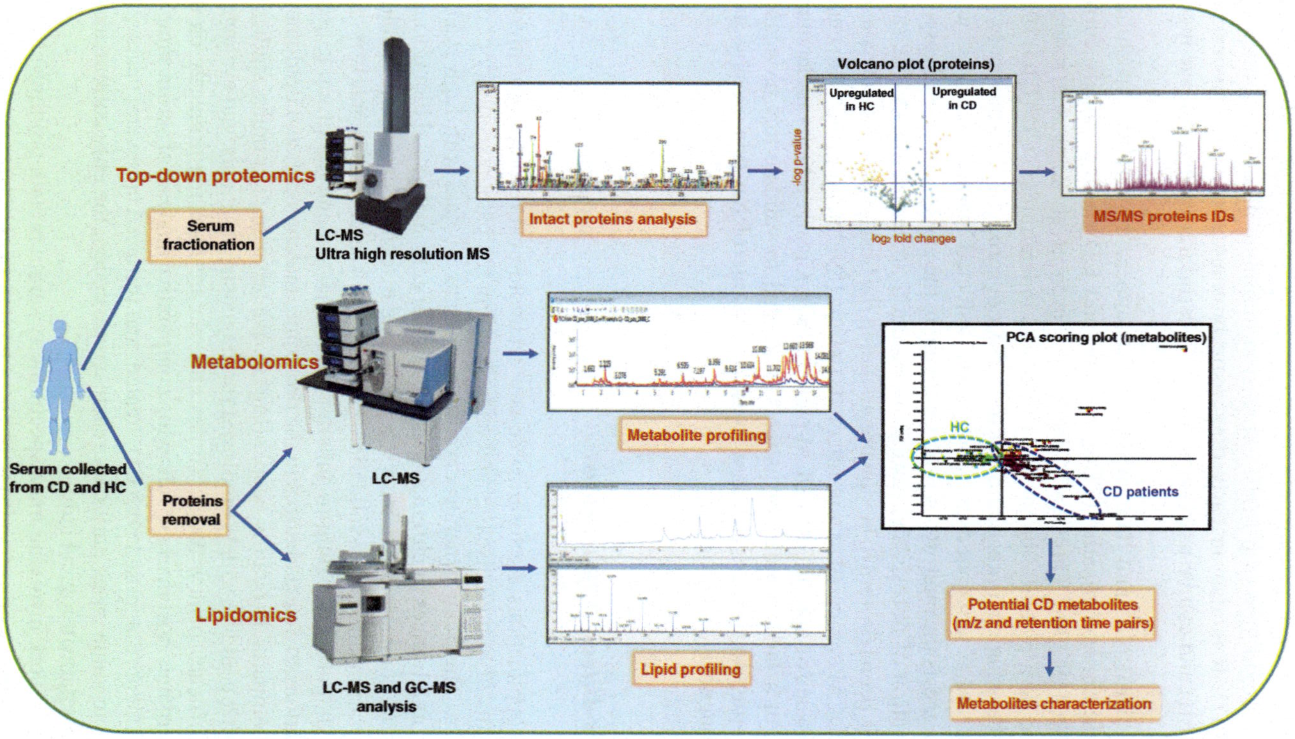

Fig. 1 Omics-based applications for CD biomarkers discovery. Serum proteins, all metabolites or only lipids from Chagas Disease (CD) and Healthy Control (HC) populations are analyzed by different LC-MS systems. Features (Intact proteins in *top-down proteomics*, candidates metabolites in *metabolomics* or candidate lipids in *lipidomics*) profiles are compared between population to find those that can discriminate CD of HC. Rigorous statistical analysis is done in all cases and candidates are further study by LC-MS/(MS^n) for identification and verification

QTOF) tandem mass spectrometer coupled to ultrahigh-performance liquid chromatography (HPLC) for the same purpose. Despite the differences between MS platforms and sample processing, we have been able to reproduce our previous findings, reinforcing the robustness of our data.

In order to validate some of the protein biomarkers identified (ApoA1 and FBN fragments) using our new MS platform, we recently attempted to correlate the presence/absence of these fragments with seroreversion, which is currently the gold standard of parasitological cure in a cohort study of CD children treated with BZN. Compared to adults, seronegative conversion in children occurs in a few months to a few years, which made children's serum samples ideal for the validation of these biomarkers. Our MS analysis and specific immunoblot results showed these fragments were absent in serum at seroreversion in the entire CD pediatric population and in some cases at the end of BZN treatment even when children remained seropositive [93]. Although still preliminary, these data suggest that ApoA1 and FBN fragments could be used as biomarkers of parasitological cure and predict cure earlier than serology, which will make them better endpoint surrogates. Additional studies are needed to further explore the real potential of these new biomarkers, for instance, an evaluation of fragment disappearance kinetics following treatment until seroreversion.

While the proteome is of great importance, the metabolome can also provide an excellent pathophysiological understanding of disease, as proteins have functions in a range of complex metabolic reactions and their activity ultimately affects the phenotype. Metabolites (low molecular weight organic and inorganic chemicals) are simpler to study compared to proteins, are the final downstream products, and give a sensitive and rapid measurement of the phenotype. Various disease states may be characterized by a specific metabolite or a pattern of metabolite changes. Metabolomics has been successfully applied to clinical conditions including inborn errors of metabolism, cardiovascular disease, and cancer to identify biomarkers related to diagnosis, assessment of disease severity, or drug toxicity/efficacy—for a review, see [94]. Changes in the human metabolome due to *T. cruzi* infection are as yet unknown, and metabolites have not been explored as biomarkers for CD. Taking advantage of our high-resolution platform, we are currently studying the serum metabolite profile of CD patients, looking for biological differences after treatment and compared to a healthy population.

Finally, considering the evidence that associates *T. cruzi* with the adipose tissue, ApoA1 and with HDL and LDL modifications, as well as its interaction with LDL receptors [95], the serum lipid profile is an interesting and important metabolome component in the search for possible CD biomarkers. Lipidomics has been defined as "the full characterization of lipid molecular species and of their biological roles with respect to expression of proteins involved in lipid metabolism and function, including gene regulation." Although lipids are not used in clinical applications yet, many individual lipids have been associated with the evolution of different cancers, cardiovascular, neuropsychiatric, respiratory, and kidney diseases [96] and can provide information on disease status. Together with our metabolomics study, we are presently characterizing the lipidome in CD patients using an untargeted approach.

We expect these holistic approaches lead to the identification of analytes that can discriminate between infected and non-infected populations.

Taken together, data issued from these different technologies will hopefully speed up the identification of markers suitable for clinical settings and proof of concept clinical studies and during drug development.

6.5 Other Technologies

Other no less important approaches to finding potential biomarkers rely on the identification of specific antibodies and the detection of anti-live *T. cruzi* antibodies by flow cytometry. To discover pathogen-specific linear B-cell epitopes from clinical samples, Carmona et al. used a highly multiplexed platform based on next-generation high-density peptide microarrays to map antibody specificities in CD [97]. In this approach, individual peptides (~180,000) are synthesized in situ on a glass slide at high densities, which reduces cost and allows a high-throughput and precise mapping of antibodies. After screening the arrays with antibodies purified from CD patients and HC, 2031 disease-specific peptides and 97 novel parasite antigens were identified, together with their linear B-cell epitopes [97]. Recently, Mucci et al. assessed the serological performance of 27 of these epitopes and their use in a multipeptide-based diagnostic method [98]. Seven peptides were evaluated in ELISA against 199 serum samples from CD and HC, including samples from leishmaniasis subjects. The assay showed a sensitivity of 96.3% and a specificity of 99.15% for CD, which suggests that the peptides could be used in CD diagnosis; however, their usefulness in treatment efficacy evaluation needs to be further studied.

As mentioned in Sect. 5.1, the detection of lytic antibodies against live parasites is an alternative to the evaluation of specific antibodies. In a double-blinded study with 94 coded samples, Martins-Filho et al. found that anti-live trypomastigote antibody (ALTA) measured by flow cytometry (FC) was able to discriminate not treated (NT), treated but not cured (TNC), and treated and cured (TC) patients when using a 1:256 serum dilution [99]. In a larger study population with four different cohorts, the same group demonstrated that anti-fixed epimastigote antibody (AFEA) discriminates the clinical status of CD patients after treatment at higher serum dilutions (1/2048). FC-AFEA-IgG showed 100% sensitivity (80.3–100%) and specificity (85.6–100%) with positive and negative predictive values of 100%. This suggests both antibodies measured by FC are not only good enough for diagnosis and prognosis but could be useful as criteria of cure.

7 The Way Forward

The identification and validation of biomarkers is in general is a very challenging process. For Chagas disease, the definition of cure or clinical benefit following treatment is clearly another major challenge. In fact, in principle, we are looking for a surrogate of a surrogate for clinical benefit, i.e., looking for a surrogate of serological cure assuming that the latter is a surrogate marker for clinical cure or halting of progression of the disease, and searching for a surrogate marker of serological cure that allows the rapid determination of seroreversion, indicating that *T. cruzi* parasites have been eliminated from the patient's body. The identification and validation of new biomarkers for Chagas disease is, therefore, a major challenge but a serious and important one that needs to be tackled in the endeavor to develop drugs for CD. Unfortunately, no biomarkers have yet progressed to clinical validation. No correlation has been made between either the absence of *T. cruzi* DNA in blood (as assessed by PCR) or the reduction in titers of specific *T. cruzi* antigens or antibodies (as assessed using serological tests at specific timepoints after treatment) and seroreversion; which make these markers potential pharmacodynamic markers but not surrogates for parasitological cure at this point. Newly identified potential markers using different technology platforms, from either the host (−omics) or other antigens of the parasite (microarrays), are showing promise, but more work is needed to assess their validity. In particular, the development, optimization, and analytical verification and validation of tests for these new markers, either as single prototypes or in a multiplex, are needed before moving forward to clinical validation. It is also reasonable to believe that a set of biomarkers rather that a single "magic bullet" might be needed to ensure a correlation with parasitological cure.

Pending a robust validated multiplex assay with valid biomarkers, the challenge of being able to clinically validate these markers and obtain regulatory acceptance remains significant. In order to do this, a sufficient number of high-quality samples from well-defined cohorts are needed to move forward (biostatistical plan). The use of retrospective cohorts could be envisaged, but a large prospective study (clinical trial with long post-therapeutic follow-up of patients) might be required to verify that markers are surrogates for seroconversion. In either case, the entire Chagas community needs to make a concerted unified effort.

Acknowledgments The authors wish to thank Louise Burrows for the editing of this manuscript. The National Reference Centre for Parasitology is supported by Public Health Agency of Canada/National Microbiology Laboratory grant WPG-6-39147 (005), the Foundation of the Montreal General Hospital, The Foundation of the McGill University Health Centre and the Research Institute of the McGill University Health Centre. The DND*i* is grateful for its donors, public and private, who have provided funding for the DND*i* since its inception in 2003. A full list of DND*i*'s donors can be found at http://www.dndi.org/donors/donors/. The DND*i* received financial support from the following donors: the Department for International Development (DFID), UK; Reconstruction Credit Institution-Federal Ministry of Education and Research (KfW-BMBF), Germany; Directorate-General for International Cooperation (DGIS), the Netherlands; Swiss Agency for Development and Cooperation (SDC), Switzerland; and Médecins Sans Frontières (Doctors Without Borders), international. The donors had no role in the study design, data collection and analysis, decision to publish, or preparation of the manuscript.

References

1. Pinazo M-J, Thomas M-C, Bustamante J, ICd A, Lopez M-C, Gascon J. Biomarkers of therapeutic responses in chronic Chagas disease: state of the art and future perspectives. Mem Inst Oswaldo Cruz. 2015;110:422–32.
2. Pinazo MJ, Thomas MC, Bua J, Perrone A, Schijman AG, Viotti RJ, Ramsey JM, Ribeiro I, Sosa-Estani S, Lopez MC, Gascon J. Biological markers for evaluating therapeutic efficacy in

Chagas disease, a systematic review. Expert Rev Anti-Infect Ther. 2014;12:479–96. https://doi.org/10.1586/14787210.2014.899150.
3. Pinho RT, Waghabi MC, Cardillo F, Mengel J, Antas PR. Scrutinizing the biomarkers for the neglected Chagas disease: how remarkable! Front Immunol. 2016;7:306. https://doi.org/10.3389/fimmu.2016.00306.
4. Chagas C. Nova tripanozomiaze humana: estudos sobre a morfolojia e o ciclo evolutivo do Schizotrypanum cruzi n. gen., n. sp., ajente etiolojico de nova entidade morbida do homem. Mem Inst Oswaldo Cruz. 1909;1:159–218.
5. Schmunis GA, Yadon ZE. Chagas disease: a Latin American health problem becoming a world health problem. Acta Trop. 2010;115:14–21.
6. Meymandi SK, Hernandez S, Forsyth CJ. A community-based screening program for Chagas disease in the USA. Trends Parasitol. 2017;33:828–31.
7. Requena-Méndez A, Aldasoro E, de Lazzari E, Sicuri E, Brown M, Moore DA, Gascon J, Muñoz J. Prevalence of Chagas disease in Latin-American migrants living in Europe: a systematic review and meta-analysis. PLoS Negl Trop Dis. 2015;9:e0003540.
8. Jackson Y, Pinto A, Pett S. Chagas disease in Australia and New Zealand: risks and needs for public health interventions. Tropical Med Int Health. 2014;19:212–8.
9. Moolani Y, Bukhman G, Hotez PJ. Neglected tropical diseases as hidden causes of cardiovascular disease. PLoS Negl Trop Dis. 2012;6:e1499.
10. WHO. Chagas disease. Geneva: World Health Organization; 2018. www.who.int/mediacentre/factsheets/fs340/en/. Accessed 16 Feb 2018
11. Cucunubá ZM, Okuwoga O, Basáñez M-G, Nouvellet P. Increased mortality attributed to Chagas disease: a systematic review and meta-analysis. Parasit Vectors. 2016;9:42.
12. Urbina JA. The long road towards a safe and effective treatment of chronic Chagas disease. Lancet Infect Dis. 2018;18:363–5.
13. Ayub-Ferreira SM, Mangini S, Issa VS, Cruz FD, Bacal F, Guimarães GV, Chizzola PR, Conceição-Souza GE, Marcondes-Braga FG, Bocchi EA. Mode of death on Chagas heart disease: comparison with other etiologies. a subanalysis of the REMADHE prospective trial. PLoS Negl Trop Dis. 2013;74:e2176.
14. Brum-Soares L, Cubides J-C, Burgos I, Monroy C, Castillo L, González S, Viñas PA, Urrutia PPP. High seroconversion rates in Trypanosoma cruzi chronic infection treated with benznidazole in people under 16 years in Guatemala. Rev Soc Bras Med Trop. 2016;49:721–7.
15. Olivera MJ, Cucunubá ZM, Álvarez CA, Nicholls RS. Safety profile of nifurtimox and treatment interruption for chronic Chagas disease in Colombian adults. Am J Trop Med Hyg. 2015;93:1224–30.
16. Cançado JR. Criteria of Chagas disease cure. Mem Inst Oswaldo Cruz. 1999;94:331–5.
17. de Lana M, Martins-Filho OA. Revisiting the posttherapeutic cure criterion in Chagas disease: time for new methods, more questions, doubts, and polemics or time to change old concepts? Biomed Res Int. 2015;2015:652985. https://doi.org/10.1155/2015/652985.
18. Temple R. A regulatory authority's opinion about surrogate endpoints. In: Nimmo WS, Tuck GT, editors. Clinical measurement in drug evaluation. New York, NY: Wiley; 1995. p. 1–22.
19. Balouz V, Agüero F, Buscaglia CA. Chagas disease diagnostic applications: present knowledge and future steps. In: Advances in parasitology, vol. 97. New York City, NY: Elsevier; 2017. p. 1–45.
20. FDA. FDA approves first U.S. treatment for Chagas disease. www.fda.gov/NewsEvents/Newsroom/PressAnnouncements/ucm573942.htm. Accessed 16 Feb 2018.
21. Sosa ES. Soluble platelet selectin (sP-selectin) and soluble vascular cell adhesion molecule-1 (sVCAM-1) decrease during therapy with benznidazole in children with indeterminate form of Chagas' disease. Clin Exp Immunol. 1999;118:423–7.
22. de Andrade ALSS, Zicker F, de Oliveira RM, e Silva SA, Luquetti A, Travassos LR, Almeida IC, de Andrade SS, de Andrade JG, Martelli CM. Randomised trial of efficacy of benznidazole in treatment of early Trypanosoma cruzi infection. Lancet. 1996;348:1407–13.
23. Henao-Martínez AF, Schwartz DA, Yang IV. Chagasic cardiomyopathy, from acute to chronic: is this mediated by host susceptibility factors? Trans R Soc Trop Med Hyg. 2012;106:521–7.

24. Deng X, Sabino EC, Cunha-Neto E, Ribeiro AL, Ianni B, Mady C, Busch MP, Seielstad M. Genome wide association study (GWAS) of Chagas cardiomyopathy in Trypanosoma cruzi seropositive subjects. PLoS One. 2013;8:e79629.
25. Del Puerto F, Nishizawa JE, Kikuchi M, Roca Y, Avilas C, Gianella A, Lora J, Velarde FUG, Miura S, Komiya N. Protective human leucocyte antigen haplotype, HLA-DRB1* 01-B* 14, against chronic Chagas disease in Bolivia. PLoS Negl Trop Dis. 2012;6:e1587.
26. Clipman SJ, Henderson-Frost J, Fu KY, Bern C, Flores J, Gilman RH. Genetic association study of NLRP1, CARD, and CASP1 inflammasome genes with chronic Chagas cardiomyopathy among Trypanosoma cruzi seropositive patients in Bolivia. PLoS One. 2018;13:e0192378.
27. Nohaile M. The biomarker is not the end. Drug Discov Today. 2011;16:878–83.
28. Porrás AI, Yadon ZE, Altcheh J, Britto C, Chaves GC, Flevaud L, Martins-Filho OA, Ribeiro I, Schijman AG, Shikanai-Yasuda MA. Target product profile (TPP) for Chagas disease point-of-care diagnosis and assessment of response to treatment. PLoS Negl Trop Dis. 2015;9:e0003697.
29. Tuck MK, Chan DW, Chia D, Godwin AK, Grizzle WE, Krueger KE, Rom W, Sanda M, Sorbara L, Stass S. Standard operating procedures for serum and plasma collection: early detection research network consensus statement standard operating procedure integration working group. J Proteome Res. 2008;8:113–7.
30. Omenn GS, States DJ, Adamski M, Blackwell TW, Menon R, Hermjakob H, Apweiler R, Haab BB, Simpson RJ, Eddes JS. Overview of the HUPO Plasma Proteome Project: Results from the pilot phase with 35 collaborating laboratories and multiple analytical groups, generating a core dataset of 3020 proteins and a publicly-available database. Proteomics. 2005;5:3226–45.
31. Requena-Méndez A, López MC, Angheben A, Izquierdo L, Ribeiro I, Pinazo M-J, Gascon J, Muñoz J. Evaluating Chagas disease progression and cure through blood-derived biomarkers: a systematic review. Expert Rev Anti-Infect Ther. 2013;11:957–76.
32. Altcheh J, Corral R, Biancardi M, Freilij H. Anti-F2/3 antibodies as cure marker in children with congenital Trypanosoma cruzi infection. Medicina. 2003;63:37–40.
33. Cancado JR. Long term evaluation of etiological treatment of chagas disease with benznidazole. Rev Inst Med Trop Sao Paulo. 2002;44:29–37.
34. de Apparecida Levy AM, Boainain E, Kloetzel JK. In situ indirect fluorescent antibody: a new specific test to detect ongoing chagasic infections. J Clin Lab Anal. 1996;10:98–103.
35. Andrade ALS, Martelli CM, Oliveira RM, Silva SA, Aires AI, Soussumi LM, Covas DT, Silva LS, Andrade JG, Travassos LR. Benznidazole efficacy among Trypanosoma cruzi-infected adolescents after a six-year follow-up. Am J Trop Med Hyg. 2004;71:594–7.
36. Pinazo M-J, de Jesus Posada E, Izquierdo L, Tassies D, Marques A-F, de Lazzari E, Aldasoro E, Muñoz J, Abras A, Tebar S. Altered hypercoagulability factors in patients with chronic Chagas disease: potential biomarkers of therapeutic response. PLoS Negl Trop Dis. 2016;10:e0004269.
37. Almeida IC, Covas DT, Soussumi LM, Travassos LR. A highly sensitive and specific chemiluminescent enzyme-linked immunosorbent assay for diagnosis of active Trypanosoma cruzi infection. Transfusion. 1997;37:850–7.
38. Negrette OS, Valdéz FJS, Lacunza CD, Bustos MFG, Mora MC, Uncos AD, Basombrío MÁ. Serological evaluation of specific-antibody levels in patients treated for chronic Chagas' disease. Clin Vaccine Immunol. 2008;15:297–302.
39. Fabbro DL, Olivera V, Bizai ML, Denner S, Diez C, Mancipar I, Streiger M, Arias E, del Barco M, Mendicino D. Humoral immune response against P2β from Trypanosoma cruzi in persons with chronic Chagas disease: its relationship with treatment against parasites and myocardial damage. Am J Trop Med Hyg. 2011;84:575–80.
40. Fernández-Villegas A, Pinazo MJ, Marañón C, Thomas MC, Posada E, Carrilero B, Segovia M, Gascon J, López MC. Short-term follow-up of chagasic patients after benznidazole treatment using multiple serological markers. BMC Infect Dis. 2011;11:206.
41. Flechas ID, Cuellar A, Cucunubá ZM, Rosas F, Velasco V, Steindel M, del Carmen Thomas M, López MC, González JM, Puerta CJ. Characterising the KMP-11 and HSP-70 recombinant antigens' humoral immune response profile in chagasic patients. BMC Infect Dis. 2009;9:186.
42. Krettli AU. The utility of anti-trypomastigote lytic antibodies for determining cure of Trypanosoma cruzi infections in treated patients: an overview and perspectives. Mem Inst Oswaldo Cruz. 2009;104:142–51.

43. Krautz GM, Galvão L, Cancado JR, Guevara-Espinoza A, Ouaissi A, Krettli AU. Use of a 24-kilodalton Trypanosoma cruzi recombinant protein to monitor cure of human Chagas' disease. J Clin Microbiol. 1995;33:2086–90.
44. Estani SS, Segura EL, Ruiz AM, Velazquez E, Porcel BM, Yampotis C. Efficacy of chemotherapy with benznidazole in children in the indeterminate phase of Chagas' disease. Am J Trop Med Hyg. 1998;59:526–9.
45. Fabbro D, Velazquez E, Bizai ML, Denner S, Olivera V, Arias E, Pravia C, Ruiz AM. Evaluation of the ELISA-F29 test as an early marker of therapeutic efficacy in adults with chronic Chagas disease. Rev Inst Med Trop Sao Paulo. 2013;55:167–72.
46. Cooley G, Etheridge RD, Boehlke C, Bundy B, Weatherly DB, Minning T, Haney M, Postan M, Laucella S, Tarleton RL. High throughput selection of effective serodiagnostics for Trypanosoma cruzi infection. PLoS Negl Trop Dis. 2008;2:e316.
47. Viotti R, Vigliano C, Álvarez MG, Lococo B, Petti M, Bertocchi G, Armenti A, De Rissio AM, Cooley G, Tarleton R. Impact of aetiological treatment on conventional and multiplex serology in chronic Chagas disease. PLoS Negl Trop Dis. 2011;5:e1314.
48. Meira WS, Galvão LM, Gontijo ED, Machado-Coelho GL, Norris KA, Chiari E. Use of the Trypanosoma cruzi recombinant complement regulatory protein to evaluate therapeutic efficacy following treatment of chronic chagasic patients. J Clin Microbiol. 2004;42:707–12.
49. Zrein M, Granjon E, Gueyffier L, Caillaudeau J, Liehl P, Pottel H, Cardoso CS, Oliveira CDL, de Oliveira LC, Lee T-H. A novel antibody surrogate biomarker to monitor parasite persistence in Trypanosoma cruzi-infected patients. PLoS Negl Trop Dis. 2018;12:e0006226.
50. Ndao M, Spithill TW, Caffrey R, Li H, Podust VN, Perichon R, Santamaria C, Ache A, Duncan M, Powell MR, Ward BJ. Identification of novel diagnostic serum biomarkers for Chagas' disease in asymptomatic subjects by mass spectrometric profiling. J Clin Microbiol. 2010;48:1139–49. doi:JCM.02207-09
51. Santamaria C, Chatelain E, Jackson Y, Miao Q, Ward BJ, Chappuis F, Ndao M. Serum biomarkers predictive of cure in Chagas disease patients after nifurtimox treatment. BMC Infect Dis. 2014;14:302. https://doi.org/10.1186/1471-2334-14-302.
52. Pinazo M-J, Tassies D, Muñoz J, Fisa R, de Jesús Posada E, Monteagudo J, Ayala E, Gállego M, Reverter J-C, Gascon J. Hypercoagulability biomarkers in Trypanosoma cruzi-infected patients. Thromb Haemost. 2011;106:617–23.
53. Laucella SA, Mazliah DP, Bertocchi G, Alvarez MG, Cooley G, Viotti R, Albareda MC, Lococo B, Postan M, Armenti A. Changes in Trypanosoma cruzi-specific immune responses after treatment: surrogate markers of treatment efficacy. Clin Infect Dis. 2009;49:1675–84.
54. Dutra WO, Cançado JR, Pereira ME, Brígido-Nunes R, Galvão L, Colley DG, Brener Z, Gazzinelli G, Carvalho-Parra JF. Influence of parasite presence on the immunologic profile of peripheral blood mononuclear cells from chagasic patients after specific drug therapy. Parasite Immunol. 1996;18:579–85.
55. Sathler-Avelar R, Vitelli-Avelar DM, Massara RL, de Lana M, Dias JCP, Teixeira-Carvalho A, Elói-Santos SM, Martins-Filho OA. Etiological treatment during early chronic indeterminate Chagas disease incites an activated status on innate and adaptive immunity associated with a type 1-modulated cytokine pattern. Microbes Infect. 2008;10:103–13.
56. Argüello RJ, Albareda MC, Alvarez MG, Bertocchi G, Armenti AH, Vigliano C, Meckert PC, Tarleton RL, Laucella SA. Inhibitory receptors are expressed by Trypanosoma cruzi-specific effector T cells and in hearts of subjects with chronic Chagas disease. PLoS One. 2012;7:e35966.
57. Sguassero Y, Cuesta CB, Roberts KN, Hicks E, Comandé D, Ciapponi A, Sosa-Estani S. Course of chronic Trypanosoma cruzi infection after treatment based on parasitological and serological tests: a systematic review of follow-up studies. PLoS One. 2015;10:e0139363.
58. Britto CC. Usefulness of PCR-based assays to assess drug efficacy in Chagas disease chemotherapy: value and limitations. Mem Inst Oswaldo Cruz. 2009;104:122–35.
59. Wei B, Chen L, Kibukawa M, Kang J, Waskin H, Marton M. Development of a PCR assay to detect low level Trypanosoma cruzi in blood specimens collected with PAXgene blood DNA tubes for clinical trials treating Chagas disease. PLoS Negl Trop Dis. 2016;10:e0005146.

60. Morillo CA, Waskin H, Sosa-Estani S, del Carmen Bangher M, Cuneo C, Milesi R, Mallagray M, Apt W, Beloscar J, Gascon J. Benznidazole and posaconazole in eliminating parasites in asymptomatic T. cruzi carriers: the STOP-CHAGAS trial. J Am Coll Cardiol. 2017;69:939–47.
61. Eiro C, Melisa D, Alvarez MG, Cooley G, Viotti RJ, Bertocchi GL, Lococo B, Albareda MC, De Rissio AM, Natale MA. The significance of discordant serology in Chagas disease: enhanced T-cell immunity to Trypanosoma cruzi in serodiscordant subjects. Front Immunol. 2017;8:1141.
62. Moure Z, Angheben A, Molina I, Gobbi F, Espasa M, Anselmi M, Salvador F, Tais S, Sánchez-Montalvá A, Pumarola T. Serodiscordance in chronic Chagas disease diagnosis: a real problem in non-endemic countries. Clin Microbiol Infect. 2016;22:788–92.
63. Moure Z, Sulleiro E, Iniesta L, Guillen C, Molina I, Alcover MM, Riera C, Pumarola T, Fisa R. The challenge of discordant serology in Chagas disease: the role of two confirmatory techniques in inconclusive cases. Acta Trop. 2018;185:144–8. https://doi.org/10.1016/j.actatropica.2018.05.010.
64. Cardoso CS, Sabino EC, Oliveira CDL, de Oliveira LC, Ferreira AM, Cunha-Neto E, Bierrenbach AL, Ferreira JE, Haikal DSA, Reingold AL. Longitudinal study of patients with chronic Chagas cardiomyopathy in Brazil (SaMi-Trop project): a cohort profile. BMJ Open. 2016;6:e011181.
65. Galvão L, Nunes R, Cançado J, Brener Z, Krettli A. Lytic antibody titre as a means of assessing cure after treatment of Chagas disease: a 10 years follow-up study. Trans R Soc Trop Med Hyg. 1993;87:220–3.
66. Ndao M, Rainczuk A, Rioux M-C, Spithill TW, Ward BJ. Is SELDI-TOF a valid tool for diagnostic biomarkers? Trends Parasitol. 2010;26:561–7.
67. Ndao M. Biomarker discovery in serum/plasma using surface enhanced laser desorption ionization time of flight (SELDI-TOF) mass spectrometry. In: Clarke CH, Bankert McCarthy DL, editors. SELDI-TOF mass spectrometry. New York, NY: Springer; 2012. p. 67–79.
68. Gomes J, Bahia-Oliveira L, Rocha M, Martins-Filho O, Gazzinelli G, Correa-Oliveira R. Evidence that development of severe cardiomyopathy in human Chagas' disease is due to a Th1-specific immune response. Infect Immun. 2003;71:1185–93.
69. Bustamante JM, Bixby LM, Tarleton RL. Drug-induced cure drives conversion to a stable and protective CD8+ T central memory response in chronic Chagas disease. Nat Med. 2008;14:542.
70. Bustamante JM, Craft JM, Crowe BD, Ketchie SA, Tarleton RL. New, combined, and reduced dosing treatment protocols cure Trypanosoma cruzi infection in mice. J Infect Dis. 2013;209:150–62.
71. Gold L, Ayers D, Bertino J, Bock C, Bock A, Brody EN, Carter J, Dalby AB, Eaton BE, Fitzwater T. Aptamer-based multiplexed proteomic technology for biomarker discovery. PLoS One. 2010;5:e15004.
72. Morris KN, Jensen KB, Julin CM, Weil M, Gold L. High affinity ligands from in vitro selection: complex targets. Proc Natl Acad Sci. 95(6):2902–7.
73. Shamah SM, Healy JM, Cload ST. Complex target SELEX. Acc Chem Res (1998). 2008;41:130–8.
74. Nagarkatti R, Bist V, Sun S, de Araujo FF, Nakhasi HL, Debrabant A. Development of an aptamer-based concentration method for the detection of Trypanosoma cruzi in blood. PLoS One. 2012;7:e43533.
75. Nagarkatti R, de Araujo FF, Gupta C, Debrabant A. Aptamer based, non-PCR, non-serological detection of Chagas disease biomarkers in Trypanosoma cruzi infected mice. PLoS Negl Trop Dis. 2014;8:e2650.
76. de Araujo FF, Nagarkatti R, Gupta C, Marino AP, Debrabant A. Aptamer-based detection of disease biomarkers in mouse models for chagas drug discovery. PLoS Negl Trop Dis. 2015;9:e3451.
77. Sin ML, Mach KE, Wong PK, Liao JC. Advances and challenges in biosensor-based diagnosis of infectious diseases. Expert Rev Mol Diagn. 2014;14:225–44.
78. Rocha-Gaso M-I, Villarreal-Gómez L-J, Beyssen D, Sarry F, Reyna M-A, Ibarra-Cerdeña C-N. Biosensors to diagnose Chagas disease: a brief review. Sensors. 2017;17:2629.

79. Pereira SV, Bertolino FA, Fernández-Baldo MA, Messina GA, Salinas E, Sanz MI, Raba J. A microfluidic device based on a screen-printed carbon electrode with electrodeposited gold nanoparticles for the detection of IgG anti-Trypanosoma cruzi antibodies. Analyst. 2011;136:4745–51.
80. Regiart M, Pereira SV, Bertolino FA, Garcia CD, Raba J, Aranda PR. An electrochemical immunosensor for anti-T. cruzi IgM antibodies, a biomarker for congenital Chagas disease, using a screen-printed electrode modified with gold nanoparticles and functionalized with shed acute phase antigen. Microchim Acta. 2016;183:1203–10.
81. Marcipar IS, Lagier CM. Advances in serological diagnosis of Chagas' disease by using recombinant proteins. In: Current topics in tropical medicine. London: InTechOpen; 2012. p. 273–92. https://doi.org/10.5772/28100.
82. Cortina ME, Melli LJ, Roberti M, Mass M, Longinotti G, Tropea S, Lloret P, Serantes DAR, Salomón F, Lloret M. Electrochemical magnetic microbeads-based biosensor for point-of-care serodiagnosis of infectious diseases. Biosens Bioelectron. 2016;80:24–33.
83. Luz JG, Souto DE, Machado-Assis GF, de Lana M, Kubota LT, Luz RC, Damos FS, Martins HR. Development and evaluation of a SPR-based immunosensor for detection of anti-Trypanosoma cruzi antibodies in human serum. Sensors Actuators B Chem. 2015;212:287–96.
84. Phillips KS, Cheng Q. Recent advances in surface plasmon resonance based techniques for bioanalysis. Anal Bioanal Chem. 2007;387:1831–40.
85. Janissen R, Sahoo PK, Santos CA, da Silva AM, von Zuben AA, Souto DE, Costa AD, Celedon P, Zanchin NI, Almeida DB. InP nanowire biosensor with tailored biofunctionalization: ultrasensitive and highly selective disease biomarker detection. Nano Lett. 2017;17:5938–49.
86. Wang J, Chen J, Sen S. MicroRNA as biomarkers and diagnostics. J Cell Physiol. 2016;231:25–30.
87. Satoh M, Minami Y, Takahashi Y, Tabuchi T, Nakamura M. Expression of microRNA-208 is associated with adverse clinical outcomes in human dilated cardiomyopathy. J Card Fail. 2010;16:404–10.
88. Wang C, Jing Q. Non-coding RNAs as biomarkers for acute myocardial infarction. Acta Pharmacol Sin. 2018; https://doi.org/10.1038/aps.2017.205.
89. Ferreira LRP, Ferreira FM, Laugier L, Cabantous S, Navarro IC, Cândido DS, Rigaud VC, Real JM, Pereira GV, Pereira IR. Integration of miRNA and gene expression profiles suggest a role for miRNAs in the pathobiological processes of acute Trypanosoma cruzi infection. Sci Rep. 2017;7:17990.
90. Ferreira LRP, Frade AF, Santos RHB, Teixeira PC, Baron MA, Navarro IC, Benvenuti LA, Fiorelli AI, Bocchi EA, Stolf NA. MicroRNAs miR-1, miR-133a, miR-133b, miR-208a and miR-208b are dysregulated in chronic Chagas disease cardiomyopathy. Int J Cardiol. 2014;175:409–17.
91. Linhares-Lacerda L, Granato A, Gomes-Neto JF, Conde L, Freire-de-Lima L, de Freitas EO, Freire-de-Lima CG, Coutinho Barroso SP, Jorge de Alcântara Guerra R, Pedrosa RC. Circulating plasma microRNA-208a as potential biomarker of chronic indeterminate phase of Chagas disease. Front Microbiol. 2018;9:269.
92. Matthews H, Hanison J, Nirmalan N. "Omics"-informed drug and biomarker discovery: opportunities, challenges and future perspectives. Proteomes. 2016;4:28.
93. Ruiz-Lancheros E, Chatelain E, Bournissen F, Moroni S, Moscatelli G, Altcheh J, Ndao M. Surrogate biomarkers of cure for Chagas' disease in children treated with benznidazole. Open Forum Infect Dis. In press 2018. Doi: https://doi.org/10.1093/ofid/ofy236
94. Mamas M, Dunn WB, Neyses L, Goodacre R. The role of metabolites and metabolomics in clinically applicable biomarkers of disease. Arch Toxicol. 2011;85:5–17.
95. Miao Q, Ndao M. *Trypanosoma cruzi* infection and host lipid metabolism. Mediat Inflamm. 2014;2014:902038. https://doi.org/10.1155/2014/902038.
96. Zhao Y-Y, Cheng X-L, Lin R-C. Lipidomics applications for discovering biomarkers of diseases in clinical chemistry. Int Rev Cell Mol Biol. 2014;313:1–26.

97. Carmona SJ, Nielsen M, Schafer-Nielsen C, Mucci J, Altcheh J, Balouz V, Tekiel V, Frasch AC, Campetella O, Buscaglia CA. Towards high-throughput immunomics for infectious diseases: use of next-generation peptide microarrays for rapid discovery and mapping of antigenic determinants. Mol Cell Proteomics. 2015;14:1871–84.
98. Mucci J, Carmona SJ, Volcovich R, Altcheh J, Bracamonte E, Marco JD, Nielsen M, Buscaglia CA, Agüero F. Next-generation ELISA diagnostic assay for Chagas disease based on the combination of short peptidic epitopes. PLoS Negl Trop Dis. 2017;11:e0005972.
99. Martins-Filho OA, Eloi-Santos SM, Carvalho AT, Oliveira RC, Rassi A, Luquetti AO, Rassi GG, Brener Z. Double-blind study to evaluate flow cytometry analysis of anti-live trypomastigote antibodies for monitoring treatment efficacy in cases of human Chagas' disease. Clin Diagn Lab Immunol. 2002;9:1107–13.

Index

J. M. Altcheh, H. Freilij (eds.), *Chagas Disease*, Birkhäuser Advances in Infectious Diseases, https://doi.org/10.1007/978-3-030-00054-7

The manufacturer's authorised representative in the EU is Springer Nature Customer Service Centre GmbH, Europaplatz 3, 69115 Heidelberg, Germany. If you have any concerns regarding our products, please contact ProductSafety@springernature.com

Printed and bound by CPI Group (UK) Ltd, Croydon, CR0 4YY
22/09/2026
02221065-0005